T0289847

Satellite Radar Altimetry Applications in Earth Analysis

Satellite Radar Altimetry Applications in Earth Analysis

Edited by Israel Morris

SYRAWOOD
PUBLISHING HOUSE

New York

Published by Syrawood Publishing House,
750 Third Avenue, 9th Floor,
New York, NY 10017, USA
www.syrawoodpublishinghouse.com

Satellite Radar Altimetry Applications in Earth Analysis
Edited by Israel Morris

International Standard Book Number: 978-1-64740-407-9 (Hardback)

Cataloging-in-publication Data

Satellite radar altimetry applications in earth analysis / edited by Israel Morris.
    p. cm.
Includes bibliographical references and index.
ISBN 978-1-64740-407-9
1. Earth sciences--Remote sensing. 2. Earth (Planet)--Observations. 3. Remote sensing. I. Morris, Israel.
QE33.2.R4 .S28 2023
550.28--dc23

TABLE OF CONTENTS

PREFACE

Satellite radar altimetry refers to a technique that is used for calculating the vertical distance between a satellite and the nadir surface of the Earth by monitoring the time it takes for radar pulses to travel from the satellite antenna to the surface and back to the altimeter. It enables accurate remote sensing of the topography of the ocean's surface. Satellite altimetry is utilized in geodesy to investigate sea-level variability, tectonic plate motion, natural hazards, size and form of the Earth, inland water-related occurrences and Earth's gravity field over oceans. Satellite radar altimetry is also used widely in Earth analysis for regional and global geoid modeling, monitoring ice sheet melting, unifying vertical height systems, and observing the effects of sea level rise. This book aims to shed light on the applications of satellite radar altimetry in Earth analysis. A number of latest researches have been included to keep the readers up-to-date with the global concepts in this area of study.

All of the data presented henceforth, was collaborated in the wake of recent advancements in the field. The aim of this book is to present the diversified developments from across the globe in a comprehensible manner. The opinions expressed in each chapter belong solely to the contributing authors. Their interpretations of the topics are the integral part of this book, which I have carefully compiled for a better understanding of the readers.

At the end, I would like to thank all those who dedicated their time and efforts for the successful completion of this book. I also wish to convey my gratitude towards my friends and family who supported me at every step.

Editor

Monitoring Reservoir Drought Dynamics with Landsat and Radar/Lidar Altimetry Time Series in Persistently Cloudy Eastern Brazil

Jamon Van Den Hoek [1,*], Augusto Getirana [2,3], Hahn Chul Jung [3,4], Modurodoluwa A. Okeowo [5] and Hyongki Lee [5]

[1] Geography Program, College of Earth, Ocean, and Atmospheric Sciences, Oregon State University, Corvallis, OR 97331, USA
[2] Earth System Science Interdisciplinary Center, University of Maryland, College Park, MD 20740, USA; augusto.getirana@nasa.gov
[3] Hydrological Sciences Laboratory, NASA Goddard Space Flight Center, Greenbelt, MD 20771, USA; hahnchul.jung@nasa.gov
[4] Science Systems and Applications, Inc., Lanham, MD 20706, USA
[5] Department of Civil and Environmental Engineering, University of Houston, Houston, TX 77005, USA; maokeowo@uh.edu (M.A.O.); hlee@uh.edu (H.L.)
* Correspondence: vandenhj@oregonstate.edu

Abstract: Tropical reservoirs are critical infrastructure for managing drinking and irrigation water and generating hydroelectric power. However, long-term spaceborne monitoring of reservoir storage is challenged by data scarcity from near-persistent cloud cover and drought, which may reduce volumes below those in the observational record. In evaluating our ability to accurately monitor long-term reservoir volume dynamics using spaceborne data and overcome such observational challenges, we integrated optical, lidar, and radar time series to estimate reservoir volume dynamics across 13 reservoirs in eastern Brazil over a 12-year (2003–2014) period affected by historic drought. We (i) used 1560 Landsat images to measure reservoir surface area; (ii) built reservoir-specific regression models relating surface area and elevation from ICESat GLAS and Envisat RA-2 data; (iii) modeled volume changes for each reservoir; and (iv) compared modeled and in situ reservoir volume changes. Regression models had high goodness-of-fit (median RMSE = 0.89 m and r = 0.88) across reservoirs. Even though 88% of an average reservoir's volume time series was based on modeled area–elevation relationships, we found exceptional agreement (RMSE = 0.31 km^3 and r = 0.95) with in situ volume time series, and accurately captured seasonal recharge/depletion dynamics and the drought's prolonged drawdown. Disagreements in volume dynamics were neither driven by wet/dry season conditions nor reservoir capacity, indicating analytical efficacy across a range of monitoring scenarios.

Keywords: reservoir volume; clouds; Envisat; ICESat; time series

1. Introduction

Reservoirs are critical global infrastructure that occupy at least 26×10^4 km^2 (0.2%) of global land surface area and contribute 6×10^3 km^3 (0.0004%) of global freshwater storage [1]. The freshwater held by reservoirs is essential for meeting global demand for drinking water, irrigation water for agriculture [2,3], and hydroelectric power generation [4]. Recent research has identified the dual roles of reservoirs in atmospheric carbon dynamics being large-scale sites of both carbon sequestration as well as greenhouse gas emission [5–11]. Tropical reservoirs only make up 15% (1040) of the

6824 Global Reservoir and Dam (GRanD) Database reservoirs [12] but play a relatively outsized role in atmospheric carbon flux compared to temperate reservoirs [13–18]. Consistent, systematic, and long-term monitoring of reservoir storage throughout drought as well as high precipitation periods is essential to understanding the role of tropical reservoirs on society and environment [8,16]).

Reservoir storage is temporally dynamic and satellite remote sensing time series from optical (e.g., Landsat, MODIS, Sentinel-2) and radar or lidar (e.g., Envisat, GLAS/ICESat) systems support systematic monitoring of changes in reservoir volume at low cost per reservoir [19–27]. Spaceborne observation of reservoir dynamics are especially valuable for geographically remote or high-elevation reservoirs where consistent and long-term monitoring are a challenge, or in regions where the cost of constructing or maintaining a hydrologic gauge network is prohibitively expensive [28–30]. However, optical remote sensing of tropical reservoir dynamics is complicated by persistent cloud cover [31–33]. Indeed, as shown in Figure 1, the 178 Brazilian reservoirs included in GRanD are on average obscured by clouds 54% of days between 2003 and 2014.

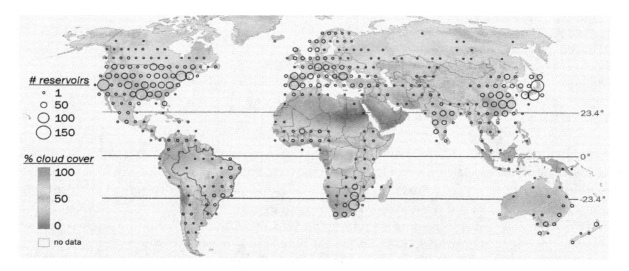

Figure 1. Global distribution of Global Reservoir and Dam (GRanD) Database reservoirs [12]. Reservoir counts are summarized at 5° resolution), and percent daily cloud cover is based on MODIS MOD09GA [34] (2003–2014, 1 km resolution, Robinson projection).

In addition to the challenges of persistent cloud cover, monitoring long-term tropical reservoir dynamics is complicated by sensor spatial resolution, satellite revisit period, and, in some cases, a limited observational record. MODIS has regularly been used to monitor reservoir dynamics at sub-monthly time scales (e.g., [23,25,35]). While spectral unmixing of MODIS imagery has been effective at measuring sub-pixel fractional water area [36,37], the relatively coarse 250 m spatial resolution impedes detection of the reservoir edge and quantification of fine-scale surface area changes (e.g., [38]) with heightened consequences for reservoirs with shallow sloped near-surface bathymetry [25]. Landsat's 30 m spatial resolution, over 40 years of global coverage makes it well suited for long-term, high spatial resolution monitoring of reservoir surface water dynamics [39–43]. Unfortunately, Landsat's 16-day revisit period limits the opportunities to mitigate cloud cover or haze compared to 8- or 16-day MODIS temporal composites [44], which further restricts opportunities for pairing surface area estimates with concurrent altimetry elevation observations. While spaceborne radar is broadly robust to atmospheric effects, altimeters on TOPEX/Poseidon and Jason-1, 2, and 3 satellites with 3–10 km spatial resolutions are not well suited for monitoring smaller reservoirs [45,46]. More recent platforms such as and Cryosat-2 and Sentinel 3 have higher spatial resolution but lack long-term coverage [24,47,48]. However, Envisat RA-2 (Radar Altimeter 2) radar (2002–2012) and ICESat GLAS (Geoscience Laser Altimeter System) lidar (2003–2010) altimeters lend themselves to reservoir monitoring applications given their sub-kilometer resolution and many years of coverage [28,35,49–52].

Using a case study of 13 reservoirs in eastern Brazil, the goal of this study is to accurately model tropical reservoir volume dynamics from 2003 to 2014 despite near-persistent cloud cover and reservoir depletion due to a historic drought in 2014. To achieve our goal, we estimated surface area and elevation across study reservoirs by integrating Landsat, Envisat RA-2, and ICESat GLAS time series, modeling surface area–elevation relationship, calculating volume changes over the study period, and assessing agreement between modeled and in situ volume time series data per date, month, and year with attention to differences in agreement before and during the drought.

2. Materials and Methods

2.1. Study Area

Brazil is the largest (8.5 million km^2) and most populous country (approximately 209 million residents in 2017) in South America. The country is dominated by tropical rain forest in its west and north, has tropical semideciduous forest along its southeastern Atlantic coast with tropical savannas (cerrado) stretching between the Brazilian Highlands to the Mato Grosso Plateau, and is home to major urban population centers including São Paulo, Rio de Janeiro, and the capital, Brasilia, in the eastern half of the country. Brazil has the largest reserve of renewable surface water in the world, approximately 32% of which is used for agricultural production [53]. Beginning in 2012, changing atmospheric circulation patterns, declining rainfall, and increased temperatures culminated in the summer of 2014 being the warmest and driest since 1951 [54] and a series of historically intense droughts that parched extensive cropland, dwindled drinking water supply, and shrank Brazil's rivers and reservoirs [55,56].

During the peak of the drought in the summer of 2014, national hydroelectric production declined by approximately 20% compared to average 2000–2010 levels, and the Cantareira reservoir system relied upon by São Paulo saw nearly 11% reduction in its total capacity [53]. Over half of Brazil's largest reservoirs (by hydroelectric production) are located in the drought-affected eastern region of the country. Thirteen reservoirs were included in this study that span southeast Brazil (Figure 2) with at least three dates of observation between Envisat and GLAS. Four study reservoirs–Agua Vermelha, Furnas, Marimbondo, and Tres Marias—with a range in nominal storage capacity from 1.3 to 34.1 km^3 and a diversity of inter-annual storage dynamics were selected for focused data presentation below with the remaining reservoirs' data included in the Appendix A.

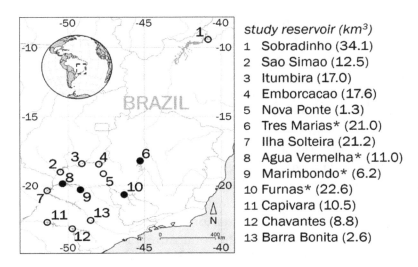

Figure 2. Geographic distribution and storage capacities (km^3) of study reservoirs in eastern Brazil [12]. Four focus reservoirs whose volume dynamics are illustrated in the manuscript text are identified with a black point on the map and an asterisk (*) in the reservoir list; all reservoirs' dynamics are illustrated in Appendix A.

2.2. Modeling Framework

The volumetric modeling framework (Figure 3) for a given reservoir comprised five main objectives: (1) generate reservoir surface area time series; (2) generate surface elevation time series; (3) build and apply a linear regression model relating surface area and elevation; (4) estimate volumetric change over the study period; and (5) assess agreement between modeled and in situ volumes.

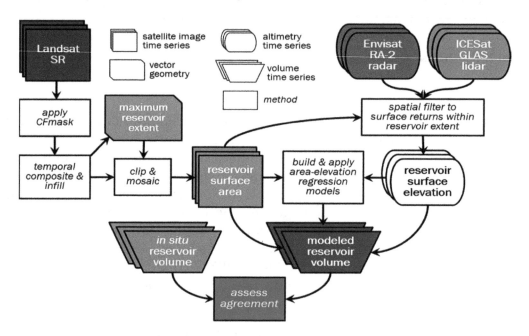

Figure 3. Overview of analytical framework to generate surface area, elevation, and volumetric time series for a given reservoir. Colors shown for each dataset are consistent throughout all figures.

2.3. Surface Area Time Series Generation

The 13 study reservoirs collectively span 15 Landsat WRS-2 path-row footprints, with three reservoirs stretching across 2–4 footprints, respectively. A total of 1460 Landsat 5, 7, and 8 surface reflectance (SR)-corrected scenes with 20% or less cloud cover were collected from 2003 through 2014; this cloud cover threshold reduced the likelihood of missing data due to cloud cover while still supporting an intra-annually dense time series. We measured surface water time series with the commonly used Modified Normalized Difference Water Index (MNDWI), which is effective at discriminating water from non-water features [57]. We included all pixels that have a MDWI value larger than 0.1 as 'water' for each image date. Missing pixels associated with clouds, shadows, and Landsat 7's Scan Line Corrector (SLC) Error data gaps were identified and removed using Landsat's CFmask product. For each image's missing pixels, pixels from images collected within a 36-day window before/after (e.g., Figure 4a–c) were infilled into the image, prioritized by the nearest date of observation (e.g., Figure 4d). Infilling using a 36-day temporal window ensured inclusion of images from up to two along-track (every 16 days) and up to four across-track (7 or 9 days) observations and reduced the amount of missing data within an average reservoir from 8.6% to 1.2% (Figure 5). On some dates, an image's missing pixels could not be directly infilled through temporal compositing; in this case, if these missing pixels were in locations classified as 'water' on at least the median frequency for a given reservoir, the missing pixels were automatically included in the surface water extent. A reservoir's water frequency map showing the total number of days when 'water' was detected (Figure 4e) was also used to generate a consistent spatial extent within which surface water area was measured: pixels classified as 'water' on at least two dates made up an 'ever-water' envelope for this purpose (shown in Figure 4). Finally, each reservoir's surface area time series was clipped by an integrated (i.e., upstream to dam wall) reservoir region that was visually interpreted using very high-resolution imagery hosted by Google Earth.

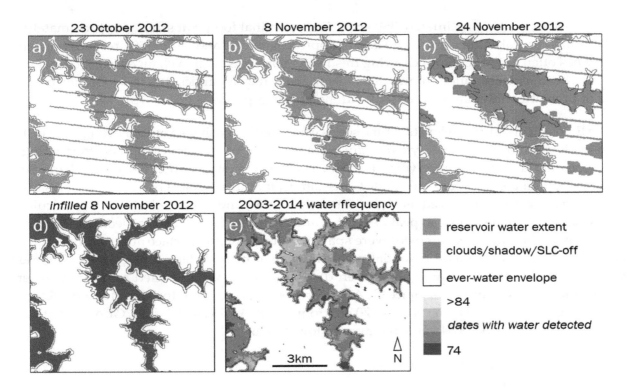

Figure 4. Example generation of temporally composited, infilled surface water extent maps showing southeastern Chavantes reservoir. (**a–c**) Surface water coverage from three near-date images (23 October, 8 November, and 24 November 2012) with data gaps. (**d**) Coverage maps are composited to create an infilled surface water map for 8 November 2012. (**e**) A time series of infilled water maps are used to generate a water frequency map for 2003–2014.

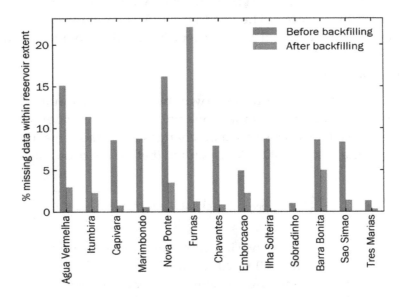

Figure 5. Reductions in average percentage of missing data across all image dates and reservoirs after infilling.

2.4. Surface Elevation Time Series Generation

Separate surface water elevation time series were built using Envisat RA-2 (2002–2010) and ICESat GLAS (2003–2009) altimetry data, respectively. The long wavelength of Envisat's RA-2 (Ku-band) radar altimetry provides resilience to cloud cover effects while the vertical profiling capabilities of ICESat's GLAS support the discrimination of ground-level conditions in all but the densest cloud cover [31,58]. Both RA-2 and GLAS datasets are suitable for reservoir monitoring with along-track sampling distances of approximately 350 and 172 m, nominal footprint sizes of approximately 2 km and 70 m, and vertical accuracies (i.e., RMSE) of approximately 50 and 15 cm [59], respectively. Envisat altimetry has often been used for measuring freshwater surface elevation changes (e.g., [60–67] though ICESat has been used less often for monitoring surface water dynamics due to its erratic operational history of premature laser failures see [68] with notable exceptions (e.g., [69–73]).

The Envisat RA-2 surface elevation time series was generated using an automated return clustering algorithm detailed in [74], which yielded an average RMSE of 48.9 cm compared to in situ gauge observation data across study reservoirs. ICESat GLAS GLA14 product (Level-2 Global Land Surface Altimetry; [75]) data were filtered to single Gaussian peak returns typical of surface water, saturated returns were corrected or removed using the saturation index, and surface elevation was calculated at the centroid of the single Gaussian peak. All RA-2 and GLAS surface returns collected within 15 days of a Landsat surface area measurement were spatially filtered by the surface area perimeter buffered inwards by 100 m to eliminate potential measurement of near-surface topography. The median elevation over a reservoir's extent on a given date was measured and used to generate reservoir elevation and volume time series. Since GLAS' reference ellipsoid (equatorial radius = 6378.1363 km; flattening coefficient = 1/298.257) is offset 70 cm from RA-2's WGS 84 ellipsoid, GLAS elevation data were adjusted to fit the WGS 84 vertical datum following [76].

2.5. Surface Area–Elevation Model Generation

Surface area and elevation time series were used to estimate changes in reservoir volumes. However, since the vast majority of surface area measurements lacked same-day measurements of surface elevation, regression models were built for each reservoir to relate surface area measurements to altimeter elevations collected within 15 days of the surface area measurement; linear regression was used following successful regional application (e.g., [23,25,49]). Modeled surface elevation values based on measured surface areas were used to build an elevation time series with the same temporal sampling as the surface area time series. The goodness-of-fit of each reservoir's area-elevation model was measured using the Pearson correlation coefficient (r) and the root mean square error (RMSE).

2.6. Volume Time Series Generation

As the volume of a reservoir with unknown bathymetry cannot be remotely measured, the reservoir was rather conceived as a circular cone with a series of water layers as detailed in [77]. Reservoir volume on a given date, t, was based on changes in surface area and elevation time series data measured relative to the median 2003–2011 volume (before the drought):

$$V_{med-t} = \frac{1}{3}(z_t - z_{med}) * \left(A_{med} + A_t + \sqrt{(A_{med} * A_t)} \right) \tag{1}$$

where, V_{med-t} represents the relative difference in volume, A_{med} and A_t and z_{med} and z_t represent surface areas and elevations based on median 2003–2011 values and the date of observation, respectively. The median 2003–2011 in situ volume was measured for each reservoir to serve as a pre-drought baseline, and the relative volume difference for each date of in situ observation was measured.

2.7. Comparison between In Situ and Modeled Volumetric Time Series

Changes in modeled reservoir volumes were compared to a median 4383 daily in situ reservoir volume measurements collected from 2003–2014 by the Brazilian Electric Sector available at http://www.ons.org.br/. However, study reservoir boundaries deviated from (unavailable) management boundaries used by the Brazilian Electric Sector to collect in situ volume and modeled and in situ volume time series could not be directly compared. To mitigate this disparity, both time series were standardized to preserve relative variations with the following equation:

$$S_t = \frac{(V_t - \overline{V})}{\sigma} \qquad (2)$$

where S_t and V_t are standardized and modeled volume changes for a given reservoir on date, t, respectively, and $\overline{V}$ and σ are the mean and standard deviation of modeled volume changes over the full time series, respectively. Using standardized (unitless) values, the agreement between modeled and in situ volume dynamics time series were assessed using linear regression, r, and RMSE, and annual and monthly agreement between modeled and in situ data were evaluated.

3. Results

3.1. Surface Area and Elevation Time Series

Over the 12-year study period, a median 216 Landsat image dates were used to generate each reservoir surface area time series (Figures 6 and A1). Prior to the drought's intensification in 2014, reservoirs typically reached their lowest extent in December or January before recharging to the maximum surface area between March and June following the end of the rainy season. Surface water expansion and contraction was highly dynamic with a median annual range (i.e., difference between maximum and minimum) of reservoir surface areas representing 31.3% of the maximum surface area in the pre-drought 2003–2011 period. During the drought period (2012–2014), this range was nominally consistent at 30.7% though the minimum and maximum surface areas declined relative to the pre-drought period by median 4.5% and 3.6%, respectively. Seasonal dynamics shifted under the drought, as well, as early year recharge was typically reduced in 2012 and 2013 and absent by 2014. Reservoirs such as Furnas and Tres Marias contracted during the drought so much that low surface areas (below 900 km^2 and 350 km^2, respectively) were without precedent in the pre-drought time series.

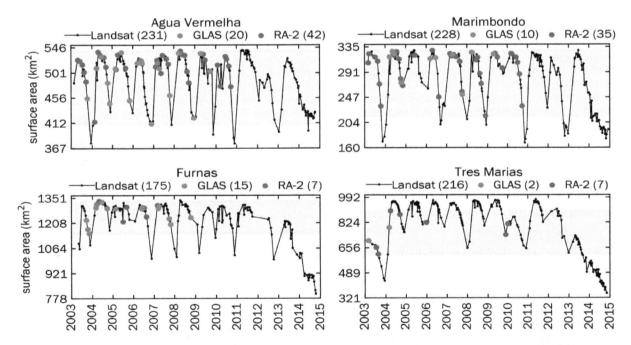

Figure 6. Landsat-derived surface area time series for four reservoirs with dates of ICESat GLAS (green dot) and Envisat RA-2 (red dot) measurement. The gray background indicates the range of surface areas measured on altimetry data collection dates. Time series for all 13 reservoirs are shown in Figure A1.

While each reservoir's seasonal dynamics are apparent, there was often a disproportionate exclusion of cloudy wet season imagery (October–March; Figures 7a and A2). Conversely, since cloud-free observations are more likely in eastern Brazil during winter months, there is a relative abundance of imagery available in July and August when reservoirs tend to be near capacity. Fewer images and generally more cloud cover during the wet season meant fewer near-date images for infilling and compositing, which may yield a larger amount of residual missing data and low biased surface area estimates [33], as well as fewer surface area measurements with which to couple elevation data. While the collection of surface elevation values was not affected by atmospheric conditions in the same way as surface area measurements, having a larger surface area (unobstructed by clouds) meant that more altimetric returns could be collected over the reservoir body (Figures 7b and A3).

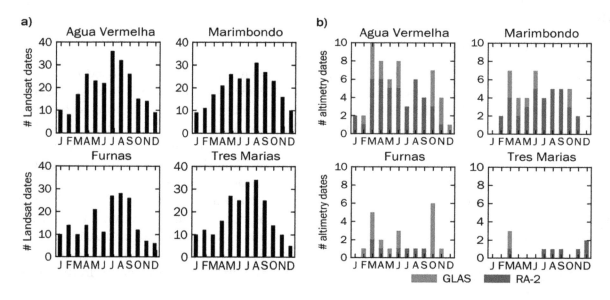

Figure 7. Monthly frequency of (**a**) Landsat images and (**b**) GLAS and RA-2 surface elevation observations for four selected reservoirs from 2003–2014. Monthly frequency of Landsat images and surface elevation observations for all 13 reservoirs are shown in Figures A2 and A3, respectively.

3.2. Modeled Surface Area–Elevation Relationships

Linear regression models relating Landsat-derived surface area and altimeter-derived surface elevation measurements were built for the 13 study reservoirs (Figures 8 and A4). The median number of area–elevation pairs for "combined" regression models was 21, with 13 and 17 measurements for GLAS- and RA-2-specific models, respectively. Combined linear models showed high goodness-of-fit with a median RMSE of 0.89 m and r of 0.88 across reservoirs. A poor model may reflect a combination of seasonal variation in surface area and elevation, differences in Landsat and altimetry acquisition dates, variation in bathymetric slope along a reservoir's perimeter and across its depth, as well as lingering cloud cover effects. Though GLAS and RA-2 elevations span a median 74.9% of reservoir surface area range (highlighted in grey in Figures 6 and A1), the elevation distribution is often skewed towards including higher elevations. This skew is most pronounced for reservoirs with fewer dates of observation during the wet season as well as those with a smaller surface area, which limits the area within which altimetry data could be collected. The contraction of surface areas during the drought also limited potential for coupling surface area-elevation values as very low surface areas during the drought had no paired elevation values.

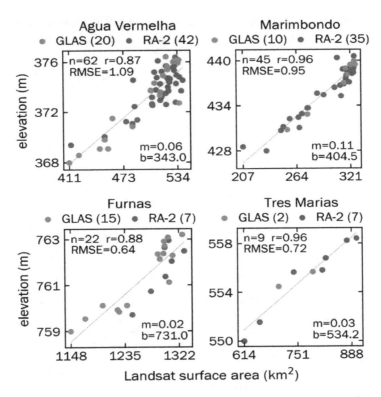

Figure 8. Derived surface area–elevation linear regressions for four selected reservoirs based on Landsat surface area and combined GLAS and RA-2 elevation data. n = number of area–elevation pairs, r = Pearson correlation coefficient, RMSE = root mean square error, m = linear slope, and b = linear intercept. Area-elevation models for all 13 reservoirs are in Figure A4.

GLAS-specific models provided a slightly better fit with median RMSE = 0.68 m and r = 0.94 compared to RA-2's RMSE = 0.94 m and r = 0.90 (see Table A1). The better fit provided by GLAS is expected given its higher vertical accuracy and smaller footprint. For the 8 of 11 reservoirs with both GLAS and RA-2 coverage over the study period, GLAS also captured a larger surface area range despite GLAS having fewer samples for these reservoirs (see Figure A4). Further, GLAS alone provided coverage for two reservoirs, Ilha Solteira and Sao Simao. Despite the driving role of GLAS data in shaping surface area–elevation models, there remains considerable value in combining altimeter measurements into a single area–elevation model. Since GLAS and RA-2 captured reservoirs at different stages over the time series, combining altimetric observations in a single model expanded the seasonality of coverage, which helped compensate for irregularity of observations at a given reservoir. Combining altimetric data also increased the average range of observed surface elevations across reservoirs by at least 1.3 m compared to the elevation range observed by either GLAS or RA-2 alone. This expanded range supports characterizing reservoir dynamics from the drawdown through recharge conditions, which is especially valuable for reservoirs with a large annual range, e.g., Agua Vermelha. 3.3. Volume Time Series and In Situ Comparison.

Across reservoirs, the median modeled volume time series is composed of 216 time steps (Figures 9a and A5a). Of these, 88% of time series values were based on reservoir-specific surface area–elevation regression models (i.e., Figure 8) while the remaining 12% were based on direct observations. While some reservoirs such as Agua Vermelha have a clear, consistent seasonality with a pronounced annual recharge and depletion, other reservoirs such as Chavantes or Sao Simao (see Figure A5) are much less dynamic. Linear regression models showed exceptionally high agreement between modeled and in situ volumetric changes with median RMSE = 0.31 km^3 and r = 0.95 (Figure 9b). The low median bias of 0.11 km^3 indicates an underestimation of in situ volume, and, indeed, each reservoir's maximum modeled volume was consistently less than the maximum in situ volume. This divergence likely results from the mismatch between reservoir boundaries used to generate modeled values and boundaries considered in in situ data collection, differing 2003–2011 reference volumes, and spuriously low modeled volumes that most often occur during the wet season.

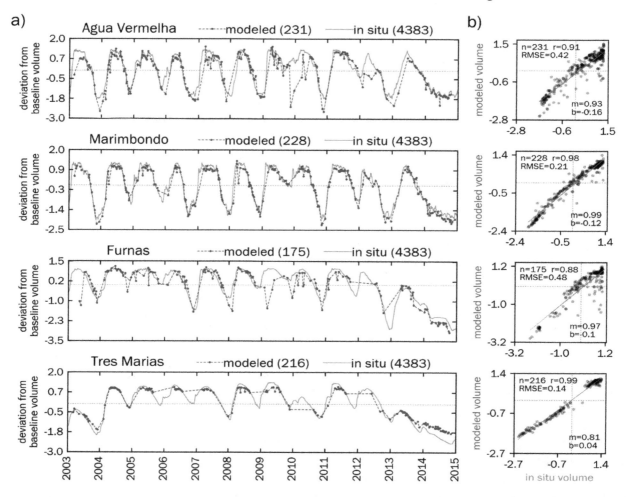

Figure 9. (a) Time series of difference between standardized reservoir volumes and each reservoir's baseline, pre-drought (i.e., median 2003–2011) volume for modeled (purple) and in situ (green) data, respectively; for dates without altimetry data, regression-based elevation values are used in modeling volume. (b) Linear regressions relating standardized in situ and modeled volumes on dates of mutual observation. Light grey lines in (a) and (b) indicate zero values on respective axes. All volume values are in km³. Time series and regressions for all 13 reservoirs are in Figure A5.

To examine intra-annual variation in agreement, the mean absolute difference between modeled and in situ volumetric change was estimated for each month (Figures 10 and A6). Wet season months (October–March) showed higher disagreement than winter months (July–August) in 5 of 13 study reservoirs. Since reservoirs tend to have their lowest volumes during wet months of November–January, accurately modeling reservoir volumes during these low volume months would improve modeling of anomalously low volumes during drought periods. Considering inter-annual variation, the mean annual flux (i.e., the range of standardized volumes within a calendar year) of modeled and in situ time series only differed between 0.004 and 0.041 in 2003–2010 across reservoirs (Figure 11). With the drought came a larger disagreement of 0.62 in 2012 but was reduced to a very low disagreement of only 0.04 in 2014. The divergence between modeled and in situ flux in 2011–2013 was likely influenced by fewer Landsat images being collected after Landsat 5 operational imaging ended in November 2011, 17 months before Landsat 8's operational imaging began in April 2013. The agreement in 2014 benefitted from Landsat 8's data collection, a low annual flux due to the drought, and good fitting area–elevation regression models that allowed for estimating elevations beyond the range of historical direct observation.

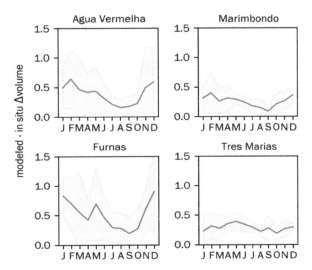

Figure 10. Mean monthly absolute difference between modeled and in situ standardized (unitless) volume changes (2003–2014; red line) with standard deviation range (red field). Monthly comparisons for all 13 reservoirs are in Figure A6.

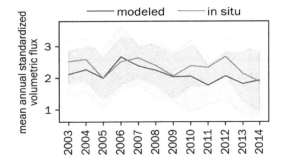

Figure 11. Comparison between mean annual flux of standardized (unitless) modeled (purple line) and in situ (green line) volume dynamics across all study reservoirs with ±1 standard deviation ranges (purple and green fields, respectively).

4. Discussion

Accurate long-term monitoring of reservoir dynamics achieved by this study is of paramount interest for assessing United Nations Development Program Sustainable Development Goal (SDG) 6 (Clean Water and Sanitation) and 7 (Affordable and Clean Energy) [78], managing national-level water resources, examining reservoir dynamics and hydroelectric power generation (e.g., [3]), assessing the impact of changing precipitation and temperature regimes on reservoir storage [54,56,79,80], and as input to or validation of hydrodynamic models (e.g., [81–83]. This study's high agreement between remote sensing and in situ volumetric flux (i.e., RMSE = 0.31 km^3) illustrates the potential for effective monitoring in tropical regions with persistent cloud cover or even those affected by historic drought.

Volumetric modeling is an essential tool in monitoring climate effects on global surface water dynamics especially in regions with unavailable or scarce in situ data [43]. This study builds on research by [19,20,22–25,65,84,85] in pursuing remote sensing-based estimation of reservoir volume dynamics. However, this research alongside [49] extends previous efforts by maintaining high accuracy in modeled volume changes outside the historical range of altimetric observations, in this case due to historic drought. Though the drought began in 2012, the reduction of reservoir surface areas, elevations, and volumes was broadly delayed across study reservoirs until 2013 as is evident in modeled and in situ volumetric change time series (e.g., Figure A5). In 2013, surface elevations at eight of the study reservoirs fell below the observed minimum surface elevation from 2000–2009; by 2014, 12 of the 13 study reservoirs had done so. Despite extrapolating surface area–elevation relationships to capture volumetric change in 2013–2014, the mean standardized modeled and in situ volumetric fluxes (e.g., Figure 11) were in agreement, thereby demonstrating the effectiveness of monitoring. Volume estimate accuracies for Tres Marias, Marimbondo, and others are comparable to those achieved by [19] (i.e., r = 0.99) despite this study's challenges of persistent cloud cover and extreme volumetric loss.

Limited observational frequency remained a challenge in this study. The largest difference between mean annual standardized modeled and in situ volumetric fluxes was 0.62 (unitless) and measured in 2012. This did not result from drought-induced depletion but was rather associated with reduced surface area measurements since only Landsat 7 was operational in 2012. Nonetheless, despite intra-annual variability of surface area measurements, limited observational data, and drought-induced depletion of reservoir volume beyond the range of observations, agreement between modeled and in situ volume dynamics was high. This is perhaps best exemplified by Tres Marias, which lacked any altimetry observations for six months of the calendar year yet shows better agreement with in situ volume data (i.e., RMSE = 0.14 km^3) than any other reservoir. Future volumetric modeling efforts will benefit from the increased observational frequency by including additional surface area time series from Sentinel-1, Sentinel-2, as well as the upcoming Surface Water and Ocean Topography (SWOT) mission [86]. Similarly, while no current hydrodynamic models accurately simulates reservoirs and lakes, recent developments on hydrological modeling have taken into account reservoir operation (e.g., [87–89]). Results presented in this study can therefore be used for further model evaluation and improvement of model parameters to better model human activities.

Errors inherent to the volumetric modeling approach include Landsat-derived surface area resolution, altimetry-derived vertical resolution, area–elevation linear regression modeling, and unavailable reservoir boundaries used for in situ data collection. Moreover, referencing the median water frequency coverage for infilling remaining pixels that could not be directly infilled with temporal compositing lead to overestimating surface area and volume on low volume days and underestimating surface area and volume on high volume days; this, in turn, degraded surface area-elevation correlations. While a linear area–elevation relationship that assumes a consistent bathymetric (i.e., hypsographic) profile across all depths and along the entire boundary of a given shoreline has been used effectively by other researchers (e.g., [23]), non-linear models have also been used to model sections or the entirety of a reservoir's hypsographic profile (e.g., [36,85]). A more extensive examination of the relationship between reservoir dynamics and bathymetry would have added valuable information. Unfortunately, there were no available reference bathymetry datasets with which to evaluate this relationship, nor were there surface area validation time series data with which to evaluate the accuracy of surface area dynamics for any reservoir in the study. Including information from a high-resolution bathymetric model would have improved volumetric modeling effort [84].

There are several opportunities for improving the water classification through use of a more advanced classification approach sensitive to reservoir-specific hydrologic conditions such as clarity, turbidity, and bathymetry, e.g., [90]. The results of this study point towards more effective monitoring using recently launched sensors such as Sentinel 1 and 2 [91–93]. The use of 20 m spatial resolution Sentinel 2A and 2B time series data with a nominal 5-day temporal resolution rather than 30 m, 16-day repeat Landsat data have the potential to estimate much more detailed volumetric changes, especially for reservoirs with complex shorelines that cannot be readily captured with Landsat, let alone MODIS imagery. Use of imaging radars such Sentinel 1, SWOT, and RADARSAT-2 [94] with <100 m horizontal resolution and <10 cm vertical resolution as well as passive microwave sensors [95] offer data fusion opportunities that make the most of available optical and passive time series data.

5. Conclusions

This study presented a remote sensing-based approach for monitoring reservoir dynamics over a 12-year period (2003–2014), which included a historic two-year long drought. Using the case study of regularly clouded eastern Brazil, Landsat 5, 7, 8, ICESat GLAS, and Envisat RA-2 time series data were used to measure surface area, elevation, and volume time series for 13 study reservoirs of varying size and intra- and inter-annual dynamics. Volumetric changes have high overall agreement with in situ volumetric time series data as well as intra- and inter-annual agreement throughout the study period including the drought. The high agreement despite regular cloud obfuscation and drought depletion points the way towards routine, very high-resolution monitoring of reservoir dynamics in regions that lack existing or available in situ monitoring. While this study benefits from reservoir management data, anticipated applications extend to regularly clouded regions without existing or otherwise inaccessible data on reservoir dynamics.

Author Contributions: Conceptualization, J.V.D.H., A.G., and H.C.J.; Formal analysis, J.V.D.H., M.O., and H.L.; Methodology, J.V.D.H., A.G., and H.C.J.; Visualization, J.V.D.H.; Writing—original draft, J.V.D.H.; Writing—review & editing, J.V.D.H.

Appendix A

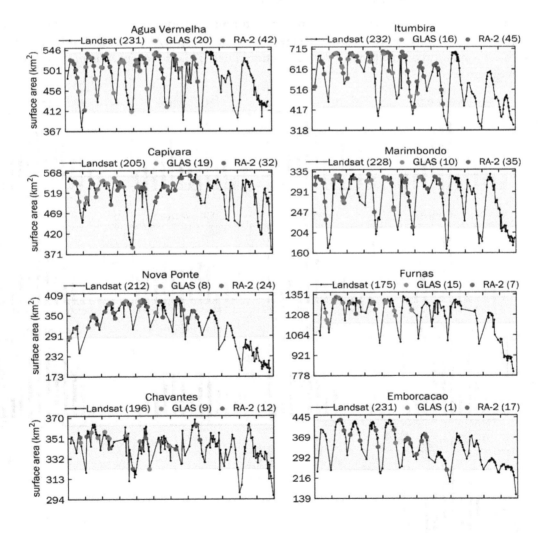

Figure A1. *Cont.*

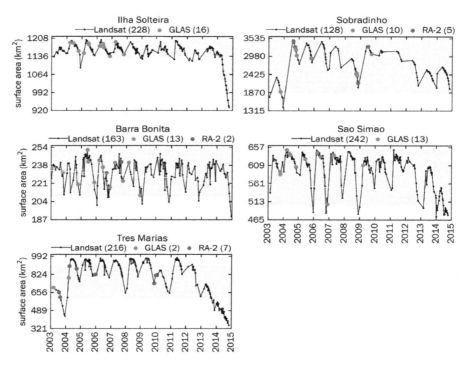

Figure A1. Landsat-derived surface area time series for all 13 study reservoirs with dates of ICESat GLAS (green dot) and Envisat RA-2 (red dot) measurement. The gray background indicates the range of surface areas measured on altimetry data collection dates. Time series for select four reservoirs are shown in Figure 6.

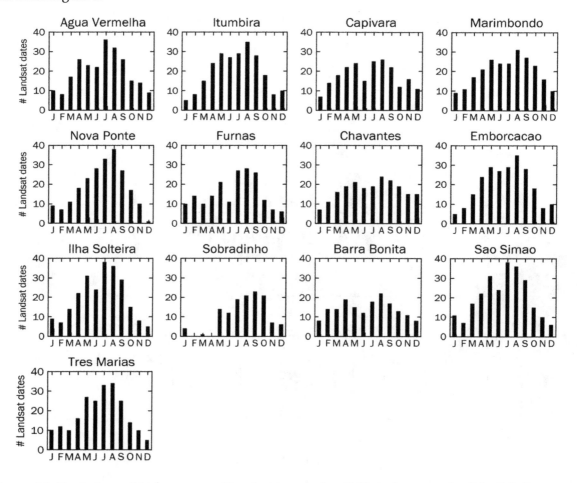

Figure A2. Stacked monthly frequency of Landsat images for all 13 study reservoirs. Monthly frequency information for four selected reservoirs is in Figure 7.

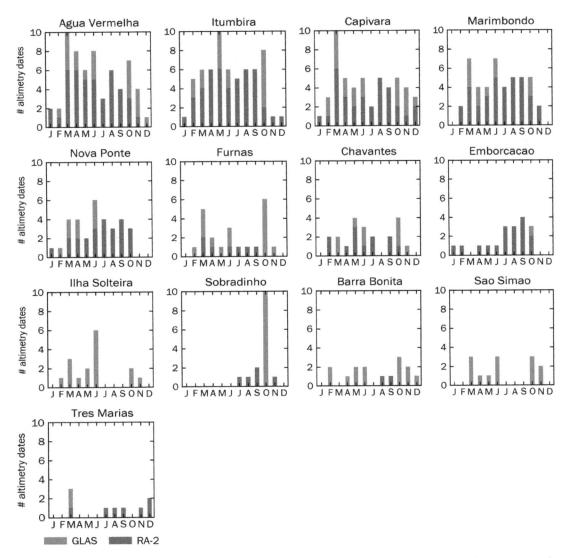

Figure A3. Stacked monthly frequency of GLAS and RA-2 surface elevation observations for all 13 study reservoirs. Monthly frequency information for four selected reservoirs is in Figure 7.

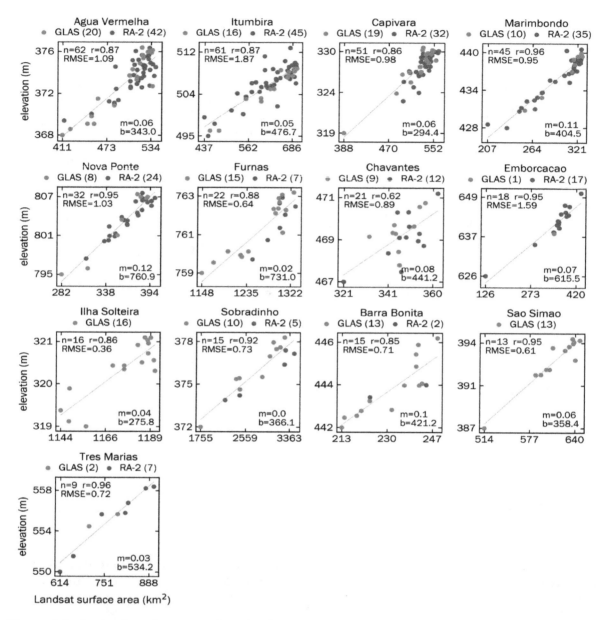

Figure A4. Derived surface area–elevation linear regressions for all 13 study reservoirs based on Landsat surface area and combined GLAS and RA-2 elevation data. n = number of area–elevation pairs, r = Pearson correlation coefficient, RMSE = root mean square error, m = linear slope, and b = linear intercept. Area–elevation models for select four reservoirs are in Figure 8.

Table A1. List of reservoir-specific linear regression model parameters and goodness-of-fit values for GLAS- or RA-2-specific models as well as 'combined' models. n = number of altimeter surface elevation measurements, m = linear slope, b = linear intercept, r = Pearson correlation coefficient, and RMSE = root mean square error.

Reservoir Name	Altimeter	n	m	b	r	RMSE
Agua Vermelha	GLAS	20	0.070	339.10	0.933	1.054
	RA-2	42	0.054	347.03	0.777	1.055
	combined	62	0.062	343.01	0.867	1.088
Barra Bonita	GLAS	13	0.109	419.13	0.884	0.683
	RA-2	2	0.097	422.63	0.839	0.745
	combined	15	0.100	421.21	0.855	0.710
Capivara	GLAS	19	0.063	295.13	0.937	0.864
	RA-2	32	0.075	288.46	0.773	0.979
	combined	51	0.064	294.41	0.862	0.982
Chavantes	GLAS	9	0.077	443.26	0.548	0.799
	RA-2	12	0.087	439.30	0.681	0.902
	combined	21	0.082	441.21	0.622	0.887
Emborcacao	GLAS	0	–	–	–	–
	RA-2	17	0.074	615.69	0.951	1.534
	combined	17	0.074	615.69	0.951	1.534
Furnas	GLAS	15	0.024	730.67	0.921	0.567
	RA-2	7	0.038	712.25	0.863	0.559
	combined	22	0.024	731.00	0.883	0.641
Ilha Solteira	GLAS	16	0.039	275.78	0.863	0.362
	RA-2	0	–	–	–	–
	combined	16	0.039	275.78	0.863	0.362
Itumbira	GLAS	16	0.054	471.90	0.977	0.861
	RA-2	45	0.045	478.62	0.840	1.994
	combined	61	0.048	476.73	0.872	1.869
Marimbondo	GLAS	10	0.127	398.20	0.985	0.504
	RA-2	35	0.102	405.62	0.954	0.995
	combined	45	0.106	404.51	0.958	0.947
Nova Ponte	GLAS	8	0.129	757.66	0.968	1.210
	RA-2	24	0.113	762.89	0.942	0.887
	combined	32	0.119	760.86	0.952	1.028
Sao Simao	GLAS	13	0.057	358.41	0.955	0.612
	RA-2	0	–	–	–	–
	combined	13	0.057	358.41	0.955	0.612
Sobradinho	GLAS	10	0.004	364.81	0.963	0.544
	RA-2	5	0.003	367.01	0.974	0.370
	combined	15	0.004	366.13	0.925	0.729
Tres Marias	GLAS	2	0.029	535.31	0.963	0.743
	RA-2	7	0.029	533.10	0.977	0.658
	combined	9	0.028	534.18	0.964	0.719

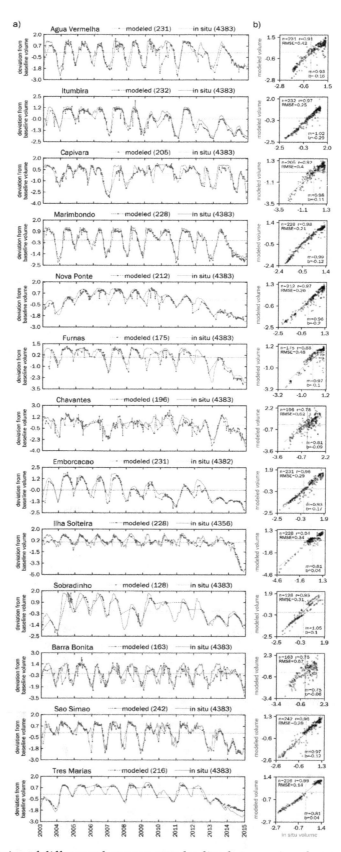

Figure A5. (a) Time series of difference between standardized reservoir volumes and each reservoir's baseline, pre-drought (i.e., median 2003-2011) volume for modeled (purple) and in situ (green) data, respectively; for dates without altimetry data, regression-based elevation values are used in modeling volume. **(b)** Linear regressions relating standardized in situ and modeled volumes on dates of mutual observation. Light grey lines in **(a)** and **(b)** indicate zero values on respective axes. All volume values are in km^3. Time series and regressions for four select reservoirs are in Figure 9.

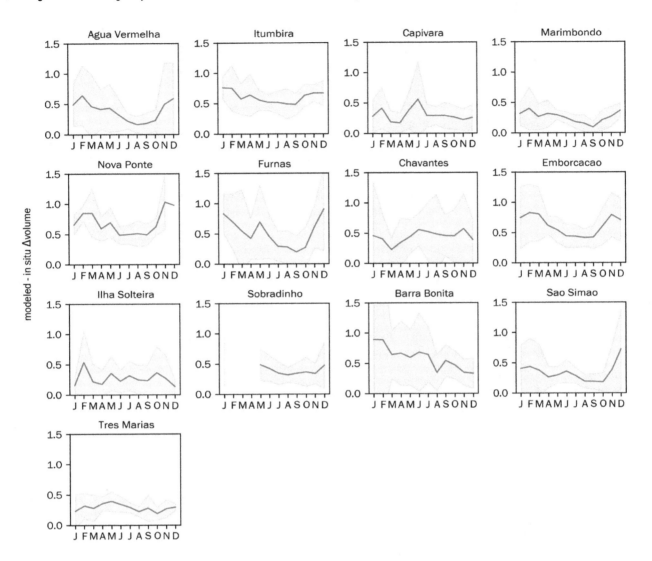

Figure A6. Mean monthly absolute difference between modeled and in situ standardized (unitless) volume changes (2003–2014; red line) with standard deviation range depicted (red field). Monthly comparisons for four select study reservoirs are in Figure 10.

References

1. Messager, M.L.; Lehner, B.; Grill, G.; Nedeva, I.; Schmitt, O. Estimating the volume and age of water stored in global lakes using a geo-statistical approach. *Nat. Commun.* **2016**, *7*, 13603. [CrossRef] [PubMed]

2. García, L.; Rodríguez, D.; Wijnen, M.; Pakulski, I. (Eds.) *Earth Observation for Water Resources Management: Current Use and Future Opportunities for the Water Sector*; The World Bank: Washington, DC, USA, 2016; ISBN 978-1-4648-0475-5.

3. Vörösmarty, C.J.; Sahagian, D. Anthropogenic Disturbance of the Terrestrial Water Cycle. *Bioscience* **2006**, *50*, 753. [CrossRef]

4. Zambon, R.C.; Barros, M.T.L.; Yeh, W.W.G. Impacts of the 2012–2015 Drought on the Brazilian Hydropower System. In Proceedings of the World Environmental and Water Resources Congress 2016, West Palm Beach, FL, USA, 22–26 May 2016; American Society of Civil Engineers: Reston, VA, USA, 2016; pp. 82–91.

5. Bastviken, D.; Tranvik, L.J.; Downing, J.A.; Crill, P.M.; Enrich-Prast, A. Freshwater methane emissions offset the continental carbon sink. *Science* **2011**, *331*. [CrossRef] [PubMed]

6. Fearnside, P.M. Do hydroelectric dams mitigate global warming? The case of Brazil's Curuá-Una Dam. *Mitig. Adapt. Strateg. Glob. Chang.* **2005**, *10*, 675–691. [CrossRef]

7. Kosten, S.; van den Berg, S.; Mendonça, R.; Paranaíba, J.R.; Roland, F.; Sobek, S.; Van Den Hoek, J.; Barros, N. Extreme drought boosts CO2 and CH4 emissions from reservoir drawdown areas. *Inland Waters* **2018**, *8*, 329–340. [CrossRef]

8. Mendonça, R.; Kosten, S.; Sobek, S.; Barros, N.; Cole, J.J.; Tranvik, L.; Roland, F. Hydroelectric carbon sequestration. *Nat. Geosci.* **2012**, *5*, 838–840. [CrossRef]
9. Rudd, J.W.M.; Hecky, R.E.; Harris, R.; Kelly, C.A. Are hydroelectric reservoirs significant sources of greenhouse gases. *Ambio* **1993**, *22*, 246–248.
10. Duchemin, É.; Rudd, J.W.M.; St. Louis, V.L.; Kelly, C.A.; Rosenberg, D.M. Reservoir Surfaces as Sources of Greenhouse Gases to the Atmosphere: A Global Estimate. *Bioscience* **2006**, *50*, 766.
11. Knoll, L.B.; McKnight, D.M.; Larsen, S.; Dillon, P.; Fortino, K.; Cotner, J.B.; Striegl, R.G.; Prairie, Y.; Finlay, K.; Sherman, B.S.; et al. Lakes and reservoirs as regulators of carbon cycling and climate. *Limnol. Oceanogr.* **2011**, *54*, 2298–2314.
12. Lehner, B.; Liermann, C.R.; Revenga, C.; Vörömsmarty, C.; Fekete, B.; Crouzet, P.; Döll, P.; Endejan, M.; Frenken, K.; Magome, J.; et al. High-resolution mapping of the world's reservoirs and dams for sustainable river-flow management. *Front. Ecol. Environ.* **2011**, *9*, 494–502. [CrossRef]
13. Barros, N.; Cole, J.J.; Tranvik, L.J.; Prairie, Y.T.; Bastviken, D.; Huszar, V.L.M.; del Giorgio, P.; Roland, F. Carbon emission from hydroelectric reservoirs linked to reservoir age and latitude. *Nat. Geosci.* **2011**, *4*, 593–596. [CrossRef]
14. Deshmukh, C.; Serça, D.; Delon, C.; Tardif, R.; Demarty, M.; Jarnot, C.; Meyerfeld, Y.; Chanudet, V.; Guédant, P.; Rode, W.; et al. Physical controls on CH4 emissions from a newly flooded subtropical freshwater hydroelectric reservoir: Nam Theun 2. *Biogeosciences* **2014**, *11*, 4251–4269. [CrossRef]
15. Fearnside, P.M. Greenhouse gas emissions from a hydroelectric reservoir (Brazil's Tucuruídam) and the energy policy implications. *Water Air Soil Pollut.* **2002**, *133*, 69–96. [CrossRef]
16. Galy-Lacaux, C.; Delmas, R.; Kouadio, G.; Richard, S.; Gosse, P. Long-term greenhouse gas emissions from hydroelectric reservoirs in tropical forest regions. *Glob. Biogeochem. Cycles* **1999**, *13*, 503–517. [CrossRef]
17. Lima, I.B.T.; Ramos, F.M.; Bambace, L.A.W.; Rosa, R.R. Methane emissions from large dams as renewable energy resources: A developing nation perspective. *Mitig. Adapt. Strateg. Glob. Chang.* **2008**, *13*, 193–206. [CrossRef]
18. Deemer, B.R.; Harrison, J.A.; Li, S.; Beaulieu, J.J.; DelSontro, T.; Barros, N.; Bezerra-Neto, J.F.; Powers, S.M.; dos Santos, M.A.; Arie Vonk, J. Greenhouse Gas Emissions from Reservoir Water Surfaces: A New Global Synthesis. *Bioscience* **2016**, *66*, 949–964. [CrossRef]
19. Baup, F.; Frappart, F.; Maubant, J. Combining high-resolution satellite images and altimetry to estimate the volume of small lakes. *Hydrol. Earth Syst. Sci.* **2014**, *18*, 2007–2020. [CrossRef]
20. Crétaux, J.F.; Jelinski, W.; Calmant, S.; Kouraev, A.; Vuglinski, V.; Bergé-Nguyen, M.; Gennero, M.C.; Nino, F.; Abarca Del Rio, R.; Cazenave, A.; et al. SOLS: A lake database to monitor in the Near Real Time water level and storage variations from remote sensing data. *Adv. Space Res.* **2011**, *47*, 1497–1507. [CrossRef]
21. Ding, X.W.; Li, X.F. Monitoring of the water-area variations of Lake Dongting in China with ENVISAT ASAR images. *Int. J. Appl. Earth Obs. Geoinf.* **2011**, *13*, 894–901. [CrossRef]
22. Duan, Z.; Bastiaanssen, W.G.M. Estimating water volume variations in lakes and reservoirs from four operational satellite altimetry databases and satellite imagery data. *Remote Sens. Environ.* **2013**, *134*, 403–416. [CrossRef]
23. Gao, H.; Birkett, C.; Lettenmaier, D.P. Global monitoring of large reservoir storage from satellite remote sensing. *Water Resour. Res.* **2012**, *48*. [CrossRef]
24. Jiang, L.; Nielsen, K.; Andersen, O.B.; Bauer-Gottwein, P. CryoSat-2 radar altimetry for monitoring freshwater resources of China. *Remote Sens. Environ.* **2017**, *200*, 125–139. [CrossRef]
25. Smith, L.C.; Pavelsky, T.M. Remote sensing of volumetric storage changes in lakes. *Earth Surf. Process. Landf.* **2009**, *34*, 1353–1358. [CrossRef]
26. Tseng, K.H.; Shum, C.K.; Kim, J.W.; Wang, X.; Zhu, K.; Cheng, X. Integrating Landsat Imageries and Digital Elevation Models to Infer Water Level Change in Hoover Dam. *IEEE J. Sel. Top. Appl. Earth Obs. Remote Sens.* **2016**, *9*, 1696–1709. [CrossRef]
27. Zhang, J.; Xu, K.; Yang, Y.; Qi, L.; Hayashi, S.; Watanabe, M. Measuring Water Storage Fluctuations in Lake Dongting, China, by Topex/Poseidon Satellite Altimetry. *Environ. Monit. Assess.* **2006**, *115*, 23–37. [CrossRef]
28. Liebe, J.; van de Giesen, N.; Andreini, M. Estimation of small reservoir storage capacities in a semi-arid environment. *Phys. Chem. Earth* **2005**, *30*, 448–454. [CrossRef]

29. Ogilvie, A.; Belaud, G.; Massuel, S.; Mulligan, M.; Le Goulven, P.; Calvez, R.; Ogilvie, A.; Belaud, G.; Massuel, S.; Mulligan, M.; et al. Assessing Floods and Droughts in Ungauged Small Reservoirs with Long-Term Landsat Imagery. *Geosciences* **2016**, *6*, 42. [CrossRef]

30. Singh, A.; Kumar, U.; Seitz, F. Remote Sensing of Storage Fluctuations of Poorly Gauged Reservoirs and State Space Model (SSM)-Based Estimation. *Remote Sens.* **2015**, *7*, 17113–17134. [CrossRef]

31. Wang, L.; Dessler, A.E. Instantaneous cloud overlap statistics in the tropical area revealed by ICESat/GLAS data. *Geophys. Res. Lett.* **2006**, *33*. [CrossRef]

32. Rodrigues, L.N.; Sano, E.E.; Steenhuis, T.S.; Passo, D.P. Estimation of small reservoir storage capacities with remote sensing in the Brazilian Savannah region. *Water Resour. Manag.* **2012**, *26*, 873–882. [CrossRef]

33. Sano, E.E.; Ferreira, L.G.; Asner, G.P.; Steinke, E.T. Spatial and temporal probabilities of obtaining cloud-free Landsat images over the Brazilian tropical savanna. *Int. J. Remote Sens.* **2007**, *28*, 2739–2752. [CrossRef]

34. Vermote, E.; Wolfe, R. *MOD09GA MODIS/Terra Surface Reflectance Daily L2G Global 1kmand 500m SIN Grid V006 [Data set]*; NASA EOSDIS Land Processes DAAC: Sioux Falls, SD, USA, 2015.

35. Zhang, S.; Gao, H.; Naz, B.S. Monitoring reservoir storage in South Asia from multisatellite remote sensing. *Water Resour. Res.* **2014**, *50*, 8927–8943. [CrossRef]

36. Chipman, J.W. A Multisensor Approach to Satellite Monitoring of Trends in Lake Area, Water Level, and Volume. *Remote Sens.* **2019**, *11*, 158. [CrossRef]

37. Hamilton, S.K.; Sippel, S.J.; Melack, J.M. Seasonal inundation patterns in two large savanna floodplains of South America: The Llanos de Moxos (Bolivia) and the Llanos del Orinoco (Venezuela and Colombia). *Hydrol. Proc.* **2004**, *18*, 2103–2116. [CrossRef]

38. Kang, S.; Hong, S.Y. Assessing Seasonal and Inter-Annual Variations of Lake Surface Areas in Mongolia during 2000-2011 Using Minimum Composite MODIS NDVI. *PLoS ONE* **2016**, *11*, e0151395. [CrossRef]

39. Donchyts, G.; Baart, F.; Winsemius, H.; Gorelick, N.; Kwadijk, J.; Van De Giesen, N. Earth's surface water change over the past 30 years. *Nat. Clim. Chang.* **2016**, *6*, 810–813. [CrossRef]

40. Feyisa, G.L.; Meilby, H.; Fensholt, R.; Proud, S.R. Automated Water Extraction Index: A new technique for surface water mapping using Landsat imagery. *Remote Sens. Environ.* **2014**, *140*, 23–35. [CrossRef]

41. Feng, M.; Sexton, J.O.; Channan, S.; Townshend, J.R. A global, high-resolution (30-m) inland water body dataset for 2000: First results of a topographic–spectral classification algorithm. *Int. J. Digit. Earth* **2016**, *9*, 113–133. [CrossRef]

42. Mueller, N.; Lewis, A.; Roberts, D.; Ring, S.; Melrose, R.; Sixsmith, J.; Lymburner, L.; McIntyre, A.; Tan, P.; Curnow, S.; et al. Water observations from space: Mapping surface water from 25years of Landsat imagery across Australia. *Remote Sens. Environ.* **2016**, *174*, 341–352. [CrossRef]

43. Pekel, J.F.; Cottam, A.; Gorelick, N.; Belward, A.S. High-resolution mapping of global surface water and its long-term changes. *Nature* **2016**, *540*, 418–422. [CrossRef]

44. Alsdorf, D.E.; Rodríguez, E.; Lettenmaier, D.P. Measuring surface water from space. *Rev. Geophys.* **2007**, *45*. [CrossRef]

45. Avisse, N.; Tilmant, A.; François Müller, M.; Zhang, H. Monitoring small reservoirs' storage with satellite remote sensing in inaccessible areas. *Hydrol. Earth Syst. Sci.* **2017**, *21*, 6445–6459. [CrossRef]

46. Uebbing, B.; Kusche, J.; Forootan, E. Waveform retracking for improving level estimations from TOPEX/Poseidon, Jason-1, and Jason-2 altimetry observations over African lakes. *IEEE Trans. Geosci. Remote Sens.* **2015**, *53*, 2211–2224. [CrossRef]

47. Jiang, L.; Schneider, R.; Andersen, O.B.; Bauer-Gottwein, P. CryoSat-2 altimetry applications over rivers and lakes. *Water* **2017**, *9*, 211. [CrossRef]

48. Politi, E.; MacCallum, S.; Cutler, M.E.J.; Merchant, C.J.; Rowan, J.S.; Dawson, T.P. Selection of a network of large lakes and reservoirs suitable for global environmental change analysis using Earth Observation. *Int. J. Remote Sens.* **2016**, *37*, 3042–3060. [CrossRef]

49. Busker, T.; de Roo, A.; Gelati, E.; Schwatke, C.; Adamovic, M.; Bisselink, B.; Pekel, J.F.; Cottam, A. A global lake and reservoir volume analysis using a surface water dataset and satellite altimetry. *Hydrol. Earth Syst. Sci.* **2019**, *23*, 669–690. [CrossRef]

50. Narasimha Rao, P.V.; Sesha Sai, M.V.R. Cover: Clouds over land in Envisat ASAR C-band image. *Int. J. Remote Sens.* **2006**, *27*, 833–834. [CrossRef]

51. Pipitone, C.; Maltese, A.; Dardanelli, G.; Brutto, M.L.; Loggia, G.L. Monitoring water surface and level of a reservoir using different remote sensing approaches and comparison with dam displacements evaluated via GNSS. *Remote Sens.* **2018**, *10*, 71. [CrossRef]

52. Ye, Z.; Liu, H.; Chen, Y.; Shu, S.; Wu, Q.; Wang, S. Analysis of water level variation of lakes and reservoirs in Xinjiang, China using ICESat laser altimetry data (2003–2009). *PLoS ONE* **2017**, *12*, e0183800. [CrossRef]

53. Getirana, A. Extreme Water Deficit in Brazil Detected from Space. *J. Hydrometeorol.* **2016**, *17*, 591–599. [CrossRef]

54. Nobre, C.A.; Marengo, J.A.; Seluchi, M.E.; Cuartas, L.A.; Alves, L.M. Some Characteristics and Impacts of the Drought and Water Crisis in Southeastern Brazil during 2014 and 2015. *J. Water Resour. Prot.* **2016**, *08*, 252–262. [CrossRef]

55. Marengo, J.A.; Alves, L.M.; Soares, W.R.; Rodriguez, D.A. Two Contrasting Severe Seasonal Extremes in Tropical South America in 2012: Flood in Amazonia and Drought in Northeast Brazil. *J. Clim.* **2013**, *26*, 9137–9154. [CrossRef]

56. Marengo, J.A.; Torres, R.R.; Alves, L.M. Drought in Northeast Brazil—Past, present, and future. *Theor. Appl. Climatol.* **2017**, *129*, 1189–1200. [CrossRef]

57. Xu, H. Modification of normalised difference water index (NDWI) to enhance open water features in remotely sensed imagery. *Int. J. Remote Sens.* **2006**, *27*, 3025–3033. [CrossRef]

58. Wang, X.; Cheng, X.; Gong, P.; Huang, H.; Li, Z.; Li, X. Earth science applications of ICESat/GLAS: A review. *Int. J. Remote Sens.* **2011**, *32*, 8837–8864. [CrossRef]

59. Phan, V.H.; Lindenbergh, R.; Menenti, M. ICESat derived elevation changes of Tibetan lakes between 2003 and 2009. *Int. J. Appl. Earth Obs. Geoinf.* **2012**, *17*, 12–22. [CrossRef]

60. Birkett, C.; Reynolds, C.; Beckley, B.; Doorn, B. From research to operations: The USDA global reservoir and lake monitor. In *Coastal Altimetry*; Springer-Verlag Berlin Heidelberg: Berlin, Germany, 2011; pp. 19–50. ISBN 9783642127953.

61. da Silva, J.S.; Calmant, S.; Seyler, F.; Rotunno Filho, O.C.; Cochonneau, G.; Mansur, W.J. Water levels in the Amazon basin derived from the ERS 2 and ENVISAT radar altimetry missions. *Remote Sens. Environ.* **2010**, *114*, 2160–2181. [CrossRef]

62. Frappart, F.; Calmant, S.; Cauhopé, M.; Seyler, F.; Cazenave, A. Preliminary results of ENVISAT RA-2-derived water levels validation over the Amazon basin. *Remote Sens. Environ.* **2006**, *100*, 252–264. [CrossRef]

63. Lee, H.; Shum, C.K.; Tseng, K.H.; Guo, J.Y.; Kuo, C.Y. Present-day lake level variation from envisat altimetry over the northeastern qinghai-tibetan plateau: Links with precipitation and temperature. *Terr. Atmos. Ocean. Sci.* **2011**, *22*, 169–175. [CrossRef]

64. Medina, C.E.; Gomez-Enri, J.; Alonso, J.J.; Villares, P. Water level fluctuations derived from ENVISAT Radar Altimeter (RA-2) and in-situ measurements in a subtropical waterbody: Lake Izabal (Guatemala). *Remote Sens. Environ.* **2008**, *112*, 3604–3617. [CrossRef]

65. Medina, C.; Gomez-Enri, J.; Alonso, J.J.; Villares, P. Water volume variations in Lake Izabal (Guatemala) from in situ measurements and ENVISAT Radar Altimeter (RA-2) and Advanced Synthetic Aperture Radar (ASAR) data products. *J. Hydrol.* **2010**, *382*, 34–48. [CrossRef]

66. Schwatke, C.; Dettmering, D.; Bosch, W.; Seitz, F. DAHITI—An innovative approach for estimating water level time series over inland waters using multi-mission satellite altimetry. *Hydrol. Earth Syst. Sci.* **2015**, *19*, 4345–4364. [CrossRef]

67. Siddique-E-Akbor, A.H.M.; Hossain, F.; Lee, H.; Shum, C.K. Inter-comparison study of water level estimates derived from hydrodynamic-hydrologic model and satellite altimetry for a complex deltaic environment. *Remote Sens. Environ.* **2011**, *115*, 1522–1531. [CrossRef]

68. Abshire, J.B.; Sun, X.; Riris, H.; Sirota, J.M.; McGarry, J.F.; Liiva, P.; Palm, S.; Yi, D. Geoscience Laser Altimeter System (GLAS) on the ICESat Mission: On-orbit measurement performance. *Geophys. Res. Lett.* **2005**, *32*. [CrossRef]

69. Hall, A.C.; Schumann, G.J.P.; Bamber, J.L.; Bates, P.D.; Trigg, M.A. Geodetic corrections to Amazon River water level gauges using ICESat altimetry. *Water Resour. Res.* **2012**, *48*. [CrossRef]

70. O'Loughlin, F.E.; Neal, J.; Yamazaki, D.; Bates, P.D. ICESat-derived inland water surface spot heights. *Water Resour. Res.* **2016**, *52*, 3276–3284. [CrossRef]

71. Swenson, S.; Wahr, J. Monitoring the water balance of Lake Victoria, East Africa, from space. *J. Hydrol.* **2009**, *370*, 163–176. [CrossRef]

72. Wang, X.; Cheng, X.; Li, Z.; Huang, H.; Niu, Z.; Li, X.; Gong, P. Lake water footprint identification from time-series ICESat/GLAS data. *IEEE Geosci. Remote Sens. Lett.* **2012**, *9*, 333–337. [CrossRef]

73. Zhang, G.Q.; Xie, H.J.; Yao, T.D.; Kang, S.C. Water balance estimates of ten greatest lakes in China using ICESat and Landsat data. *Chin. Sci. Bull.* **2013**, *58*, 3815–3829. [CrossRef]

74. Okeowo, M.A.; Lee, H.; Hossain, F.; Getirana, A. Automated Generation of Lakes and Reservoirs Water Elevation Changes from Satellite Radar Altimetry. *IEEE J. Sel. Top. Appl. Earth Obs. Remote Sens.* **2017**, *10*, 3465–3481. [CrossRef]

75. Zwally, H.J.; Schutz, R.; Bentley, C.; Bufton, J.; Herring, T.; Minster, J.; Spinhirne, J.; Thomas, R. *GLAS/ICESat L2 Global Land Surface Altimetry Data, Version 34. [GLA14]*; NASA National Snow and Ice Data Center: Boulder, CO, USA, 2014. [CrossRef]

76. Frequently Asked Questions. Available online: https://nsidc.org/data/icesat/faq.html#alt7 (accessed on 18 May 2018).

77. Taube, C.M. Three Methods for Computing the Volume of a Lake. In *Manual of Fisheries Survey Methods II: With Periodic Updates*; Schneider, J.C., Ed.; Michigan Department of Natural Resources: Lansing, MI, USA, 2000.

78. Ait-Kadi, M. Water for Development and Development for Water: Realizing the Sustainable Development Goals (SDGs) Vision. *Aquat. Procedia* **2016**, *6*, 106–110. [CrossRef]

79. Chadwick, R.; Good, P.; Martin, G.; Rowell, D.P. Large rainfall changes consistently projected over substantial areas of tropical land. *Nat. Clim. Chang.* **2016**, *6*, 177–181. [CrossRef]

80. Feng, X.; Porporato, A.; Rodriguez-Iturbe, I. Changes in rainfall seasonality in the tropics. *Nat. Clim. Chang.* **2013**, *3*, 811–815. [CrossRef]

81. De Paiva, R.C.D.; Collischonn, W.; Calmant, S.; Bulhões Mendes, C.A.; Bonnet, M.P.; Frappart, F.; Buarque, D.C. Large-scale hydrologic and hydrodynamic modeling of the Amazon River basin. *Water Resour. Res.* **2013**, *49*, 1226–1243. [CrossRef]

82. Getirana, A.C.; Boone, A.; Yamazaki, D.; Decharme, B.; Papa, F.; Mognard, N. The Hydrological Modeling and Analysis Platform (HyMAP): Evaluation in the Amazon Basin. *J. Hydrometeorol.* **2012**, *13*, 1641–1665. [CrossRef]

83. Jarihani, A.A.; Callow, J.N.; McVicar, T.R.; Van Niel, T.G.; Larsen, J.R. Satellite-derived Digital Elevation Model (DEM) selection, preparation and correction for hydrodynamic modelling in large, low-gradient and data-sparse catchments. *J. Hydrol.* **2015**, *524*, 489–506. [CrossRef]

84. Lu, S.; Ouyang, N.; Wu, B.; Wei, Y.; Tesemma, Z. Lake water volume calculation with time series remote-sensing images. *Int. J. Remote Sens.* **2013**, *34*, 7962–7973. [CrossRef]

85. Pillco Zolá, R.; Bengtsson, L. Three methods for determining the area-depth relationship of Lake Poopó, a large shallow lake in Bolivia. *Lakes Reserv.* **2007**, *12*, 275–284. [CrossRef]

86. Solander, K.C.; Reager, J.T.; Famiglietti, J.S. How well will the Surface Water and Ocean Topography (SWOT) mission observe global reservoirs? *Water Resour. Res.* **2016**, *52*, 2123–2140. [CrossRef]

87. Doll, P.; Fiedler, K.; Zhang, J. Global-scale analysis of river flow alterations due to water withdrawals and reservoirs. *Hydrol. Earth Syst. Sci.* **2009**, *13*, 2413–2432. [CrossRef]

88. Haddeland, I.; Skaugen, T.; Lettenmaier, D.P. Anthropogenic impacts on continental surface water fluxes. *Geophys. Res. Lett.* **2006**, *33*. [CrossRef]

89. Mateo, C.M.; Hanasaki, N.; Komori, D.; Tanaka, K.; Kiguchi, M.; Champathong, A.; Sukhapunnaphan, T.; Yamazaki, D.; Oki, T. Assessing the impacts of reservoir operation to floodplain inundation by combining hydrological, reservoir management, and hydrodynamic models. *Water Resour. Res.* **2014**, *50*, 7245–7266. [CrossRef]

90. Ji, L.; Zhang, L.; Wylie, B. Analysis of Dynamic Thresholds for the Normalized Difference Water Index. *Photogramm. Eng. Remote Sens.* **2013**, *75*, 1307–1317. [CrossRef]

91. Amitrano, D.; di Martino, G.; Iodice, A.; Mitidieri, F.; Papa, M.N.; Riccio, D.; Ruello, G. Sentinel-1 for monitoring reservoirs: A performance analysis. *Remote Sens.* **2014**, *6*, 10676–10693. [CrossRef]

92. Du, Y.; Zhang, Y.; Ling, F.; Wang, Q.; Li, W.; Li, X. Water bodies' mapping from Sentinel-2 imagery with Modified Normalized Difference Water Index at 10-m spatial resolution produced by sharpening the swir band. *Remote Sens.* **2016**, *8*, 354. [CrossRef]

93. Rucci, A.; Ferretti, A.; Monti Guarnieri, A.; Rocca, F. Sentinel 1 SAR interferometry applications: The outlook for sub millimeter measurements. *Remote Sens. Environ.* **2012**, *120*, 156–163. [CrossRef]

94. Eilander, D.; Annor, F.O.; Iannini, L.; van de Giesen, N. Remotely sensed monitoring of small reservoir dynamics: A Bayesian approach. *Remote Sens.* **2014**, *6*, 1191–1210. [CrossRef]

95. Sippel, S.J.; Hamilton, S.K.; Melack, J.M.; Novo, E.M.M. Passive microwave observations of inundation area and the area/stage relation in the Amazon River floodplain. *Int. J. Remote Sens.* **1998**, *19*, 3055–3074. [CrossRef]

2

On the Contribution of Satellite Altimetry-Derived Water Surface Elevation to Hydrodynamic Model Calibration in the Han River

Youjiang Shen [1,2], Dedi Liu [1,2,*], Liguang Jiang [3]🆔, Jiabo Yin [1,2], Karina Nielsen [4]🆔, Peter Bauer-Gottwein [3]🆔, Shenglian Guo [1]🆔 and Jun Wang [1]

1   State Key Laboratory of Water Resources and Hydropower Engineering Science, Wuhan University, Wuhan 430072, China; yjshen@whu.edu.cn (Y.S.); jboyn@whu.edu.cn (J.Y.); slguo@whu.edu.cn (S.G.); wangjwd@whu.edu.cn (J.W.)
2   Hubei Province Key Lab of Water System Science for Sponge City Construction, Wuhan University, Wuhan 430072, China
3   Department of Environmental Engineering, Technical University of Denmark, 2800 Kgs. Lyngby, Denmark; ljia@env.dtu.dk (L.J.); pbau@env.dtu.dk (P.B.-G.)
4   DTU Space, National Space Institute, Technical University of Denmark, 2800 Kgs. Lyngby, Denmark; karni@space.dtu.dk
*   Correspondence: dediliu@whu.edu.cn

Abstract: Satellite altimetry can fill the spatial gaps of in-situ gauging networks especially in poorly gauged regions. Although at a generally low temporal resolution, satellite altimetry has been successfully used for water surface elevation (WSE) estimation and hydrodynamic modeling. This study aims to investigate the contribution of WSE from both short-repeat and geodetic altimetry to hydrodynamic model calibration, and also explore the contribution of the new Sentinel-3 mission. Two types of data sources (i.e., in-situ and satellite altimetry) are investigated together with two roughness cases (i.e., spatially variable and uniform roughness) for calibration of a hydrodynamic model (DHI MIKE 11) with available bathymetry. A 150 km long reach of Han River in China with rich altimetry and in-situ gauging data is selected as a case study. Results show that the performances of the model calibrated by satellite altimetry-derived datasets are acceptable in terms of Root Mean Square Error (RMSE) of simulated WSE. Sentinel-3A can support hydrodynamic model calibration even though it has a relatively low temporal resolution (27-day repeat cycle). The CryoSat-2 data with a higher spatial resolution (7.5 km at the Equator) are proved to be more valuable than the Sentinel-3A altimetry data with a low spatial resolution (104 km at the Equator) for hydrodynamic model calibration in terms of RMSE values of 0.16 and 0.18 m, respectively. Moreover, the spatially variable roughness can also improve the model performance compared to the uniform roughness case, with decreasing RMSE values by 2–14%. Our finding shows the value of satellite altimetry-derived datasets for hydrodynamic model calibration and therefore supports flood risk assessment and water resources management.

Keywords: satellite altimetry; hydrodynamic model; Sentinel-3A; CryoSat-2; roughness parameters

1. Introduction

Climate change and human activities have altered the river flow regimes and led to more frequent and severe extreme natural disasters, such as floods and droughts [1,2]. Monitoring hydraulic variables and relevant hydrological-hydraulic processes of rivers usually play an essential role in flood risk assessment and water resources management [3,4]. One of the key hydraulic variables is the water surface elevation (WSE), which refers to the height of surface water above a given datum (i.e., 1985 National Height Datum of China for in-situ records, and EGM2008 datum for satellite WSE) [5]. Reliable and timely estimation of WSE is important for flood and water resources management tasks [6,7]. Traditionally, WSE measurements relied on in-situ gauging stations and often suffered from insufficient spatial coverage and data access constraints [8,9]. Hydraulic infrastructure and engineering projects are often designed in data sparsely or ungauged river reaches, which hampers the stakeholders to make decisions about river system management [10]. Hence, it is difficult for engineers to find an efficient way for WSE estimation along the entire river reach. Currently, a practical way to achieve this goal is the comprehensive estimation of WSE and other hydraulic variables on a whole river reach by using hydrodynamic models [4,11].

A hydrodynamic model is an efficient tool for flood forecasting and water resources management, as it can simulate the spatio-temporal dynamics of hydraulic variables given the boundary and initial conditions [12,13]. Observed data including WSE is often employed to calibrate the parameters of the hydrodynamic model and thus influences the model performance of simulated hydraulic variables [14]. However, observations from in-situ gauging stations are often sparse, and model parameters (e.g., spatially variable roughness parameters) probably cannot be estimated well from in-situ data through a calibration procedure [15]. The hydrodynamic model application is constrained in flood and water resources management, especially in ungauged or poorly gauged basins [16–18].

With the rapid advances in remote sensing sciences and technologies, space-borne sensors have provided valuable estimates of hydrological-hydraulic variables such as river extents and WSE [19–22]. WSE can be freely obtained from satellite altimetry missions (e.g., Envisat and Jason-2/-3), while river extents can be derived from optical or SAR (synthetic aperture radar) imagery satellite missions (e.g., Landsat and Sentinel-1/-2) [23,24]. These data from a satellite mission called satellite observations have been taken as a key supplement to in-situ data for hydrodynamic modeling. There have been many studies focusing on integrating satellite observations in facilitating hydrodynamic modeling. For example, satellite imagery observations have been adopted to calibrate two-dimensional (2-D) hydrodynamic models [25,26], to validate the efficiency of hydrodynamic models [27,28], and even integrated into model structures for data assimilation [29]. Although promising results have been shown by integrating satellite imagery observations in hydrodynamic modeling, there are also a lot of challenges. For example, optical imagery data are easily affected by cloud conditions while SAR imagery data always suffer the low spatial resolution [30], inundation extents are hard to be interpreted due to the noise of the backscatter and are also unavailable during the flood period.

WSE characterizes the surface water variations [31]. Satellite altimetry-derived WSE has been widely used to calibrate hydrodynamic models as an alternative to in-situ records, especially in ungauged areas [32,33]. For example, short-repeat missions such as Envisat and ERS-2 have been adopted in calibrating a quasi-two-dimensional hydrodynamic model in Italy [30]. Recently, CryoSat-2 was used to calibrate a one-dimensional hydrodynamic model in the Brahmaputra River [34].

Hydrodynamic models can benefit from satellite altimetry data in model calibration, improving the skills of simulating hydraulic variables such as WSE and discharge. However, the contributions of various altimetry-derived WSE datasets to hydrodynamic model calibration are different. The data-quality and sampling density are two potential pivotal factors influencing the contribution of satellite WSE on hydrodynamic model calibration. Previous works have revealed the importance of altimetry spatial sampling density to hydrodynamic model calibration through providing a high number of virtual river stage gauging stations from satellite altimetry [35,36], exploring the value of CryoSat-2 altimetry with dense sampling patterns [34], or applying various altimetry-derived datasets to hydrodynamic model calibration [37]. However, few studies have simultaneously evaluated the contributions of WSE from both short-repeat and geodetic altimetry missions on hydrodynamic model calibration, and also little is known about the contribution of the new Sentinel-3 mission [38].

Short-repeat missions such as Sentinel-3 have a repeat cycle of 10 to 35 days with a relatively sparse ground track pattern, while geodetic missions such as CryoSat-2 have dense point observations of the river longitudinal profile due to their drifting ground track pattern [39,40]. As a new generation satellite altimetry mission with a 27-day repeat cycle, Sentinel-3 is likely to deliver high-quality WSE observation for its new open-loop tracking and SAR modes [38]. The CryoSat-2 was launched in July 2010 to an orbit with a 369-day full repeat cycle and dense ground tracks [41]. Sentinel-3 is operating in a constellation of two-satellites (Sentinel-3A and Sentinel-3B). The Sentinel-3A (S3A) and Sentinel-3B (S3B) were launched in February 2016 and April 2018, respectively.

This study aims to investigate the contribution of altimetry-derived WSE from short-repeat and geodetic missions (Sentinel-3A and CryoSat-2) to hydrodynamic model calibration. The 150 km long reach of the Han River with rich data in China was selected as a case study area. Model calibration and validation are carried out under fourteen scenarios by combining different data sources (i.e., in-situ and satellite altimetry) and two roughness cases (i.e., spatially variable and uniform roughness). Therefore, the results of our study will help to implement satellite altimetry-derived WSE into hydrodynamic model calibration and therefore support flood forecasting and water resources management.

2. Materials and Methods

2.1. Study Area

The Han River offers a large amount of biological and hydropower resources and serves as one of the major water sources for the South-to-North Water Diversion in China [42]. Concerning the availability of in-situ hydrometric and satellite altimetry data, a 150 km long reach of Han River in China (Figure 1) is selected to explore the contributions of satellite altimetry-derived datasets on hydrodynamic model calibration. The case area belongs to the middle stream of the Han River between Huangjiagang and Yicheng hydrometric stations. Three main tributaries are located in the study area, and two hydraulic projects, the Wangbuzhou and Cuijiaying reservoirs, are also located with the functions of flood control and hydropower production (Figure 1).

2.2. Data Description

2.2.1. Hydrometric Data

Daily discharge of four hydrometric stations and daily water level of five hydrometric stations from 01/01/2016 to 12/31/2018 are collected from the Bureau of Hydrology, Changjiang Water Resources Commission, affiliated to the Ministry of Water Resources of China (Table 1). The main-channel geometry of the reach is described by the 94 surveyed cross-section profiles, which are surveyed by the Bureau of Hydrology in 2016. The information on the two hydraulic projects can be found in Table 1 and Figure 2. The operational rules applied for the two projects are dependent on their water levels and water inflows. Specifically, the outflow equals the inflow when the inflow is lower than the maximum allowable release under the normal water level, otherwise, the outflow equals the maximum allowable release under its corresponding water level (Figure 2c,d).

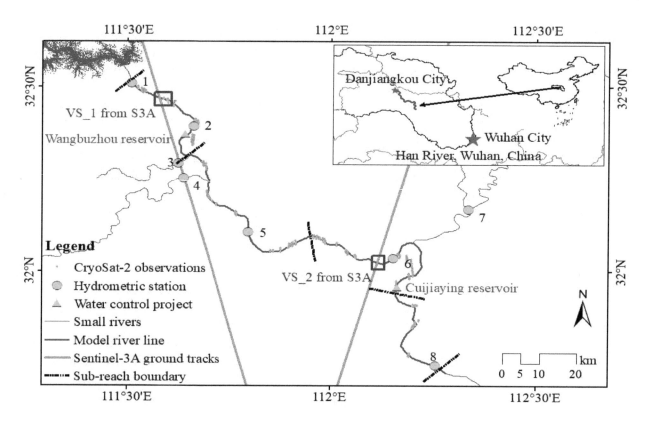

Figure 1. Map of the study site. The study reach (from Huangjiagang to Yicheng stations) belongs to the middle stream of the Han River, in China. Four sub-reaches based on the distribution of hydrometric stations and river morphology are established for a spatially variable roughness scheme. Numbers indicate the hydrometric stations: 1—Huangjiagang, 2—Laohekou, 3—Bei, 4—Gucheng, 5—Miaogang, 6—Xiangyang, 7—Dongcheng, 8—Yicheng.

Table 1. Summary of the hydrometric stations. Q denotes discharge; W denotes water level.

ID	Station Name	Chainage (m)	River Width (m)	Data Type
1	Huangjiagang (HJG)	0	438	Q/W
2	Laohekou (LHK)	20,655	441	W
3	Bei (B)	31,196	101	Q
4	Gucheng (GC)	37,867	182	Q
5	Miaogang (MG)	60,599	280	W
6	Xiangyang (XY)	98,950	534	W
7	Dongcheng (DC)	107,667	351	Q
8	Yicheng (YC)	150,379	650	W
9	Wangbuzhou (WBZ)	28,763	1351	Water control
10	Cuijiaying (CJY)	113,819	1213	project

Note: Chainage refers to the distance along the river centerline between points of interest. River width indicates the bank's full river width from surveyed cross-section river profiles.

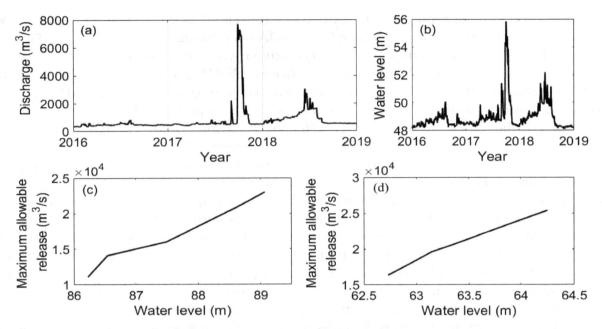

Figure 2. Flow regimes of the case study river. (**a**) Daily discharge at Huangjiagang station from 2016 to 2018. (**b**) Daily mean water level at Yicheng station from 2016 to 2018. (**c**) Maximum allowable release-water level relationship applied for Wangbuzhou water control project. (**d**) Maximum allowable release-water level relationship applied for Cuijiaying water control project.

2.2.2. Satellite Altimetry Data

The altimetry-derived WSE are retrieved from Sentinel-3A and CryoSat-2. Sentinel-3A data are collected from the ESA level 2 product "Enhanced measurements", which are downloaded from the Copernicus Open Access Hub (https://scihub.copernicus.eu/dhus/). The ESA level 2 product contains 20Hz measurements including the waveform, tracker range, satellite altitude, and corrections. The waveforms are retracked with a traditional offset center of gravity retracker [7]. For CryoSat-2, we use the ESA level 1b product from baseline C (https://science-pds.cryosat.esa.int/). The ESA level 1b product contains 20Hz measurements including the waveform, tracker range, satellite altitude, and corrections. The CryoSat-2 waveforms were retracked with a narrow primary peak threshold retracker [43].

The water surface elevations WSE, are constructed via the following equation

$$WSE = h - R - N \tag{1}$$

where h is the altitude of the satellite, and N is the geoid height, which is the EGM2008 geoid model [44]. R is the range that is the distance from the satellite to the surface, which can be expressed as

$$R = R_{trac} + R_{retrac} + R_{atm} + R_{geo} \tag{2}$$

where R_{trac} is the distance measured by the onboard tracker, hence the distance to the nominal bin in the waveform and R_{retrac} is the re-tracking correction. R_{atm} is the atmospheric corrections including the wet and dry tropospheric corrections and the ionospheric correction. R_{geo} is the geophysical corrections including solid earth tide, pole tide, and ocean loading tide. All corrections are taken from the altimetry products.

2.3. WSE Data Processing

The altimetry-derived WSE are extracted by a water mask from Global Surface Water Explorer (https://global-surface-water.appspot.com/). Specifically, the maximum extent is used to pre-select as many observations as possible [45].

The identical satellite ground tracks with a repeat cycle of 27 days allow to derive the WSE time series at two virtual stations (VSs) from Sentinel-3A over the study reach (Figure 1). Firstly, the outliers are discarded by comparing the altimetry-derived WSE with ASTER GDEM v.3 (Advanced Spaceborne Thermal Emission and Reflection Radiometer Global Digital Elevation Model) [46] with an absolute bias threshold of 20 m. The DEM datasets have been reprojected to the EGM2008 geoid at the platform of VDatum software [47]. Secondly, a filtering process is performed for each track to discard outliers based on the work by Jiang et al. [40] and finally, the median value of all WSE observations for a given track is used to derive WSE time series. The data processing for CryoSat-2 altimetry is slightly different from that of Sentinel-3A due to its drifting ground track pattern. A series of virtual stations were selected and all observations were relocated to the nearest virtual stations based on their local slopes [43]. The virtual stations are determined based on the coverages of CryoSat-2 observations along the river reach and placed on the model river line (i.e., the centerline of the river from the extracted water mask). The procedure of relocating CryoSat-2 observations onto virtual stations has been proved to be an efficient way of constructing WSE time series at VSs and strengthening the hydrologic and hydrodynamic applications of CryoSat-2 [43]. Balancing the maximization of the VS numbers and the minimization of error from projecting observations to virtual stations, 16 VSs have been figured out in the study reach (Table 2). An outlier filtering algorithm is performed to obtain the median values of each track: a measurement is discarded if $|h-u|>3\delta$ (h is the point observation, u is the mean value and δ is the standard deviation) [40]. All the median observations are relocated to the nearest virtual stations through local slope correction. It should be noted that there are often systematic biases between satellite altimetry-derived observations and in-situ data due to the unknown local vertical datums and other factors. If their mean values of every series are removed, both in-situ data and satellite altimetry-derived datasets are transformed to WSE anomalies.

Table 2. Characteristics of each virtual station from the Sentinel-3A and CryoSat-2 missions.

ID	Lon (degree)	Lat (degree)	Chainage (m)	Width (m)	Number
\multicolumn{6}{c}{VS Platform: CryoSat-2}
1	111.539	32.485	3681	1060	3
2	111.605	32.459	114,59	1494	3
3	111.666	32.401	193,53	1351	5
4	111.688	32.286	32,570	575	3
5	111.704	32.191	42,175	604	3
6	111.768	32.155	51,246	348	3
7	111.905	32.069	69,674	563	5
8	111.981	32.086	80,502	862	4
9	112.053	32.042	87,091	749	2
10	112.18	32.037	92,876	1290	3
11	112.194	31.987	104,022	1068	3
12	112.16	31.955	113,369	1206	1
13	112.205	31.907	121,534	776	3
14	112.208	31.857	129,284	434	2
15	112.192	31.772	140,443	403	2
16	112.241	31.750	146,961	571	4
\multicolumn{6}{c}{VS Platform: Sentinel-3A}
1	111.579	32.467	7654	775	36
2	112.109	32.025	94,622	1253	38

Note: VS indicates the virtual stations. Width indicates the bank full river width from surveyed cross-section river profiles. The number indicates the number of processed median observations at each virtual station applied for hydrodynamic modeling.

2.4. Hydrodynamic Model in the Study Area

A one-dimension (1D) hydrodynamic model is used to simulate flow regimes and water levels of the study reach. The model was running at the DHI MIKE 11 software [48]. The software solves the

Saint–Venant equations with an implicit 6-point finite-difference scheme. The model is directly forced by the observed daily discharge data from the main upstream and three tributaries, while the daily water level of Yicheng hydrometric station is served as the downstream boundary condition (shown in Figure 1). The study reach is sketched by 94 cross-section profiles, and the profiles are derived from the detailed bathymetric map with an interval of around 50 m and thus can capture the variations of riverbed geometry. The operational rules of Wangbuzhou and Cuijiaying reservoirs are also simulated in the model as it is described in Section 2.2.

2.5. Hydrodynamic Model Calibration

The hydrodynamic model calibration focuses on the estimation of the channel roughness, i.e., the Strickler coefficient Ks (the reciprocal of the Manning coefficient, see in Manning's Equation (3)).

$$Q = Ar^{\frac{2}{3}}S^{\frac{1}{2}}K_s \tag{3}$$

With given bathymetry data, the cross-section area (A), the hydraulic radius (r), the slope of the hydraulic gradient line (S) at the discharge (Q) can be described as a function of water depth relating to WSE. Only Ks is a parameter that varies along the river and is often calibrated through maximizing the degree of goodness of fit between observed and simulated WSE. To figure out the effect of heterogeneity of roughness distribution on model performance, two roughness cases are assumed in the calibration process. One is that only a single Ks coefficient is assigned to the whole reach (uniform Ks). The other one is that the river reach is partitioned into sub-reaches according to the hydro-morphological characteristics of the river and the distribution of hydrometric stations (variable Ks). The variable Ks with sub-reaches has also been proved to be a feasible approach to improve the hydrodynamic model performance [34]. Specifically, four sub-reaches are used in our case study for the variable Ks. The first 29 km long reach is very gentle (0.035 m/km) and the following 44 km long reach is narrow and steep with a complex river morphology. In the following 38 km long reach, the river channel is steeper with a gradient of 0.125 m/km. The last reach is a little wider and deeper.

To investigate the efficiency and reliability of the various satellite altimetry-derived datasets, three configurations are constructed for the hydrodynamic model calibration. The first one is from the in-situ data (called Configuration A), the second one is from the altimetry-derived data from Sentinel-3A and CryoSat-2 (called Configuration B) and the last one is from the combination of in-situ and satellite altimetry-derived data (called Configuration C). Combining the three configurations and two roughness cases, there are fourteen scenarios for calibrating the hydrodynamic model as shown in Table 3.

Table 3. Summary of hydrodynamic model calibration.

Scenarios Design		Using In-Situ	Using Sentinel-3A	Using CryoSat-2
Configuration A: using in-situ data (two schemes)				
AU1	AV1	Y	-	-
Configuration B: using satellite altimetry data (six schemes)				
BU1	BV1	-	Y	-
BU2	BV2	-	-	Y
BU3	BV3	-	Y	Y
Configuration C: using both in-situ and satellite altimetry data (six schemes)				
CU1	CV1	Y	Y	-
CU2	CV2	Y	-	Y
CU3	CV3	Y	Y	Y

Note: Y indicates objective function. U/V indicate the uniform and variable roughness cases, respectively. 1, 2, 3 indicate the number of data sources in the corresponding configuration A, B, and C.

The 1D hydrodynamic model is calibrated by the fourteen scenarios from 01/01/2016 to 12/31/2017 through minimizing the Root Mean Square Error (RMSE) between simulated and observed WSE. The model is also validated by the data from 01/01/2018 to 12/31/2018. The manual trial-and-error method is adopted for testing the variable Ks case and the scenario-based method is used (running models with a range of values of Ks at an interval of 0.5/0.1 $m^{1/3}s^{-1}$) for the uniform Ks case [33,49].

$$RMSE = \sqrt{\frac{1}{n}\sum_{i=1}^{n}(\hat{p}_i - o_i)^2} \qquad (4)$$

where O_i and $\hat{p}_i$ are the observed and simulated WSE at the ith point, respectively. n is the total number of WSE observations.

3. Results

3.1. Calibrated Strickler Coefficient Ks

The calibrated uniform Ks plays a vital role in hydrodynamic simulation, and the optimal Ks value can be easily determined from the clear patterns shown in Figure 3a (e.g., 34.6 $m^{1/3}$/s as the optimal value for the scheme AU1). The model performance is sensitive to uniform Ks value (RMSE ranging from 0.105–0.450 m) and the optimal RMSE values are assigned near the high bound of the parameter range. The calibrated Ks values of four sub-reaches reflect the characteristics of each sub-reach (Figure 3b). The Ks values of the second and the final sub-reaches range from 27–33 $m^{1/3}$/s. The Ks is relatively high at the first upstream 29-km reach (approximately 37 $m^{1/3}$/s) due to the flat and straight river channel. On the contrary, the Ks value has decreased significantly in the third reach due to the increasing water depth and highly vegetated channel. Roughness is closely related to flow resistance, which is positively influenced by river depth and vegetable conditions. More specifically, flow resistance increases with a highly vegetated channel and river depth, and leads to higher roughness (smaller Ks) [49,50]. It is interesting to find that the RMSE values jump near the Strickler coefficient close to 30 in Figure 3a. As there are discharge bursts in 2017 that can challenge the calibration of parameters (Figure 2a), the RMSE value is sensitive to the bias of peak simulation. The peaks of discharge should be processed separately and the model parameters (Ks) should be chosen in a special way. There is not enough available data (in-situ records and satellite observations) to separately calibrate the parameters. In addition, all the jumping curves are from the uniform Ks scenarios, which proves that the uniform parameter can bring more uncertainties of discharge simulation and cut down the contribution of the satellite altimetry-derived WSE to hydrodynamic model calibration at the whole river reach.

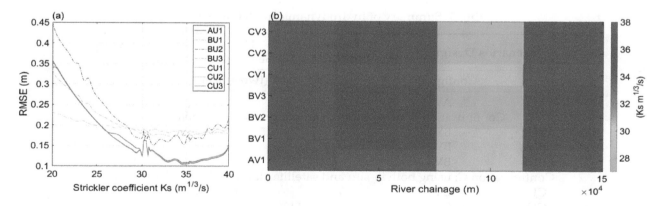

Figure 3. Calibration results of Strickler coefficients Ks for all schemes. (**a**) The sensitivity of the single Strickler coefficient Ks on model performance in terms of the root mean square error (RMSE). (**b**) The optimum values of variable Strickler coefficients Ks for different schemes.

3.2. Hydrodynamic Model Performances with Different Configurations

3.2.1. Model Calibration with the in-situ Observations

Regarding configuration A, hydrodynamic models are calibrated by the in-situ observations using the uniform and variable Ks (i.e., the schemes AU1 and AV1). The results during the calibration and validation periods reveal that the model can reproduce WSE well, where the RMSE ranging from 0.100–0.216 m (Table 4). However, there are few obvious differences in the accuracies of WSE simulation from the two roughness cases. The averaged performances from the variable Ks are slightly better than those from the uniform Ks during the calibration periods while the results are not as good as those from the uniform Ks during the validation periods. Nevertheless, the improved results from the variable Ks can be found with lower RMSE values at the specific station (Figures 4 and 5). For example, the RMSE from the variable Ks is 0.074 m at Xiangyang station, decreased by 11.5% compared to that from the uniform Ks (Figure 5). Overall, the improvements with the variable Ks highlight the potential superiority of the variable Ks in hydrodynamic model calibration.

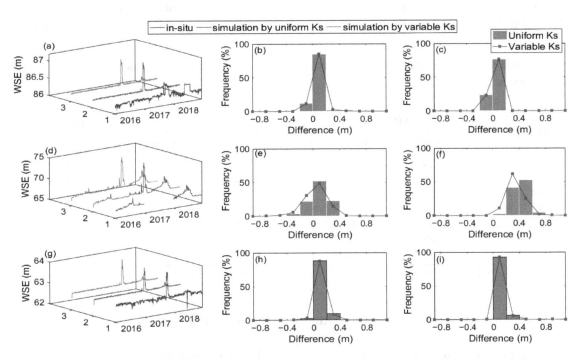

Figure 4. Validation of water surface elevation (WSE) obtained by configuration A against three in-situ gauging stations. Panel 1-3 indicate results at Laohekou, Miaogang, Xiangyang stations, respectively. (**a,d,g**) Comparisons of simulated and in-situ observations during the calibration and validation periods. (**b,e,h**) The distribution of differences between simulated and in-situ observed WSE during the calibration periods. (**c,f,i**) The distribution of differences between simulated and in-situ observed WSE during the validation periods.

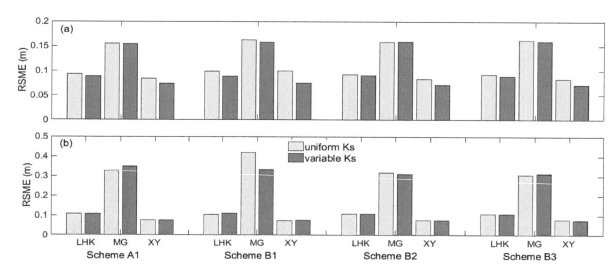

Figure 5. Statistical metrics of model performance in terms of RMSE at three in-situ stations for configuration A and B. (**a,b**) indicate results during the calibration and validation periods, respectively.

Table 4. Summary of statistical metrics of average model performance in terms of RMSE at all stations of the schemes.

Schemes	RMSE (m)		Schemes	RMSE (m)	
	Calibration	Validation		Calibration	Validation
AU1	0.105	0.204	AV1	0.100	0.216
BU1	0.179	0.146	BV1	0.177	0.148
BU2	0.154	0.313	BV2	0.153	0.316
BU3	0.172	0.227	BV3	0.168	0.234
CU1	0.108	0.202	CV1	0.104	0.210
CU2	0.106	0.206	CV2	0.102	0.204
CU3	0.109	0.205	CV3	0.105	0.205

Note: Details of calibration and validation conditions can be found in Table 3.

3.2.2. Model Calibration with the Satellite Altimetry-derived Observations

Concerning configuration B, the model is calibrated by the satellite altimetry-derived WSE. There are six model scenarios, i.e., calibration by Sentinel-3A, CryoSat-2, and the combinations of these two datasets taking the uniform and variable Ks cases (see the details in Table 3). The calibrated models show the comparable performance in comparison to those by using in-situ data in terms of RMSE values during both calibration and validation periods where the RMSE ranges 0.153–0.179 m in Table 4. Thus, the altimetry-derived datasets from Sentinel-3A and CryoSat-2 can be applied in the hydrodynamic model calibration. However, the contributions of WSE from different altimetry to hydrodynamic model calibration are different. The schemes BU(V)2 (against CryoSat-2) show the lowest values of RMSE while the schemes BU(V)1 (against Sentinel-3A) perform worst during the calibration periods. Interestingly, the model calibrated by the schemes BU(V)2 shows the best performance in the calibration period while it performs worst in the validation period probably due to their different sample size during the calibration and validation periods. The simulated WSE at virtual stations from the variable Ks consistently shows a better performance than those from the uniform Ks. The distributions of difference between model simulations and satellite altimetry-derived WSE have also been shown in Figures 6 and 7, which further proves the advantages of the variable Ks.

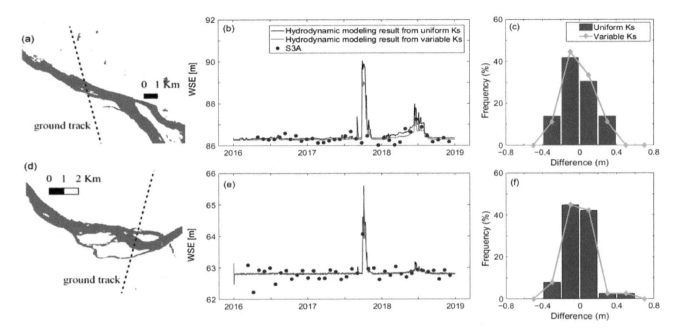

Figure 6. Validation of WSE obtained by configuration B against two virtual stations from Sentinel-3A. (**a**,**d**) Location of the two virtual stations. (**b**,**e**) Comparisons of simulated and satellite observed WSE during the calibration and validation periods. (**c**,**f**) The distribution of differences between simulated and satellite observed WSE from 2016–2018.

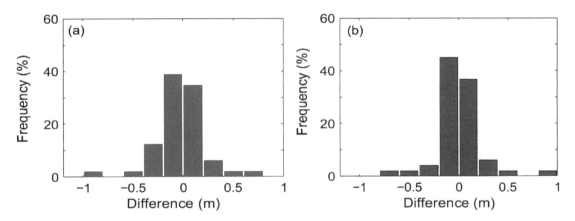

Figure 7. Validation of WSE obtained by configuration B against CryoSat-2 observations. (**a**) and (**b**) The distribution of differences between simulated and satellite WSE from 2016–2018 using uniform and variable roughness parameters, respectively.

To show the performances of hydrodynamic model calibrated by altimetry-derived datasets at three in-situ stations, their statistical metrics (RMSE) are shown in Figure 5 and their hydrographs are depicted in Figure 8. The models calibrated by satellite altimetry-derived datasets can also provide reliable WSE estimations at these in-situ stations. Moreover, the statistical metrics are different for every station due to their flow regimes and channel geometry. For example, WSE estimations at Miaogang station are not as good as those at other two in-situ stations with higher RMSE values (Figure 5). The model performance can also be dependent on the schemes with different satellite altimetry-derived datasets or roughness cases. The schemes BU(V)2 (against CryoSat-2) show better skills than the schemes BU(V)1 (against Sentinel-3A) or BU(V)3 (against all satellite altimetry-derived datasets) in simulating WSE at in-situ stations (Figure 5). For example, the RMSE values from the scheme BU2 at Miaogang station are 0.158/0.316 m for the calibration/validation periods, while the RMSE values from the schemes BU1 and BU3 are 0.162/0.420 and 0.162/0.306 m, respectively.

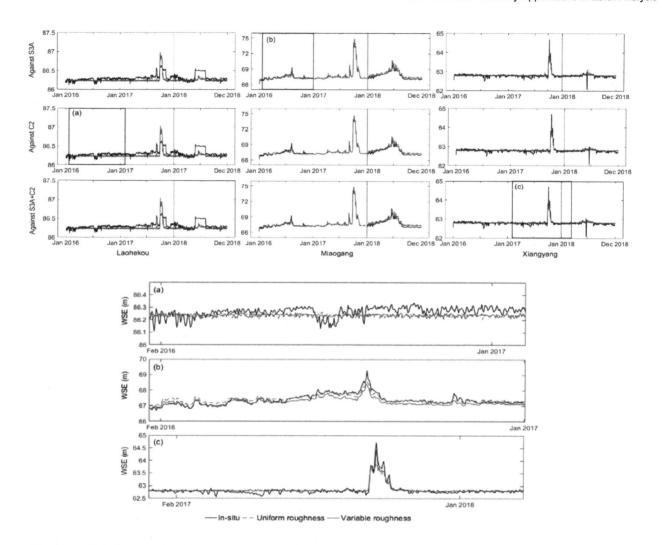

Figure 8. Validation of WSE obtained by configuration B against three in-situ gauging stations. (**a–c**) Zoom-in of the selected WSE simulations.

3.2.3. Model Calibration with Both In-Situ and Satellite Altimetry-Derived Observations

In the case of configuration C, the hydrodynamic model was calibrated by all available observations at in-situ and virtual stations. The contributions of additional satellite altimetry-derived WSE to hydrodynamic model calibration are relatively small at the evaluated in-situ and virtual stations (Table 4). This can be attributed to the fact that the temporal resolution of the available in-situ observations (daily frequency) is higher than that of satellite altimetry-derived datasets (e.g., 27-day frequency for S3A observations). Despite the little contribution, the advantages of additional satellite altimetry-derived datasets to the hydrodynamic model can still be found through the reproduction of the flood event (i.e., the flow processes with the maximum discharge at the upper boundary) (Figure 9). The WSE obtained from the model by calibration using all available observations is different from that obtained by calibration using only in-situ observations in the region where the satellite altimetry-derived observations can be found (shown in Figure 9d). With more spatially-distributed WSE observations, the hydrodynamic model provides more reliable WSE estimations on the whole river reach. The integration of altimetry-derived datasets and in-situ observations has led to the best performance for the hydrodynamic model calibration. Therefore, the satellite altimetry-derived datasets can be taken as a significant complementary to in-situ gauging networks, especially for improving the reliable simulation of WSE at the whole river reach scale.

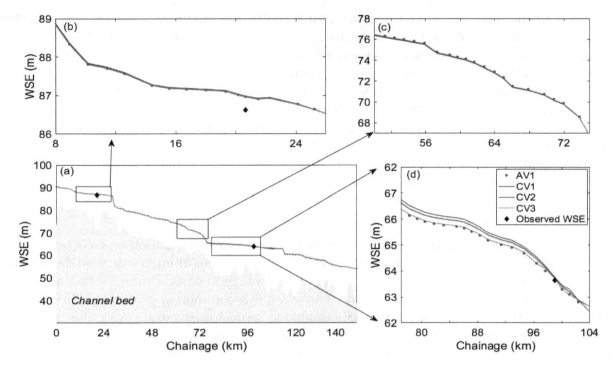

Figure 9. WSE profiles along the river reach obtained by the configuration C. (**a**) The WSE profiles at the flood event. (**b–d**) zoom-in of (**a**) focusing on the three different reaches.

4. Discussion

4.1. Effects of Satellite Altimetry-Derived WSE on Hydrodynamic Model Calibration

Two pivotal factors are influencing the performance of the hydrodynamic model calibrated by satellite altimetry-derived WSE. One is the accuracy of satellite data itself and the other one is the spatio-temporal resolution of satellite altimetry missions. The two VSs of Sentinel-3A (i.e., VS1 and VS2) are close to the Huangjiagang and Xiangyang hydrometric stations (eight km from VS1 to Huangjiagang, four km from VS2 to Xiangyang). The slopes between the two VSs and their corresponding hydrometric stations are too flat to impact the WSE. The WSE at the two VSs can replace the data observed at the Huangjiagang and Xiangyang stations. The RMSE values are 0.28 m and 0.16 m for the VS1 and VS2, respectively. In the same way, the accuracies of CryoSat-2 at VS3 and VS7 are also evaluated by the observation data from Lianghekou (one km from VS3) and Xiangyang (eight km from VS7). The RMSE values are 0.14 m and 0.71 m for the VS3 and VS7, respectively. According to the previous studies [7,51], the RMSE values for satellite altimetry below 0.3 m are taken as good, and below 0.6 m are moderate. Sentinel-3A and CryoSat-2 data are not only found to be good and moderate in the literature [7,34,43] but also have been proved to be good and moderate in our case study river, a medium river (around 400 m) with braided and single-threaded reaches. By discarding outliers based on DEM approaches, a good or moderate accuracy of altimetry observations can be obtained in the case study, indicating that the land relief issue can be well addressed by discarding invalid measurements (e.g., a careful analysis of satellite waveforms identification). Our findings suggest that a good or moderate accuracy of altimetry observations can provide satisfactory results for hydrodynamic model calibration.

The spatio-temporal resolution of satellite altimetry missions directly influences the performance of hydrodynamic model calibration. Our findings further revealed that CryoSat-2 geodetic altimetry with high spatial resolution shows better performance in hydrodynamic model calibration than Sentinel-3A altimetry with a low spatial resolution (Table 4 and Figures 3–5). It should be noted that CryoSat-2 geodetic altimetry has a full cycle of 369 days while Sentinel-3A has a 27-day cycle at the specified virtual stations. The values of the new Sentinel-3A altimetry to hydrodynamic model calibration are highlighted due to its high data-quality with a new on-board tracking mode and easier work in deriving WSE time series at virtual stations. But it is challenging to process the spatially-distributed CryoSat-2 observations and use them for hydrodynamic model calibration [34,39]. In our case study, the CryoSat-2 data are relocated onto a series of virtual stations and then transformed into anomalies for hydrodynamic model calibration [43]. The systematic bias still remains when processing such few CryoSat-2 observations into anomalies, and can be propagated into WSE simulation of the model and reduces its accuracy. This could be attributed to the fact that the results are not the best among the schemes of configuration B when pooling the Sentinel-3A and CryoSat-2 datasets for model calibration (shown in Table 4 and Figures 3–5). Further works can focus on producing blending products with more densified data from different satellite altimetry missions through the statistical and hydraulic blending approaches and use them for hydrodynamic modeling [52,53].

4.2. Comparison of Model Performance from Uniform and Variable Roughness Parameters

The effects of the two parameterizations on hydrodynamic model calibration have been reported in the literature [30,34], and are further explored on the use of various altimetry-derived datasets in our case study. The uniform *Ks* is widely used in hydrodynamic model calibration for its simplicity and promising efficiency at the evaluated stations. However, this common approach fails to characterize the variations of the riverbed and is probably unable to reproduce accurate WSE at other locations of the river under high heterogeneity [4,54]. The hydrodynamic model with uniform *Ks* calibrated by Sentinel-3A derived-WSE shows poor skills with high RMSE values at in-situ stations (Figure 5). Moreover, the uniform *Ks* can also be attributed to the equifinality issue, i.e., a model with different roughness sets performs equally well at the evaluated gauging stations (shown in Figure 3a). To find a more efficient approach for more accurate WSE estimation along the river under the heterogeneity of the channel, the variable *Ks* is recommended for hydrodynamic model calibration as it takes the spatial variability of flow resistance into account [34]. The model with the variable *Ks* can simulate WSE more accurately with lower RMSE values compared to that with the uniform *Ks*.

Two pivotal factors are influencing the hydrodynamic model calibration with the variable *Ks*. One is the availability of data for model calibration and the other is the density of resolved variable roughness parameters. With satellite altimetry-derived data available, the variable *Ks* could be an efficient approach to facilitate hydrodynamic modeling. The model performance is also closely dependent on the satellite altimetry datasets with different spatio-temporal resolution. The CryoSat-2 geodetic mission with dense observations is more helpful than those with a low spatial resolution (Sentinel-3A) in hydrodynamic model calibration (Table 4 and Figures 3–5). The contributions of the densified roughness parameters on model calibration have been highlighted in the literature [4,34], while the improvements of the densified roughness parameters come at the expense of uncertainties and computation time. The more parameters commonly required more data for calibration [4]. Schneider et al. [34] used the 10-km variable roughness parameters to facilitate hydrodynamic modeling and found that the performances of the model were not refined well with large uncertainties. This issue is further revealed in our case study that the accuracies of WSE from the variable *Ks* are slightly better than those from the uniform *Ks* during the calibration periods with reducing RMSE values by 2–14%, while the results are not as good as those from the uniform *Ks* during the validation periods. However, the advantages of the variable *Ks* are not only to improve the accuracy of simulated WSE at the evaluated gauging stations but also to improve the hydrodynamic simulation on the entire river reach via a physically-based strategy. With more satellite altimetry-derived WSE available, the variable Ks could be estimated well and therefore increases the performance of the hydrodynamic model [55]. As the calibrated Ks values are reasonable compared to previous works [34,56], the variable *Ks* is recommended for hydrodynamic model calibration.

5. Conclusions

The contributions of current satellite altimetry missions (i.e., short-repeat and geodetic altimetry) on hydrodynamic model calibration are reviewed. The contribution of the new Sentinel-3A altimetry is also assessed. Hydrodynamic model (DHI MIKE 11) calibrations with available bathymetry data are carried out by combining seven calibration datasets (i.e., in-situ, satellite altimetry, and both of them) and two roughness cases (i.e., spatially variable roughness and uniform roughness). In the case study river reach with a river width of around 400 m, both Sentinel-3A and CryoSat-2 can deliver useful observations with the averaged RMSE values of 0.22 and 0.49 m, respectively. Our findings suggest that a good or moderate accuracy of altimetry observations (RMSE below 0.6 m) can contribute to the hydrodynamic model calibration. Specifically, calibration against altimetry-derived WSE achieves promising results in terms of RMSE as 0.100–0.216 m during the calibration periods, which are in line with previous studies [30,34]. Moreover, model performance varies with different altimetry-derived datasets due to their different characteristics such as spatio-temporal sampling patterns.

Although the satellite altimetry has a low temporal resolution, both Sentinel-3A and CryoSat-2 can provide a comparable result as those from the daily in-situ observations. CryoSat-2 with a higher spatial resolution (7.5 km at the Equator) brings better performances than short-repeat Sentinel-3A altimetry with a low resolution (104 km at the Equator) in the hydrodynamic model calibration according to their RMSE values of 0.16 and 0.18 m, respectively. Moreover, the variable Ks improves the performance of hydrodynamic simulation compared to the uniform Ks reducing the RMSE values by 2–14%, and is recommended in hydrodynamic model calibration for its physically-based strategy. Our study not only enriches the existing knowledge on integrating satellite altimetry-derived datasets into hydrodynamic model calibration but also supports the future hydrology-related satellite altimetry design.

Author Contributions: Conceptualization, Y.S. and D.L.; methodology, Y.S. and D.L.; formal analysis, J.W. and S.G.; investigation, Y.S.; data curation, L.J., J.W., and K.N.; writing—original draft preparation, Y.S.; writing—review and editing, Y.S., L.J., J.Y., P.B.-G., and D.L.; visualization, Y.S.; supervision, D.L.; project administration, D.L.; funding acquisition, D.L. All authors have read and agreed to the published version of the manuscript.

References

1.	Mann, M.E.; Rahmstorf, S.; Kornhuber, K.; Steinman, B.A.; Miller, S.K.; Coumou, D. Influence of Anthropogenic climate change on planetary wave resource and extreme weather events. *Sci. Rep.* **2017**, *7*, 45242. [CrossRef]

2.	Yin, J.; Gentine, P.; Zhou, S.; Sullivan, S.C.; Wang, R.; Zhang, Y.; Guo, S. Large increase in global storm runoff extremes driven by climate and anthropogenic changes. *Nat. Commun.* **2018**, *9*, 4389. [CrossRef]

3.	Blari, P.; Buytaert, W. Socio-hydrological modelling: A review asking "why, what and how?". *Hydrol. Earth Syst. Sci.* **2016**, *20*, 443–478. [CrossRef]

4.	Jiang, L.; Bandini, F.; Smith, O.; Jensen, I.K.; Bauer-Gottwein, P. The value of distributed high-resolution UAV-Borne observations of water surface elevation for river management and hydrodynamic modeling. *Remote Sens.* **2020**, *12*, 1171. [CrossRef]

5.	Normandin, C.; Frappart, F.; Diepkilé, A.T.; Marieu, V.; Mougin, E.; Blarel, F.; Lubac, B.; Braquet, N.; Ba, A. Evaluation of the performance of radar altimetry missions from ERS-2 to Sentinel-3A over the Inner Niger Delta. *Remote Sens.* **2018**, *10*, 833. [CrossRef]

6.	Hossain, F.; Maswood, M.; Siddique-E-Akbor, A.H.; Yigzaw, W.; Mazumdar, L.C.; Ahmed, T.; Hossain, M.; Shah-Newaz, S.M.; Limaye, A.; Lee, H.; et al. A promising radar altimetry satellite system for operational flood forecasting in flood-prone Bangladesh. *IEEE Geosci. Remote Sens. Mag.* **2014**, *2*, 27–36. [CrossRef]

7.	Kittel, C.M.M.; Jiang, L.; To, C.; Bauer-Gottwein, P. Sentinel-3 radar altimetry for river monitoring—A catchment-scale evaluation of satellite water surface elevation from Sentinel-3A and Sentinel-3B. *Hydrol. Earth Syst. Sci. Discuss.* **2020**, *165*. [CrossRef]

8.	Biancamaria, S.; Hossain, F.; Lettenmaier, D.P. Forecasting transboundary river water elevations from space. *Geophys. Res. Lett.* **2011**, *38*, 1–5. [CrossRef]

9.	Da Silva, J.S.; Calmant, S.; Seyler, F.; Moreira, D.M.; Oliveira, D.; Monteiro, A. Radar altimetry aids managing gauge networks. *Water Resour. Manag.* **2014**, *28*, 587–603. [CrossRef]

10.	Tauro, F.; Selker, J.; van de Giesen, N.; Abrate, T.; Uijlenhoet, R.; Porfiri, M.; Manfreda, S.; Caylor, K.; Moramarco, T.; Benveniste, J.; et al. Measurements and observations in the XXI century (MOXXI): Innovation and multi-disciplinarily to sense the hydrological cycle. *Hydrol. Earth Syst. Sci.* **2018**, *63*, 169–196. [CrossRef]

11. Siddique-E-Akbor, A.H.M.; Hossain, F.; Lee, H.; Shum, C.K. Inter-comparison study of water level estimates derived from hydrodynamic-hydrologic model and satellite altimetry for a complex deltaic environment. *Remote Sens. Environ.* **2011**, *115*, 1522–1531. [CrossRef]

12. García-Pintado, J.; Neal, J.C.; Mason, D.C.; Dance, S.L.; Bates, P.D. Scheduling satellite-based SAR acquisition for sequential assimilation of water level observations into flood modelling. *J. Hydrol.* **2013**, *495*, 252–266. [CrossRef]

13. Tarpanelli, A.; Brocca, L.; Melone, F.; Moramarco, T. Hydraulic modelling calibration in small rivers by using coarse resolution synthetic aperture radar imagery. *Hydrol. Process.* **2013**, *27*, 1321–1330. [CrossRef]

14. de Moraes Frasson, R.P.; Wei, R.; Durand, M.; Minear, J.T.; Domeneghetti, A.; Schumann, G.; Williams, B.A.; Rodriguez, E.; Picamilh, C.; Lion, C.; et al. Automated river reach definition strategies: Applications for the surface water and ocean topography mission. *Water Resour. Res.* **2017**, *53*, 8164–8186. [CrossRef]

15. Hostache, R.; Matgen, P.; Schumann, G.; Puech, C.; Hoffmann, L.; Pfister, L. Water level estimation and reduction of hydraulic model calibration uncertainties using satellite SAR images of floods. *IEEE Trans. Geosci. Remote Sens.* **2009**, *47*, 431–441. [CrossRef]

16. Birkinshaw, S.J.; O'Donnell, G.M.; Moore, P.; Kilsby, C.G.; Fowler, H.J.; Berry, P.A.M. Using satellite altimetry data to augment flow estimation techniques on the Mekong River. *Hydrol. Process.* **2010**, *24*, 3811–3825. [CrossRef]

17. Michailovsky, C.I.; Milzow, C.; Bauer-Gottwein, P. Assimilation of radar altimetry to a routing model of the Brahmaputra River. *Water Resour. Res.* **2013**, *49*, 4807–4816. [CrossRef]

18. Schneider, R.; Godiksen, P.N.; Villadsen, H.; Madsen, H.; Bauer-Gottwein, P. Application of CryoSat-2 altimetry data for river analysis and modelling. *Hydrol. Earth Syst. Sci.* **2017**, *21*, 751–764. [CrossRef]

19. Biancamaria, S.; Frappart, F.; Leleu, A.-S.; Marieu, V.; Blumstein, D.; Desjonquères, J.-D.; Boy, F.; Sottolichio, A.; Valle-Levinson, A. Satellite radar altimetry water elevations performance over a 200 m wide river: Evaluation over the Garonne River. *Adv. Space Res.* **2017**, *59*, 128–146. [CrossRef]

20. Birkinshaw, S.J.; Moore, P.; Kilsby, C.G.; O'Donnell, G.M.; Hardy, A.J.; Berry, P.A.M. Daily discharge estimation at ungauged river sites using remote sensing. *Hydrol. Process.* **2014**, *28*, 1043–1054. [CrossRef]

21. Sulistioadi, Y.B.; Tseng, K.H.; Shum, C.K.; Hidayat, H.; Sumaryono, M.; Suhardiman, A.; Setiawan, F.; Sunarso, S. Satellite radar altimetry for monitoring small rivers and lakes in Indonesia. *Hydrol. Earth Syst. Sci.* **2015**, *19*, 341–359. [CrossRef]

22. Ablain, M.; Meyssignac, B.; Zawadzki, L.; Jugier, R.; Ribes, A.; Spada, G.; Benveniste, J.; Cazenave, A.; Picot, N. Uncertainty in satellite estimates of global mean sea-level changes, trend and acceleration. *Earth Syst. Sci. Data* **2019**, *11*, 1189–1202. [CrossRef]

23. Goumehei, E.; Tolpekin, V.; Stein, A.; Yan, W. Surface water body detection in polarimetric SAR data using contextual complex Wishart classification. *Water Resour. Res.* **2019**, *55*, 7047–7059. [CrossRef]

24. Weekley, D.; Li, X. Tracking multidecadal lake water dynamics with Landsat imagery and topography/bathymetry. *Water Resour. Res.* **2019**, *55*, 8350–8367. [CrossRef]

25. Haque, M.M.; Seidou, O.; Mohammadian, A.; Djibo, A.G. Development of a time-varying MODIS/2D hydrodynamic model relationship between water levels and flooded areas in the Inner Niger Delta, Mali, West Africa. *J. Hydrol. Reg. Stud.* **2020**, *30*, 100703. [CrossRef]

26. Mtamba, J.; Van der Velde, R.; Ndomba, P.; Zoltán, V.; Mtalo, F. Use of Radarsat-2 and Landsat TM images for spatial parameterization of Manning's roughness coefficient in hydraulic modeling. *Remote Sens.* **2015**, *7*, 836–864. [CrossRef]

27. Paiva, R.C.D.; Collischonn, W.; Buarque, D.C. Validation of a full hydrodynamic model for large-scale hydrologic modelling in the Amazon. *Hydrol. Process.* **2011**, *27*, 333–346. [CrossRef]

28. Chen, H.; Liang, Z.; Liu, Y.; Liang, Q.; Xie, S. Integrated remote sensing imagery and two-dimensional hydraulic modeling approach for impact evaluation of flood on crop yields. *J. Hydrol.* **2017**, *553*, 262–275. [CrossRef]

29. Andreadis, K.M. Data assimilation and river hydrodynamic modeling over large scales. In *Global Flood Hazard: Applications in Modeling, Mapping, and Forecasting*; American Geophysical Union: Washington, DC, USA, 2018. [CrossRef]

30. Domeneghetti, A.; Tarpanelli, A.; Brocca, L.; Barbetta, S.; Moramarco, T.; Castellarin, A.; Brath, A. The use of remote sensing-derived water surface data for hydraulic model calibration. *Remote Sens. Environ.* **2014**, *149*, 130–141. [CrossRef]

31. Dettmering, D.; Ellenbeck, L.; Scherer, D.; Schwatke, C.; Niemann, C. Potential and limitations of satellite altimetry constellations for monitoring surface water storage changes—A case study in the Mississippi Basin. *Remote Sens.* **2020**, *12*, 3320. [CrossRef]

32. Biancamaria, S.; Lettenmaier, D.P.; Pavelsky, T.M. The SWOT mission and its capabilities for land hydrology. *Surv. Geophys.* **2016**, *37*, 307–337. [CrossRef]

33. Garambois, P.A.; Calmant, S.; Roux, H.; Paris, A.; Monnier, J.; Finaud-Guyot, P.; Montazem, A.S.; da Silva, J.S. Hydraulic visibility: Using satellite altimetry to parameterize a hydraulic model of an ungauged reach of a braided river. *Hydrol. Process.* **2017**, *31*, 756–767. [CrossRef]

34. Schneider, R.; Tarpanelli, A.; Nielsen, K.; Madsen, H.; Bauer-Gottwein, P. Evaluation of multi-mode CryoSat-2 altimetry data over the Po River against in situ data and a hydrodynamic model. *Adv. Water Resour.* **2018**, *112*, 17–26. [CrossRef]

35. Liu, G.; Schwartz, F.W.; Tseng, K.-H.; Shum, C.K. Discharge and water-depth estimates for ungauged rivers: Combining hydrologic, hydraulic and inverse modeling with stage and water-area measurements from satellites. *Water Resour. Res.* **2015**, *51*, 6017–6035. [CrossRef]

36. Getirana, A.C.V.; Peters-Lidard, C. Estimating water discharge from large radar altimetry datasets. *Hydrol. Earth Syst. Sci.* **2013**, *17*, 923–933. [CrossRef]

37. Jiang, L.; Madsen, H.; Bauer-Gottwein, P. Simultaneous calibration of multiple hydrodynamic model parameters using satellite altimetry observations of water surface elevation in the Songhua River. *Remote Sens. Environ.* **2019**, *225*, 229–247. [CrossRef]

38. Donlon, C.; Berruti, B.; Buongiorno, A.; Ferreira, M.-H.; Féménias, P.; Frerick, J.; Goryl, P.; Klein, U.; Laur, H.; Mavrocordatos, C.; et al. The global monitoring for environment and security (GMES) Sentinel-3 mission. *Remote Sens. Environ.* **2012**, *120*, 37–57. [CrossRef]

39. Jiang, L.; Schneider, R.; Andersen, O.B.; Bauer-Gottwein, P. CryoSat-2 altimetry applications over rivers and lakes. *Water* **2017**, *9*, 211. [CrossRef]

40. Jiang, L.; Nielsen, K.; Andersen, O.B.; Bauer-Gottwein, P. CryoSat-2 radar altimetry for monitoring freshwater resources of China. *Remote Sens. Environ.* **2017**, *200*, 125–139. [CrossRef]

41. Wingham, D.J.; Francis, C.R.; Baker, S.; Bouzinac, C.; Brockley, D.; Cullen, R.; de Chateau-Thierry, P.; Laxon, S.W.; Mallow, U.; Mavrocordatos, C.; et al. CryoSat: A mission to determine the fluctuations in Earth's land and marine ice fields. *Adv. Space Res.* **2006**, *37*, 841–871. [CrossRef]

42. Liu, D.; Guo, S.; Shao, Q.; Liu, P.; Xiong, L.; Wang, L.; Hong, X.; Xu, Y.; Wang, Z. Assessing the effects of adaptation measures on optimal water resources allocation under varied water availability conditions. *J. Hydrol.* **2018**, *556*, 759–774. [CrossRef]

43. Villadsen, H.; Andersen, O.B.; Stenseng, L.; Nielsen, K.; Knudsen, P. CryoSat-2 altimetry for river level monitoring-Evaluation in the Ganges-Brahmaputra River basin. *Remote Sens. Environ.* **2015**, *168*, 80–89. [CrossRef]

44. Pavlis, N.K.; Holmes, S.A.; Kenyon, S.C.; Factor, J.K. The development and evaluation of the Earth Gravitational Model 2008 (EGM2008). *J. Geophys. Res. Solid Earth* **2012**, *117*. [CrossRef]

45. Pekel, J.-F.; Cottam, A.; Gorelick, N.; Belward, A.S. High-resolution mapping of global surface water and its long-term changes. *Nature* **2016**, *540*, 418–422. [CrossRef] [PubMed]

46. ASTER GDEM Validation Team. ASTER Global DEM Validation. *Summary Report*; 2009; pp. 1–28. Available online: https://ssl.jspacesystems.or.jp/ersdac/GDEM/E/image/ASTERGDEM_ValidationSummaryReport_Ver1.pdf (accessed on 11 December 2020).

47. Myers, E.; Hess, K.; Yang, Z.; Xu, J.; Wong, A.; Doyle, D.; Woolard, J.; White, S.; Le, B.; Gill, S.; et al. VDatum and strategies for national coverage. In Proceedings of the Ocean Conference Record (IEEE), Vancouver, BC, Canada, 29 September–4 October 2007. [CrossRef]

48. DHI. MIKE 11 A Modelling System for Rivers and Channels-Reference Manual, DHI: Copenhagen, Denmark. 2015. Available online: https://manuals.mikepoweredbydhi.help/2017/Water_Resources/Mike_11_ref.pdf (accessed on 11 December 2020).

49. Moramarco, T.; Singh, V.P. Formulation of the entropy parameter based on hydraulic and geometric characteristics of river cross sections. *J. Hydrol. Eng.* **2010**, *15*, 10. [CrossRef]

50. Wood, M.; Hostache, R.; Neal, J.; Wagener, T.; Giustarini, L.; Chini, M.; Corato, G.; Matgen, P.; Bates, P. Calibration of channel depth and friction parameters in the LISFLOOD-FP hydraulic model using medium-resolution SAR data and identifiability techniques. *Hydrol. Earth Syst. Sci.* **2016**, *20*, 4983–4997. [CrossRef]

51. Villadsen, H.; Deng, X.; Andersen, O.B.; Stenseng, L.; Nielsen, K.; Knudsen, P. Improved inland water levels from SAR altimetry using novel empirical and physical retrackers. *J. Hydrol.* **2016**, *537*, 234–247. [CrossRef]

52. Boergens, E.; Buhl, S.; Dettmering, D.; Kluppelberg, C.; Seitz, F. Combination of multi-mission altimetry data along the Mekong River with spatio-temporal kriging. *J. Geod.* **2017**, *91*, 519–534. [CrossRef]

53. Tourian, M.J.; Tarpanelli, A.; Elmi, O.; Qin, T.; Brocca, L.; Moramarco, T.; Sneeuw, N. Spatiotemporal densification of river water level time series by multimission satellite altimetry. *Water Resour. Res.* **2016**, *52*, 1140–1159. [CrossRef]

54. Pappenberger, F.; Beven, K.; Frodsham, K.; Romanowicz, R.; Matgen, P. Grasping the unavoidable subjectivity in calibration of flood inundation models: A vulnerability weighted approach. *J. Hydrol.* **2007**, *333*, 275–287. [CrossRef]

55. Tuozzolo, S.; Langhorst, T.; de Moraes Frasson, R.P.; Pavelsky, T.; Durand, M.; Schobelock, J.J. The impact of reach averaging Manning's equation for an in-situ datasets of water surface elevation, width, and slope. *J. Hydrol.* **2019**, *578*, 123866. [CrossRef]

56. Dung, N.V.; Merz, B.; Bardossy, A.; Thang, T.D.; Apel, H. Multi-objective automatic calibration of hydrodynamic models utilizing inundation maps and gauge data. *Hydrol. Earth Syst. Sci.* **2011**, *15*, 1339–1354. [CrossRef]

3

Evaluation of Satellite-Altimetry-Derived Pycnocline Depth Products in the South China Sea

Yingying Chen [1,2], Kai Yu [1,3], Changming Dong [1,4], Zhigang He [5], Yunwei Yan [3] and Dongxiao Wang [2,*]

[1] School of Marine Science, Nanjing University of Information Science and Technology, Nanjing 210044, China; cyy@scsio.ac.cn (Y.C.); yukai041@nuist.edu.cn (K.Y.); cmdong@gmail.com (C.D.)

[2] State Key Laboratory of Tropical Oceanography, South China Sea Institute of Oceanology, Chinese Academy of Sciences, Guangzhou 510000, China

[3] State Key Laboratory of Satellite Ocean Environment Dynamics, Second Institute of Oceanography, State Oceanic Administration, Hangzhou 310000, China; yanyunwei@sio.org.cn

[4] Department of Atmospheric and Oceanic Sciences, University of California, Los Angeles, CA 90095, USA

[5] College of Ocean and Earth Science, Xiamen University, Xiamen 361000, China; zghe@xmu.edu.cn

* Correspondence: dxwang@scsio.ac.cn

Abstract: The climatological monthly gridded World Ocean Atlas 2013 temperature and salinity data and satellite altimeter sea level anomaly data are used to build two altimeter-derived high-resolution real-time upper layer thickness products based on a highly simplified two-layer ocean model of the South China Sea. One product uses the proportional relationship between the sea level anomaly and upper layer thickness anomaly. The other one adds a modified component (η'_M) to account for the barotropic and thermodynamic processes that are neglected in the former product. The upper layer thickness, in this work, represents the depth of the main pycnocline, which is defined as the thickness from the sea surface to the 25 kg/m^3 isopycnal depth. The mean upper layer thickness in the semi-closed South China Sea is ~120 m and the mean reduced gravity is ~0.073 m/s^2, which is about one order of magnitude larger than the value obtained in the open deep ocean. The long-term temperature observations from three moored buoys, the conductivity-temperature-depth profiles from three joint cruises, and the Argo measurements from 2006 to 2015 are used to compare and evaluate these two upper layer thickness products. It shows that adding the η'_M component is necessary to simulate the upper layer thickness in some situations, especially in summer and fall in the northern South China Sea.

Keywords: upper layer thickness; satellite altimeter; two-layer ocean model; South China Sea

1. Introduction

The development of satellite remote sensing technology has enabled the retrieval of high-resolution, real-time, global information on the ocean. However, remote sensors can only "see" the sea surface. To infer underwater information using remote sensing data, one of the simplest and most used ways is to use the two-layer ocean model [1]. This model simplifies the stratified ocean to a two-layer homogeneous fluid, regarding the main pycnocline as the interface and the main pycnocline depth as the upper layer thickness (ULT). With the two-layer ocean model, one can easily derive the ULT, and the barotropic mode and first baroclinic mode of the circulation using satellite altimeter sea level anomaly (SLA) data.

The altimeter-derived ULT and circulation products are widely used in various areas for different purposes. Garzoli et al. [2] monitored the upper layer transport in the southeastern Atlantic Ocean; Sainz-Trapaga et al. [3] identified the Kuroshio Extension, its bifurcation, and its northern branch; Goni and Wainer [4] investigated the variability of the Brazil Current front. After adding satellite sea surface temperature (SST) data to the two-layer ocean model, Shay et al. [5] estimated upper layer thermal structure and upper layer heat content, and then studied the effects of a warm oceanic feature associated with Hurricane Opal. Following Shay et al. [5], Pun et al. [6] improved typhoon intensity forecasts and analyzed the spatial and temporal errors in the western North Pacific Ocean with the satellite-derived upper layer heat content. Vertical thermal structure and upper-ocean heat content from satellite remote sensing data had also been obtained by combining climatological hydrographic data and the two-layer ocean model (e.g., [7–9]). Most of the studies mentioned above were for the open ocean. Lin et al. [10] were the first to use altimeter data and the two-layer ocean model to study the spatial and temporal variation of ULT in the semi-closed South China Sea.

The SLA changes through mainly three processes: barotropic motions, the vertical Ekman pumping of the main pycnocline and the near-surface density flux above the seasonal pycnocline [11]. The Ekman pumping of the main pycnocline is the most important process, and the other two are usually one order of magnitude smaller than that of the main pycnocline fluctuations in the open ocean. The altimeter-derived ULT product based on the two-layer ocean model is mainly dependent on the proportional relationship between the SLA and main pycnocline fluctuations. As with previous studies in the open ocean, the effects of barotropic motions and near-surface density flux on the variability of SLA were not discussed in the work of Lin et al. [10].

However, these factors may play a more important role in the semi-closed South China Sea. The South China Sea is the largest marginal sea in the Northwest Pacific. Its mean depth is ~1200 m and maximum depth is ~5000 m, with a diamond-shaped deep basin oriented along the northeast-southwest direction (Figure 1). The SLA and upper layer circulation in the South China Sea have strong seasonal variability that is primarily driven by the monsoon wind [12–14]. The coastal regions show a higher seasonal sea level cycle than deep water (e.g., [15,16]). Liu et al. [17] found that the seasonal variability of the SLA is forced mainly by surface wind curl, and secondarily (about 20%) by surface net heat flux. They concluded that the surface net heat flux can obviously change the seasonal variability of the SLA, especially in the central South China Sea, by expanding/contracting the water column in the mixed layer, and the influence of the surface net heat flux is much stronger than for the open ocean. Cheng and Qi [18] argued that on seasonal time scales the baroclinic component to a great extent explains the SLA over the deep part of the South China Sea basin, whereas the barotropic component has a significant contribution to the SLA over shallow water areas in the South China Sea.

In the present study, instead of the historical ocean profiles, the gridded data is used to derive the altimeter-derived ULT products in the South China Sea. We first get the high-resolution ($0.25° \times 0.25°$) altimeter-derived ULT products and give a comprehensive assessment of the ULT products with multifarious in situ observations. A monthly varied modified component is first introduced into the two-layer ocean model to examine the effects of the barotropic component and near-surface density flux on the ULT products in the South China Sea. This paper is organized as follows: Section 2 describes the data. Section 3 introduces the methodology. Section 4 determines the product parameters. Section 5 evaluates the products with different in situ datasets. Section 6 discusses the physical mechanisms of the modified component. Finally, we give our conclusions and discussions in Section 7.

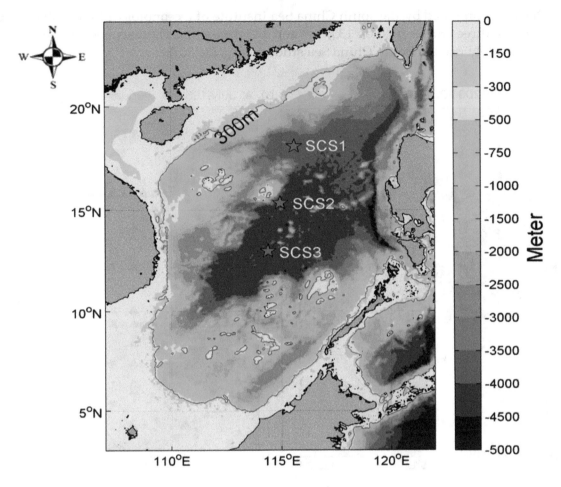

Figure 1. Bottom topography of the South China Sea. The locations of three ATLAS buoys SCS1 (18°5.9′N, 115°35.8′E), SCS2 (15°20.6′N, 114°57.3′E), SCS3 (12°58.5′N, 114°24.5′E) are indicated by red stars.

2. Data

Two datasets are used to derive the ULT product in the South China Sea. The first is the climatological monthly temperature and salinity data at standard depth levels with a horizontal resolution of 0.25° both in longitude and latitude taken from the World Ocean Atlas 2013 [19]. The second is the daily and climatological monthly delayed-time level 4 gridded SLA products from multi-satellite observation provided by Archiving, Validation and Interpretation of Satellite Oceanographic Data (AVISO [20], version 5.1). The SLA data have the same spatial resolution as the World Ocean Atlas 2013 and have been modified for various atmospheric, tidal and instrumental corrections [21,22].

Three different kinds of in situ datasets are used to validate and evaluate the ULT product. The first dataset consists of measurements from three autonomous temperature line acquisition system (ATLAS) buoys (named SCS1, SCS2 and SCS3), deployed by the Institute of Oceanography, Taiwan University during the South China Sea Monsoon Experiment, whose locations are shown in Figure 1. The temporal spans are from 17 April 1997 to 9 April 1999 for SCS1, from 11 April 1998 to 10 April 1999 for SCS2, and from 12 April 1998 to 10 April 1999 for SCS3. The temporal resolution is 10 min, and the measurement depth is 1 m, 25 m, 50 m, 75 m, 100 m, 125 m, 150 m, 200 m, 250 m, 300 m and 500 m.

The Argo measurements available in the South China Sea from 2006 to 2015 (Figure 2a) are provided by the US Global Ocean Data Assimilation Experiment (USGODAE [23]). The conductivity-temperature-depth (CTD) probe data from three joint cruises in July 1998 (from 4 June to 21 July 1998), August 2000 (from 2 August to 3 September 2000) and May 2004 (from 29 April to 25 May 2004) are provided by the South China Sea Institute of Oceanology of the Chinese Academy of Sciences. These three cruises include 133, 189 and 143 different CTD casts, respectively (Figure 2b). These data are widely used in South China Sea studies (e.g., [24,25]).

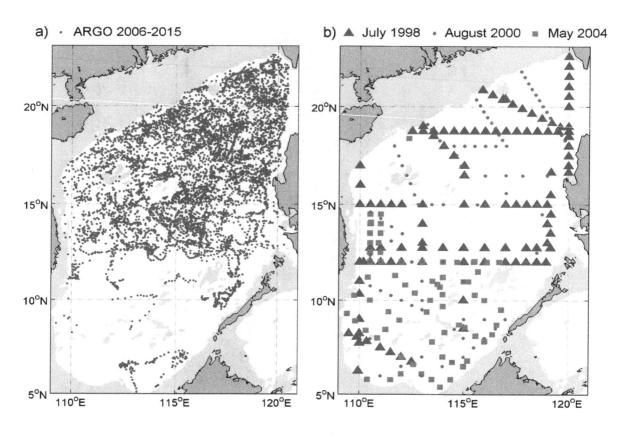

Figure 2. Distribution of (**a**) Argo measurements from 2006 to 2015 and (**b**) CTD data of three joint cruises investigation in July 1998 (blue triangles), August 2000 (red points) and May 2004 (golden squares).

3. Methods

The ocean bottom pressure (P_o) is the sum of the sea surface atmospheric pressure (P_a) and the weight per unit area of the water column:

$$P_o = P_a + g \int_{-D}^{\eta} \rho dz \tag{1}$$

where D is the mean ocean depth, η represents the sea surface height above geoid, g is the gravitational acceleration, and ρ is sea water density. According to the Boussinesq approximation, Equation (1) can be further approximated by

$$P_o = P_a + g \int_{-D}^{0} \rho dz + g\rho_0 \eta \tag{2}$$

where ρ_0 is the mean density of the ocean, which is equal to 1025 kg/m^3 in this study. The fluctuating ocean bottom pressure (P_o') is caused by the variability of atmospheric pressure, the density fluctuations under the mean sea surface and the SLA (η'):

$$P_o' = P_a' + g \int_{-D}^{0} \rho' dz + g\rho_0 \eta' \tag{3}$$

Now η' can be partitioned into barotropic (η_b') and steric components (η_s'):

$$\eta' = \eta_b' + \eta_s'. \tag{4}$$

The barotropic component changes as a result of internal mass redistribution in the ocean or water-mass flux, which are directly correlated with ocean bottom pressure, or changes as a result of the variability of atmospheric pressure:

$$\eta_b' = \frac{P_a' - P_o'}{g\rho_0}. \tag{5}$$

The steric component, also called the baroclinic component, changes depending on the seawater density of the water column but corresponds neither to ocean bottom pressure nor to sea surface atmospheric pressure:

$$\eta_s' = -\frac{1}{\rho_0} \int_{-D}^0 \rho' dz. \tag{6}$$

According to Gill and Niiler [11], the steric height anomaly, η_s', is produced mainly by the variability of vertical Ekman pumping of the main pycnocline and near-surface density flux above the seasonal pycnocline.

The ocean is continuously stratified, with a sharp vertical density gradient in the pycnocline. The two-layer model, proposed by Goni et al. [1], simplifies the stratified ocean as a two-layer homogeneous fluid, regarding the main pycnocline depth as the interface. Let the thicknesses and densities of the upper and lower layer be h_1, ρ_1, h_2, and ρ_2, respectively. The sea surface height is then

$$\eta = h_1 + h_2 - D, \tag{7}$$

and the ocean bottom pressure is then

$$P_o = P_a + g\rho_1 h_1 + g\rho_2 h_2 \tag{8}$$

Combining Equations (7) and (8) yields:

$$\eta = \varepsilon h_1 + B, \tag{9}$$

where

$$\varepsilon = (\rho_2 - \rho_1)/\rho_2, \tag{10}$$

and

$$B = \frac{P_o - P_a}{g\rho_2} - D = \frac{\rho_1}{\rho_2} h_1 + h_2 - D. \tag{11}$$

The fluctuating part of Equation (9) is

$$\eta' = \varepsilon h_1' + B'. \tag{12}$$

As one can see, B' has the same meaning as η_b' in Equation (4), and varies with the variabilities of the ocean bottom pressure and sea surface atmospheric pressure. On the other hand, the $\varepsilon h_1'$ has the same meaning as η_s' in Equation (4), and this illustrates that the steric component of the SLA in the two-layer ocean model is only caused by the fluctuations of the interface (main pycnocline). However, the effect of near-surface density flux is not considered in this model.

Generally speaking, the lower layer thickness in the deep ocean is much larger than that of the upper layer. If it is further assumed that the lower layer is infinitely deep and the fluid in the lower layer is stagnant, the model can be referred to as a 1.5-layer reduced gravity model. In the 1.5-layer reduced gravity model, the barotropic component, B', is negligible and the SLA is equal to the baroclinic component and proportional to the ULT anomaly (ULTA), h_1' [3]. Letting η_{ULT}' denote the part of SLA that is proportional to the ULTA, then we have

$$\eta'(t) = \eta_{ULT}'(t) = \varepsilon h_1'(t). \tag{13}$$

As a result, the real-time ULT product can be derived from the corresponding altimeter data:

$$h_1(t) = \overline{h}_1 + h_1'(t) = \overline{h}_1 + \frac{1}{\varepsilon}\eta'(t), \tag{14}$$

where $\overline{h}_1$ represents the climatological mean ULT. Both $\overline{h}_1$ and ε can be calculated from the World Ocean Atlas 2013 data. The 1.5-layer reduced gravity model is widely used in simulating the upper-ocean circulation in the South China Sea (e.g., [26–28]).

Equation (14) only considers the proportional relationship between the SLA and ULTA. However, the barotropic and sea surface thermodynamic processes will change the SLA, while having little effect on the main pycnocline fluctuations. They will break the proportional relationship between the SLA and ULTA and bring error into the altimeter-derived ULT product based on Equation (14). To consider the effects of these processes, we add a modified component η_M', and then η' can be divided into two parts η_{ULT}' and η_M':

$$\eta'(t) = \eta_{ULT}'(t) + \eta_M'(t). \tag{15}$$

The climatological monthly η_{ULT}' can be derived from h_1' based on Equation (13). The h_1' is calculated from the World Ocean Atlas data. Then the climatological monthly η_M' can be produced by subtracting the World Ocean Atlas-derived η_{ULT}' from the η', which is obtained from the climatological monthly altimeter data. Then the real-time ULT product can be obtained with the daily altimeter data by regarding the climatological monthly η_M' as a background field:

$$h_1(t) = \overline{h}_1 + \frac{1}{\varepsilon}\eta_{ULT}'(t) = \overline{h}_1 + \frac{1}{\varepsilon}(\eta'(t) - \eta_M'(t)). \tag{16}$$

4. Parameter Determination

To get the altimeter-derived ULT products, the first step is to determine the interface of the two-layer ocean model. Given that the effect of salinity on the density is smaller than the effect of temperature, previous investigators usually choose the depth of a certain isotherm as the interface to correspond to the main thermocline depth (e.g., [1–4]). After analyzing the in situ hydrographic data from the World Ocean Database 2005, Lin et al. [10] found the 16 °C isotherm has a better linear correlation with the SLA than those at 12 °C, 14 °C, 18 °C, or 20 °C, and thus used the depth of the 16 °C isotherm as the interface of the South China Sea. In our study, from the climatological mean temperature and density profiles over the South China Sea region derived from the World Ocean Atlas 2013 (Figure 3a), one can see that the thermocline and pycnocline depths are nearly the same in the South China Sea. Both are mainly located between 50 m and 200 m. Around this depth range, different isotherms and isopycnals depths (Figure 3a) are adopted to calculate the climatological monthly ULT and ULTA using the World Ocean Atlas 2013 data. All available spatial and temporal ULTA values in the South China Sea are compared with the corresponding SLA (Figure 3b,c). The spatial grid pixels are excluded when the water depth is shallower than 300 m (Figure 1) because of the inapplicability of the two-layer ocean model in the continental shelf area. The correlation coefficients between the SLA and ULTA reach a maximum when choosing the 20 °C isotherm or 25 kg/m^3 isopycnal. Both the maximum correlation coefficients are ~0.77 and the corresponding depths of these two choices are ~120 m (Figure 3a). The strong similarity suggests that the salinity effect is weak. The 25 kg/m^3 isopycnal depth will be used to represent the ULT in the remainder of this paper. The 20 °C isotherm depth is adopted to represent the ULT only when the ULT product is validated with the ATLAS buoys' data, because the ATLAS buoys lack salinity data.

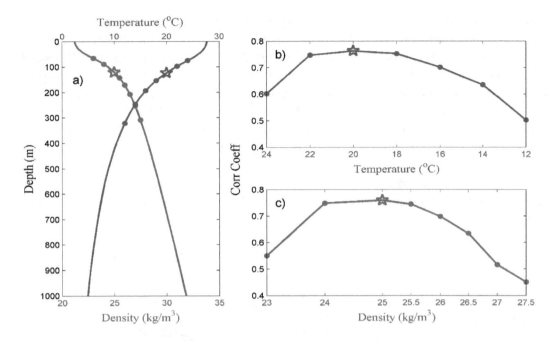

Figure 3. (a) The mean temperature (blue) and density (red) profile in the South China Sea derived from the World Ocean Atlas 2013. The correlation coefficient between the SLA and ULTA calculated by different (**b**) temperature and (**c**) density. The stars indicate where the maximum correlation coefficients appear.

Figure 4 shows the comparison of the correlation between the SLA and ULTA in different seasons. The winter, spring, summer and fall values correspond here to January, April, July and October, respectively. The correlation coefficient is high (beyond 95% t-test confidence interval) throughout the year. The mean correlation coefficient is 0.76 ± 0.07. The maximum correlation coefficient is 0.86 which happens in spring (Figure 4b), and the minimum is 0.72 in summer (Figure 4c).

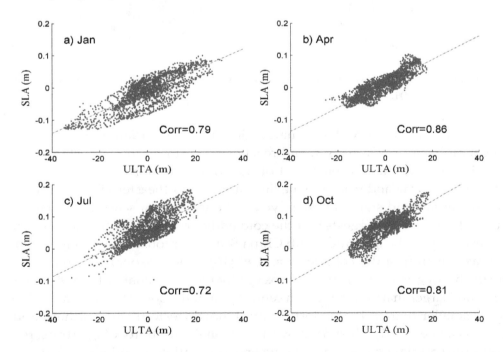

Figure 4. The scatter plots of SLA and ULTA calculated by 25 kg/m^3 isopycnal in January, April, July and October (**a–d**). The line of best fit is represented by the red dashed line.

After determining the interface of the two-layer ocean model, the climatological mean $\bar{h}_1$, ρ_1, ρ_2 and ε can be calculated from the World Ocean Atlas 2013 data. Figure 5a illustrates the climatological mean ULT, $\bar{h}_1$, in the South China Sea. The mean ULT is ~120 m and becomes thicker toward the east. There are two eddy-like thinner ULT areas at 18°N, 118°E and 14°N, 111°E, corresponding to the cold eddy northwest of the Luzon Island [13,29] and the cold eddy off the Vietnamese coast [27,30], respectively. Figure 5b shows that the distribution of ε in the South China Sea. ε increases gradually from the northwest to southeast South China Sea, with the values between 7.0×10^{-3} and 7.8×10^{-3} in most of our study area. For the regions near the continental shelf edge, the ε rapidly decreases to $~5.0 \times 10^{-3}$. Around the Nansha Islands, ε also decreases rapidly. The reduced gravity g' can be estimated by multiplying ε by the gravitational acceleration. This gives g' ~0.073 m/s^2 in the South China Sea, which is about one order of magnitude larger than the value obtained in the open deep ocean (e.g., [1,2]).

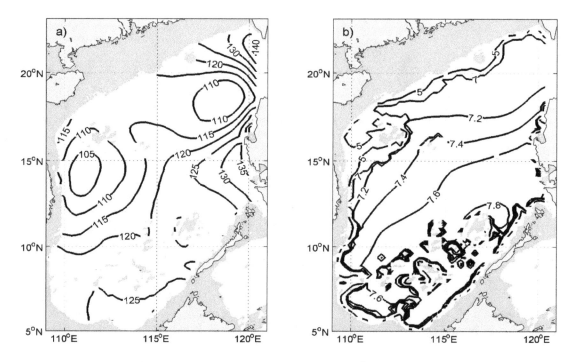

Figure 5. Climatological mean (**a**) ULT (CI: 5 m) and (**b**) ε ($\times 10^{-3}$) in the South China Sea derived from the World Ocean Atlas 2013.

The first row of Figure 6 shows the climatological seasonal variation of η', SLA. In winter, η' is negative in most of the South China Sea, and reaches a minimum in the northeastern South China Sea. This is mainly due to the reinforcement of the cold eddy northwest of the Luzon Island. η' gradually increases in spring and reaches a maximum in the northeastern South China Sea in summer. Then η' decreases in fall and starts another cycle. The maximum seasonal variability of η' in the northeastern South China Sea suggests that the cold eddy northwest of the Luzon Island dominates the seasonal variability of the η' in the South China Sea. The monsoon is thought to be the primary force for the seasonal shift of the cold eddy and η' [16]. The seasonal variability of η'_{ULT} (second row in Figure 6) shows great similarity to η', except that η'_{ULT} is smaller than η' in the northeastern South China Sea and larger than η' in the north central part of the South China Sea in winter. In summer and fall, η'_{ULT} is smaller than η' in the north central, northwestern and southern South China Sea. These differences can be clearly illustrated by the seasonal variability of η'_M (third row in Figure 6). Figure 7a shows the correlation coefficient between the monthly η' and η'_{ULT}. A significant positive correlation exists between 10°N and 15°N in the northeastern South China Sea, while in the north central, northwestern and southern South China Sea, the positive correlation is weak and even turns into a negative correlation in the north central and northwestern South China Sea. On the other hand, the correlation coefficient between η' and η'_M (Figure 7b) is relatively high where the positive correlation between η' and η'_{ULT} is lower. Interestingly, the region where the higher positive correlation between η'_M and η'_M occurs is exactly the area where one can find the larger ratio of the standard deviation between η'_M and η'_{ULT} (Figure 7c), which shows the relative importance of these two components in the seasonal variability amplitude of η'. It can be concluded that the proportional relationship between η' and h'_1 is stronger and one can get an accurate ULT product based on Equation (14) between 10°N and 15°N and in the northeastern South China Sea. However, in the north central, northwestern and southern South China Sea, η'_M becomes important, and one must derive the ULT product according to Equation (16) instead of Equation (14).

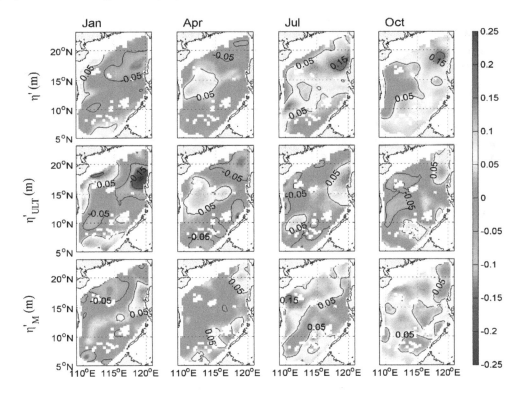

Figure 6. Seasonal variability of η', η'_{ULT} and η'_M (CI: 0.1 m).

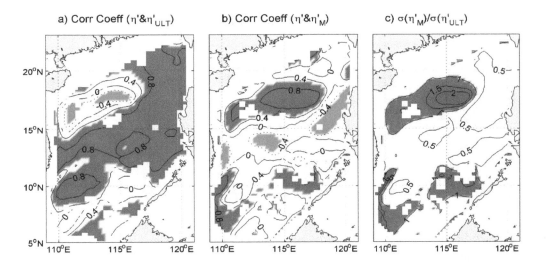

Figure 7. Spatial distributions of correlation coefficient between the monthly η' (**a**) with η'_{ULT} and (**b**) with η'_M. Shading in (**a,b**) indicates correlation beyond 95% confidence level; (**c**) The ratio of the standard deviation between η'_M and η'_{ULT}. The ratio larger than 1 is shaded in (**c**).

5. Product Evaluation

After identifying the parameters, one can easily get the altimeter-derived real-time ULT products based on Equations (14) and (16), respectively. These two products are compared and evaluated with different in situ observations.

The Argo-data-derived ULTs are directly obtained through computing the 25 kg/m^3 isopycnal depth at each Argo profile. Then the gridded altimeter-derived ULTs were interpolated to Argo profile locations to compare with the Argo-data-derived ULTs. The Argo measurements in the South China Sea are separated into the part north of 15°N and the part between 10°N and 15°N, because the relative importance of η'_{ULT} and η'_M is quite different in these two areas and the Argo data are very scarce in the southern South China Sea (Figure 2a). Figure 8 shows the comparison of the mean bias and root-mean-square error (RMSE) between the altimeter-derived and Argo-data-derived ULT in these two areas based on the two methods. Overall, the biases are positive, which indicate that the altimeter-derived products always overestimate the ULT directly calculated by the Argo observations. The overestimation can be somewhat reduced after considering η'_M. At the same time, the RMSE also achieves a certain degree of reduction. The bias and RMSE for the region north of 15°N are obviously larger than those between 10°N and 15°N. Specifically, without considering η'_M, the maximum bias (~12 m) and maximum RMSE (more than 15 m) can be found in summer and fall for the region north of 15°N. After adding η'_M, the bias is reduced to less than 3 m and the RMSE decreases by more than 27%. The situations are the same for the region between 10°N and 15°N, except that the bias and RMSE and the improvements after considering η'_M are relatively smaller than those north of 15°N. These results correspond to the large positive η'_M in summer and fall (third row in Figure 6) and prove the relative importance of η'_M in summer and fall, especially in the northern South China Sea. However, in winter and spring, the improvements are very limited after considering η'_M.

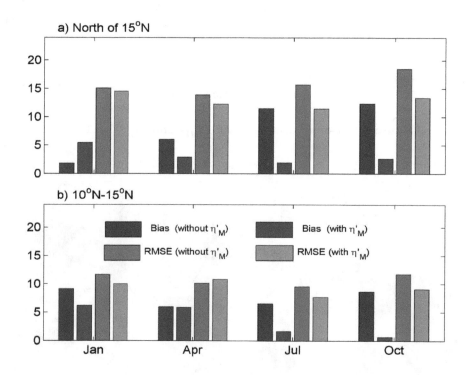

Figure 8. Bias and RMSE (Unit: m) between the altimeter-derived and Argo-data-derived ULT for the South China Sea (**a**) north of 15°N and (**b**) between 10°N and 15°N.

The CTD data collected during three joint cruises are used to validate the altimeter-derived ULT products. Figure 9 shows the differences between altimeter-derived and cruise-data-derived ULTs for these three cruises. The ULT differences at each station are derived in the same way as the

Argo profile. Results for the first two cruises are shown in the first two columns of Figure 9. They are during the summer and fall and cover nearly all the South China Sea. The altimeter-derived products, which neglect η'_M, evidently overestimate the cruise-data-derived ULT in these two cruises, especially in the north central, northwestern and southern South China Sea. After considering η'_M, the overestimations are substantially reduced, and the RMSE are reduced from 11.8 m and 14.5 m to 8.6 m and 11.7 m, respectively for the cruise in July 1998 and August 2000. The reduction rate of the RMSE for these two cruises is about 27% and 19%, respectively. However, both the altimeter-derived products underestimate the cruise-data-derived ULT off the southeast coast of Vietnam in August 2000 (second column in Figure 9). That may be because of the effects of an anticyclonic eddy, which is part of a dipole eddy pair associated with the summer eastward jet (e.g., [14,26,27,31,32]). The details still need further study. Furthermore, η'_M has a very limited influence on ULT products for the cruise in May 2004 (third column in Figure 9), and this agrees with previous studies that η'_M is very small (Figure 6) and the improvements are very limited after considering η'_M in spring (Figure 8b).

Next, we validate the two satellite-derived products with the data of the three ATLAS buoys in the South China Sea (Figure 10). The SCS1 is located in the northern South China Sea (Figure 1) where the relative importance of η'_M reaches its maximum (Figure 7c). The correlation coefficient and RMSE between the time series of ULT derived from the altimeter without considering η'_M and from the SCS1 is 0.54 m and 15.2 m. The altimeter-derived product obviously overestimates the buoy-derived ULT during the summer and fall. Consistent with our expectations, these overestimations are markedly reduced after considering η'_M. The new product fits very well with the buoy-derived ULT, especially in summer and fall. The correlation coefficient rises to 0.81, and the RMSE decreases to 9.6 m. These results confirm the important role of η'_M in simulating the ULT in the northern South China Sea in summer and fall. For the buoys SCS2 and SCS3, when η'_M is neglected, the altimeter-derived products correspond well with the buoy-derived ULT. Their correlation coefficient is 0.91 and 0.69, and their RMSE is 9.7 m and 8.1 m for SCS2 and SCS3, respectively. After considering η'_M, the time series are basically the same (Figure 10b,c), which is consistent with the result that the effects of η'_M are smaller in the middle of the South China Sea (Figure 7c).

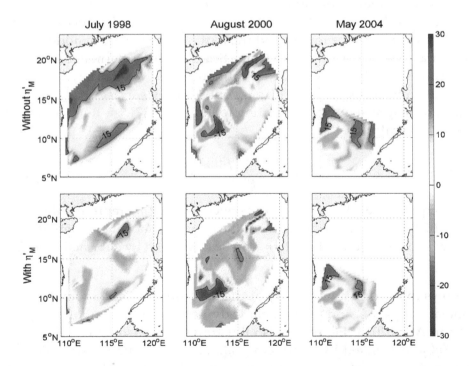

Figure 9. Spatial distribution of differences (Units: m) between altimeter-derived and cruise-data-derived ULT in July 1998 (**left**), August 2000 (**middle**) and May 2004 (**right**) without η'_M (**up**) and with η'_M (**down**).

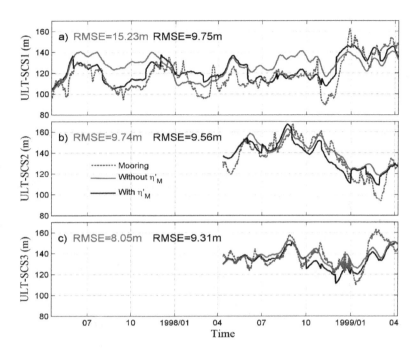

Figure 10. (a–c) Time series of buoy-data-derived ULT (blue dashed line), altimeter-derived ULT without η'_M (red solid line) and with η'_M (black solid line). The locations of three ATLAS buoys are marked in Figure 1.

6. Physical Mechanisms of the Modified Component

It is worth discussing the physical mechanisms that control η'_M. According to Gill and Niiler [11], the wind can affect η' in two ways: by changing the bottom pressure through barotropic motions and by changing the main thermocline through vertical Ekman pumping. In addition, the surface density flux can affect the near-surface steric height anomaly by changing the density in the mixed layer. Therefore, η'_{ULT} looks to be mainly under the control of baroclinic dynamics and η'_M is dominated by barotropic processes and the near-surface baroclinic thermodynamic processes. By comparing the 1.5-layer baroclinic planetary wave model and the barotropic Sverdrup model, Liu et al. [17] noted that the seasonal variability of η' in the South China Sea is mainly determined by surface wind curl on baroclinic Rossby waves and the barotropic contribution is very small. They also find that the surface steric height anomaly in the upper 50 m, which is controlled by surface heat flux forcing, contributes ~20% of the total steric height anomaly. According to Liu et al. [17], the surface steric height anomaly in the upper 50 m can be calculated as:

$$\eta'_{50} = -\frac{1}{\rho_0} \int_{-50\ m}^{0} \rho'dz, \tag{17}$$

in the same way as Equation (6). The differences between η'_s and η'_{50} represent the baroclinic dynamics component of η', which is mainly correlated with the main thermocline variability. Liu et al. [17] also mentioned that the barotropic component of η' can be calculated by subtracting η'_s from the altimeter η' if both are exact.

Figure 11 shows the seasonal variability of the total steric height anomaly (η'_{WOA}), the surface steric height anomaly (η'_{WOA50}), and the baroclinic dynamics component of η' (η'_{WOAULT}) measured by the World Ocean Atlas 2013. Figure 11 also illustrates the seasonal variability of the barotropic component of η' (η'_{BT}) calculated by subtracting η'_{WOA} from η' (first row in Figure 6). The major seasonal features of η'_{WOA} and η'_{WOAULT} agree well with η' and η'_{ULT} (Figure 6), which suggests that the seasonal variability of the SLA is dominated by the baroclinic part, especially by the baroclinic dynamic processes. The seasonal variability of η'_{WOA50} is relatively small, and mainly manifests as negative anomalies (smaller than -0.02 m) in winter and spring and positive anomalies (larger than 0.02 m) in summer and fall in the northern South China Sea. In the same way as Figure 7c, the relative importance of η'_{WOA50}, η'_{BT}, and η'_{WOAULT} is shown by comparing their standard deviation (Figure 12). From Figure 12a, η'_{WOA50} is found to be more important than η'_{WOAULT} only in the north central part of the South China Sea. This position is exactly where the buoy SCS1 is located and the maximum ratio of the standard deviation between η'_M and η'_{ULT} is found. This demonstrates that, in this region, the variability of η'_M can be partly explained as the effects of surface thermodynamic processes. The surface heat flux expands/contracts the water column in the mixed layer and brings a positive/negative η'_M in summer/winter, especially in the northern South China Sea. Comparing with η'_{WOA50}, η'_{BT} plays a more important role. η'_{BT} shows high positive values (larger than 0.05 m) in the northeastern South China Sea in winter, in the southern South China Sea in spring, and in the western South China Sea in summer and fall. The seasonal variability of η'_{BT} is the same as, but weaker than, η'_M. Besides the north central part of the South China Sea, the ratio of the standard deviation between η'_{BT} and η'_{WOAULT} is more than 1 in the northwestern and southern South China Sea (Figure 12b). After combining η'_{WOA50} and η'_{BT}, the ratio of the standard deviation between the combined effect and η'_{WOAULT} (Figure 12c) fits very well with that between η'_M and η'_{ULT} (Figure 7c). This confirms that η'_M is mainly under the control of the combined effect of surface thermodynamic and barotropic processes. However, whether the differences between η'_{WOA} and satellite-derived η' can represent the barotropic component is still open to discussion. The differences may also include the effects of the time inconsistency of these two datasets. The Gravity Recovery and Climate Experiment (GRACE) [33,34] will be included to discuss the seasonal variability of the barotropic component in future work.

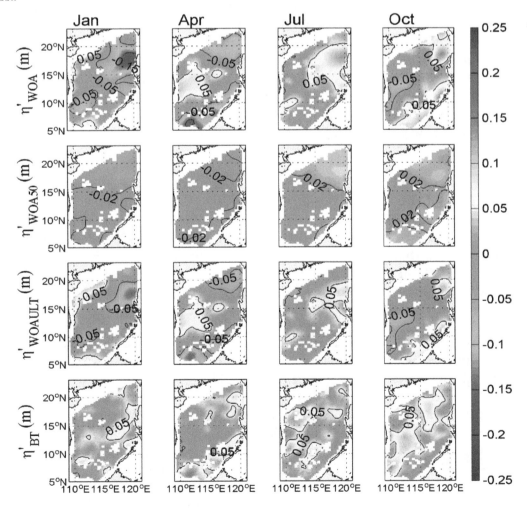

Figure 11. Seasonal variability of η'_{WOA}, η'_{WOA50}, η'_{WOAULT} and η'_{BT}.

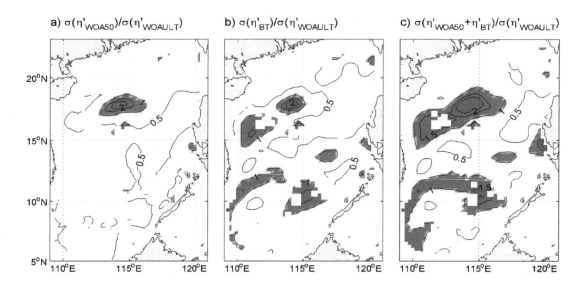

Figure 12. The ratio of the standard deviation (**a**) between η'_{WOA50} and η'_{WOAULT}; (**b**) between η'_{BT} and η'_{WOAULT}; and (**c**) between $\eta'_{WOA50} + \eta'_{BT}$ and η'_{WOAULT}. The ratio larger than 1 is shaded.

7. Conclusions and Discussion

The climatological monthly gridded temperature and salinity and SLA data are used to quantify the background field parameters of Equations (14) and (16) based on a highly simplified two-layer ocean model. The first method considers that the SLA is directly proportional to the ULTA, and the other separates SLA into two parts: the component proportional to ULTA, η'_{ULT}, and the modified component η'_M, which can correct the difference between η' and η'_{ULT} introduced by other physical processes. With these equations, one can derive two real-time ULT products using the SLA data. As far as we know, this is the first time to have obtained the altimeter-derived ULT products with gridded World Ocean Atlas 2013 data. Compared to the former products ($2° \times 2°$) derived from the site data (e.g., [10]), the new products have higher resolution ($0.25° \times 0.25°$) and thus can identify the ULT structures at smaller scales.

The $25\,\mathrm{kg/m^3}$ isopycnal depth is chosen as the representation of the ULT after comparing different choices. The alternative choice is the 20 °C isotherm depth due to the salinity effect being weak. In the South China Sea, the mean ULT is ~120 m, and there are two obvious eddy-like thinner ULT areas corresponding to two cold eddies. The mean reduced gravity is ~0.073 m/s², which is about one order of magnitude larger than the value obtained in the open deep ocean. Both η' and η'_{ULT} display a clear seasonal variability and their variability shows great similarity, except in the north central, northwestern and southern South China Sea, where the η'_M has higher correlation with η' and larger standard deviation relative to η'_{ULT}.

Three different in situ observations are used to compare and evaluate these two ULT products. Both the high-resolution altimeter-derived products satisfactorily simulate the ULT in the South China Sea. The range of the RMSE is from less than 10 m to ~15 m in different evaluations. The monthly varied η'_M is firstly introduced into the two-layer ocean model. We found that η'_M is really necessary to correct the overestimation in summer and fall reported by the altimeter-derived ULT product without considering η'_M, especially in the northern South China Sea. However, in winter and spring, and in the middle South China Sea, the corrective action of η'_M is limited. Nevertheless, it is comforting to see that the bias and RMSE is relatively small in these situations even if without η'_M. The in situ observations in the South China Sea south of 10°N were too few to evaluate the satellite-derived ULT product. Finally, the physical mechanisms of η'_M are found to be mainly under the control of the combined effect of barotropic and surface thermodynamic processes.

Acknowledgments: This research was supported by the National Key R&D Program of China (grant 2016YFA0601804 and 2016YFA0601803), National Science Foundation of China (grant 41606021, 41606017, and 41476022), Open Fund of State Key Laboratory of Satellite Ocean Environment Dynamics (grant QNHX1610), the Startup Foundation for Introducing Talent of NUIST (grant 2014r017 and 2014r072), the National Program on Global Change and Air-Sea Interaction (grant GASI-03-IPOVAI-05), and the Priority Academic Program Development of Jiangsu Higher Education Institutions (PAPD). The authors would like to thank Gregory P. King for language editing. The authors also thank South China Sea Institute of Oceanology of the Chinese Academy of Sciences for supplying the ATLAS buoys' data during the South China Sea Monsoon Experiment and the CTD probe data of the joint cruise investigations.

Author Contributions: D.W. and C.D. conceived and designed the experiments; Y.C., K.Y, Z.H. and Y.Y. performed the experiments; Y.C. and K.Y, analyzed the data; Y.C. wrote the paper. K.Y and Y.Y. helped modify the manuscript.

References

1. Goni, G.J.; Kamholz, S.; Garzoli, S.L.; Olson, D.B. Dynamics of the Brazil-Malvinas confluence based on inverted echo sounders and altimetry. *J. Geophys. Res.* **1996**, *101*, 16273–16289. [CrossRef]
2. Garzoli, S.L.; Goni, G.J.; Mariano, A.J.; Olson, D.B. Monitoring the upper southeastern Atlantic transports using altimeter data. *J. Mar. Res.* **1997**, *55*, 453–481. [CrossRef]
3. Sainz-Trapaga, S.; Goni, G.J.; Sugimoto, T. Identification of the Kuroshio Extension, its bifurcation and Northern Branch from altimetry and hydrographic data during October 1992–August 1999: Spatial and temporal variability. *Geophys. Res. Lett.* **2001**, *28*, 1759–1762. [CrossRef]
4. Goni, G.J.; Wainer, I. Investigation of the Brazil current front variability from altimeter data. *J. Geophys. Res.* **2001**, *106*, 31117–31128. [CrossRef]
5. Shay, L.K.; Goni, G.J.; Black, P.G. Effects of a warm oceanic feature on Hurricane Opal. *Mon. Weather Rev.* **2000**, *128*, 1366–1383. [CrossRef]
6. Pun, I.F.; Lin, I.I.; Wu, C.R.; Ko, D.S.; Liu, W.T. Validation and application of altimetry derived upper ocean thermal structure in the western North Pacific Ocean for typhoon intensity forecast. *IEEE Trans. Geosci. Remote Sens.* **2007**, *45*, 1616–1630. [CrossRef]
7. Polito, P.S.; Sato, O.T. Patterns of sea surface height and heat storage associated to intraseasonal Rossby waves in the tropics. *J. Geophys. Res.* **2003**, *34*, L09603. [CrossRef]
8. Lentini, C.A.D.; Goni, G.J.; Olson, D.B. Investigation of Brazil current rings in the confluence region. *J. Geophys. Res.* **2006**, *111*. [CrossRef]
9. Momin, I.M.; Sharma, R.; Basu, S. Satellite-derived heat content in the tropical Indian Ocean. *Remote Sens. Lett.* **2011**, *2*, 269–277. [CrossRef]
10. Lin, C.Y.; Ho, C.R.; Zheng, Z.W.; Kuo, N.J. Validation and variation of upper layer thickness in South China Sea from satellite altimeter data. *Sensors* **2008**, *8*, 3802–3818. [CrossRef] [PubMed]
11. Gill, A.E.; Niiler, P.P. The theory of seasonal variability in the ocean. *Deep Sea Res.* **1973**, *20*, 141–177. [CrossRef]
12. Wyrtki, K. Scientific Results of Marine Investigations of the South China Sea and the Gulf of Thailand. In *Physical Oceanography of the Southeast Asian Waters*; Scripps Institution of Oceanography: La Jolla, CA, USA, 1961; Volume 2, p. 195.
13. Qu, T. Upper-Layer Circulation in the South China Sea. *J. Phys. Oceanogr.* **2000**, *30*, 1450–1460. [CrossRef]
14. Liu, Y.; Weisberg, R.H.; Yuan, Y. Patterns of upper layer circulation variability in the South China Sea from satellite altimetry using the self-organizing map. *Acta Oceanol. Sin.* **2008**, *27*, 129–144. [CrossRef]
15. Amiruddin, A.; Haigh, I.; Tsimplis, M.; Calafat, F.; Dangendorf, S. The seasonal cycle and variability of sea level in the South China Sea. *J. Geophys. Res.* **2015**, *120*, 5490–5513. [CrossRef]
16. Cheng, Y.; Hamlington, B.; Plag, H.-P.; Xu, Q. Influence of ENSO on the variation of annual sea level cycle in the South China Sea. *Ocean Eng.* **2016**, *126*, 343–352. [CrossRef]
17. Liu, Z.; Yang, H.; Liu, Q. Regional dynamics of seasonal variability in the South China Sea. *J. Phys. Oceanogr.* **2001**, *31*, 272–284. [CrossRef]
18. Cheng, X.; Qi, Y. On steric and mass-induced contributions to the annual sea-level variations in the South China Sea. *Glob. Planet Chang.* **2010**, *72*, 227–233. [CrossRef]
19. World Ocean Atlas 2013. Available online: http://apdrc.soest.hawaii.edu (accessed on 4 March 2017).
20. Archiving, Validation and Interpretation of Satellite Oceanographic Data. Available online: ftp.aviso.altimetry.fr (accessed on 4 March 2017).
21. Le Traon, P.Y.; Ogor, F. ERS-1/2 orbit improvement using TOPEX/POSEIDON: The 2 cm challenge. *J. Geophys. Res.* **1998**, *103*, 8045–8057. [CrossRef]

22. Le Traon, P.Y.; Nadal, F.; Ducet, N. An improved mapping method of multisatellite altimeter data. *J. Atmos. Ocean. Technol.* **1998**, *15*, 522–534. [CrossRef]

23. US Global Ocean Data Assimilation Experiment. Available online: http://www.usgodae.org/cgi-bin/argo_select.pl (accessed on 4 March 2017).

24. Liu, Y.; Yuan, Y.; Su, J.; Jiang, J. Circulation in the South China Sea in summer of 1998. *Chin. Sci. Bull.* **2000**, *45*, 1648–1655. [CrossRef]

25. Zeng, L.; Wang, D.; Chen, J.; Wang, W.; Chen, R. SCSPOD14, a South China Sea physical oceanographic dataset derived from in situ measurements during 1919–2014. *Sci. Data* **2016**, *3*, 160029. [CrossRef] [PubMed]

26. Metzger, E.J.; Hurlburt, H. Coupled dynamics of the South China Sea, the Sulu Sea, and the Pacific Ocean. *J. Geophys. Res.* **1996**, *101*, 12331–12352. [CrossRef]

27. Wang, G.; Chen, D.; Su, J. Generation and life cycle of the dipole in the South China Sea summer circulation. *J. Geophys. Res.* **2006**, *111*, C06002. [CrossRef]

28. Zhuang, W.; Qiu, B.; Du, Y. Low-frequency western Pacific Ocean sea level and circulation changes due to the connectivity of the Philippine Archipelago. *J. Geophys. Res.* **2013**, *118*, 6759–6773. [CrossRef]

29. He, Y.; Cai, S.; Wang, D.; He, J. A model study of Luzon cold eddies in the northern South China Sea. *Deep Sea Res. Part I* **2015**, *97*, 107–123. [CrossRef]

30. Gan, J.; Qu, T. Coastal jet separation and associated flow variability in the southwest South China Sea. *Deep Sea Res. Part I* **2008**, *55*, 1–19. [CrossRef]

31. Wu, C.R.; Shaw, P.T.; Chao, S.Y. Assimilating altimetric data into a South China Sea model. *J. Geophys. Res.* **1999**, *104*, 29987–30005. [CrossRef]

32. Chu, X.; Xue, H.; Qi, Y.; Chen, G.; Mao, Q.; Wang, D.; Chai, F. An exceptional anticyclonic eddy in the South China Sea in 2010. *J. Geophys. Res.* **2014**, *119*, 881–897. [CrossRef]

33. Chambers, D.P.; Bonin, J.A. Evaluation of Release 05 time-variable gravity coefficients over the ocean. *Ocean Sci.* **2012**, *8*, 859–868. [CrossRef]

34. Chambers, D.P.; Willis, J.K. A Global Evaluation of Ocean Bottom Pressure from GRACE, OMCT, and Steric-Corrected Altimetry. *J. Atmos. Ocean Technol.* **2010**, *27*, 1395–1402. [CrossRef]

4

Coastal Improvements for Tide Models: The Impact of ALES Retracker

Gaia Piccioni *, Denise Dettmering, Marcello Passaro, Christian Schwatke, Wolfgang Bosch and Florian Seitz

Deutsches Geodätisches Forschungsinstitut der Technischen Universität München (DGFI-TUM), Arcisstrasse 21, 80333 München, Germany; denise.dettmering@tum.de (D.D.); marcello.passaro@tum.de (M.P.); christian.schwatke@tum.de (C.S.); wolfgang.bosch@tum.de (W.B.); florian.seitz@tum.de (F.S.)
* Correspondence: gaia.piccioni@tum.de

Abstract: Since the launch of the first altimetry satellites, ocean tide models have been improved dramatically for deep and shallow waters. However, issues are still found for areas of great interest for climate change investigations: the coastal regions. The purpose of this study is to analyze the influence of the ALES coastal retracker on tide modeling in these regions with respect to a standard open ocean retracker. The approach used to compute the tidal constituents is an updated and along-track version of the Empirical Ocean Tide model developed at DGFI-TUM. The major constituents are derived from a least-square harmonic analysis of sea level residuals based on the FES2014 tide model. The results obtained with ALES are compared with the ones estimated with the standard product. A lower fitting error is found for the ALES solution, especially for distances closer than 20 km from the coast. In comparison with in situ data, the root mean squared error computed with ALES can reach an improvement larger than 2 cm at single locations, with an average impact of over 10% for tidal constituents K_2, O_1, and P_1. For Q_1, the improvement is over 25%. It was observed that improvements to the root-sum squares are larger for distances closer than 10 km to the coast, independently on the sea state. Finally, the performance of the solutions changes according to the satellite's flight direction: for tracks approaching land from open ocean root mean square differences larger than 1 cm are found in comparison to tracks going from land to ocean.

Keywords: ocean tides; coastal altimetry; ALES retracker

1. Introduction

The ability to predict tides in coastal areas is of crucial importance for our society. In certain regions, tidal events combined with extreme meteorological conditions are responsible for severe flooding and consequent environmental issues. Another critical function of tide models is related to ocean satellite altimetry: altimetric measurements need to be corrected for tidal signal in order to separate the tidal-related variability of sea level from the anomalies coming from the ocean dynamic topography. Therefore, more accurate tide models result in more reliable altimetric sea level retrievals. During the last decades, improvements in oceanographic models and observation techniques brought remarkable results in tide monitoring and prediction. A fundamental benefit comes from satellite altimetry, which provides global-scale sea-level observations with an accuracy of few centimeters [1]. These measurements are mainly exploited in modern tide models as constraint for hydrodynamic modeling, or to empirically derive tidal information from satellite sea-level time-series. As described by [2], after the exploitation of satellite data TOPEX/Poseidon (launched in 1992) tide models showed an enhancement of approximately 5 cm over the previous models. However, significant errors for the major constituents M_2 and S_2 were found at high latitudes [3]. Also, low accuracy was observed in shallow waters, where tidal constituents are highly dependent on bathymetry and the shape of the oceanic shelf [4]. Major efforts in these areas brought a dramatic progress for shallow-water tides, with a consequent larger agreement among different models, and a clear improvement on the single constituents [5]. However, lower performances were observed in coastal regions, resulting in large discrepancies among the models. For models assimilating satellite measurements such situation may be due to a poor availability and quality of altimetric data, highly influenced by the presence of land [6], patches of water at very low sea state within the altimeter footprint [7], or ice [8]. In these areas, the returned echo assumes shapes that are considerably different from the typical open ocean radar return and therefore the signal needs to be fitted with a dedicated algorithm (called retracker). Exploiting these recent advances in data pre-processing, some dedicated coastal products are currently available [9].

The purpose of this paper is to assess the influence of a tailored coastal retracking method on the quality of an ocean tide model, which is an important step towards more oceanographic applications of coastal altimetry [10]. In other words, we want to quantify the difference at the coast between tidal constituents estimated with a dedicated coastal retracker and the same constituents derived with an ordinary open ocean retracker. The coastal retracker used for this experiment is the Adaptive Leading Edge Subwaveform (ALES) retracker. The reliability of this retracker has been proven in a number of applications such as the regional estimation of the seasonal cycle and trend of the sea level [11,12]. Moreover, improvements in areas with a complex macrotidal regimes were validated in [13]. The approach applied to derive the tidal constituents represents a prototype of the new Empirical Ocean Tide (EOT) model. This model takes advantage of the most recent altimetric products, with focus on coastal performances. The model scheme follows the former EOT11a approach [3]: residual tidal constituents are derived on a least-squares-based harmonic analysis applied to Sea Level Anomalies (SLA). In this case, an along-track solution was preferred compared to the classical grids in order to study the evolution of the performances and the impact of the retrackers with respect to the distance to the coast. The data used for the model are illustrated in Section 2 together with a brief description of the ALES retracker and the in situ dataset used for the comparison of the models (Sections 2.1 and 2.2). A more detailed explanation of the tide model approach can be found in Section 3. In Section 4 the methods used for the model comparison are shown, and in Section 5 the results are presented and discussed. Finally, in Section 6 the conclusions and future work are described.

2. Dataset Description

2.1. Altimeter Dataset

In this study, high-rate observations from Jason-1 and Jason-2 missions were used. The high-rate (20 Hz) data allow a ground spatial resolution of circa 350 m along-track, which was preferred over low-rate (1-Hz) products for this dedicated investigation over coastal areas. The data were extracted from the DGFI-TUM's Open Altimeter Database (OpenADB: https://openadb.dgfi.tum.de), which contains the original Sensor Geophysical Data Records (SGDR) and derived high-level products. Version SGDR-E is available in OpenADB for Jason-1, while for Jason-2 version SGDR-D was used. For the two missions only data provided during the reference orbit phase are included, obtaining a continuous time-series of 14 years, from January 2002 until February 2016. In order to compute the tidal constants, values of Sea Level Anomalies (SLA) are needed. At each point, SLA are calculated according to [14]:

$$SLA = H - R - h_{MSS} - h_{geo} \qquad (1)$$

where H is the orbital height of the satellite, R is the range, h_{MSS} is the height of the Mean Sea Surface (**MSS**), and h_{geo} is the sum of the heights of all the geophysical corrections. The **MSS** and the geophysical corrections applied for both missions are listed in Table 1. SLA values are additionally flagged with the following criteria:

- $-2.5 \text{ m} \le \text{SLA} \le 2.5 \text{ m}$ [3]
- SWH < 11 m [15]
- 7 dB < BS < 30 dB [15]
- Distance to coast >3 km

where SWH is the Significant Wave Height and BS is the backscatter coefficient. Note that the backscatter coefficient is commonly defined in literature as σ_0, however in this case BS is used to avoid ambiguities with the unit-weight variance (see Section 4). The tidal correction plays an important role in this investigation. The rationale behind the EOT approach consists of the following steps:

1. Application of a pre-existing tide model to correct the SLA
2. Estimation of residual periodic components associated with tides in the corrected SLA
3. Estimation of a new tide correction to adjust and improve the original FES2014 solution

Details on this procedure are given in Section 3. The pre-existing tide model used to correct SLAs is the Finite Element Solution 2014 (FES2014) and it is characterized by new high-resolution and coastal features, essential basis for a coast-dedicated tide model. According to the range used, two experiments are defined in this study: SGDR and ALES. The first uses the range obtained from the ocean retracker of the standard product. This is based on the MLE4 algorithm which adopts the Brown-Hayne (BH) functional form [16,17]. The BH models the expected reflected radar signal from the ocean surface and is considered to be suboptimal in the coastal zone. The second is based on the ALES retracker which restricts the application of BH to only a portion of the fitted radar echo (selected according to a first estimation of the sea state) in order to guarantee the precision of the measurement also in the open ocean, while avoiding spurious reflections typical of the coastal zone. The SSB correction applied depends on the range used. For both the retracking algorithms, the estimates of SSB are also provided. In particular, the ALES SSB is computed using the same SSB model of the SGDR [18] applied to the 20-Hz estimations of SWH and Wind Speed from ALES. This strategy has already been validated with in situ data in [19].

Table 1. List of corrections used to compute Sea Level Anomalies for this study.

Correction	Model	Reference
Mean Sea Surface	DTU15MSS	Andersen et al. [20]
Inverse barometer	Dynamic Atmospheric Correction (DAC)	Carrère et al. [21]
Wet and Dry troposphere	ECMWF	ECMWF [22]
Ionosphere	NOAA Ionosphere Climatology 2009 (NIC09)	Scharroo and Smith [23]
Ocean and Load tide	FES2014	Carrère et al. [24]
Solid Earth and Pole Tide	IERS Conventions 2003	McCarthy and Petit [25]
ALES Sea State Bias	ALES	Passaro et al. [19,26]
SGDR Sea State Bias	SGDR	AVISO/PODAAC [15]

2.2. Tide Gauge Dataset

The harmonic constants resulting from the along-track model are compared against in situ data at the coast. These data were taken from the Global Extreme Sea Level Analysis (GESLA) dataset, which is a unique-format collection of different datasets containing high-frequency (every one hour) sea-level measurements [27]. The harmonic constants used for the comparison were computed via the least-squares method, following e.g., [13]. Within this dataset, tide gauges were selected according to the following criteria:

- Maximum distance to satellite track: 50 km.
- GESLA data already assimilated in FES2014 model (Cancet, personal communication) are discarded.
- Stations near estuaries are discarded. Exceptions for fjords (e.g., Finnish and Canadian coasts).
- Final manual screening on the selected stations: tide gauges with timeseries shorter than one year are discarded while part of the timeseries containing doubtful offsets are not considered.

For each site, one or two crossing tracks were found, obtaining a total of 85 tracks for 70 tide gauges. Their locations are shown in Figure 1.

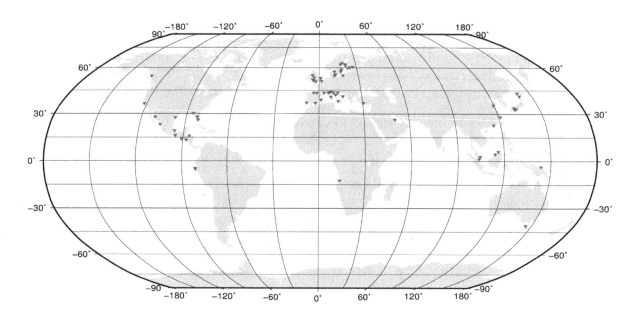

Figure 1. Location of the in situ data used in this work.

3. Tide Model Approach

The method used to compute the tidal constants is based on version 11a of DGFI-TUM's EOT model. For this work, the tidal analysis was based on Jason-1 and Jason-2 missions only, and an along-track solution was chosen. The approach is described in detail hereafter.

3.1. Selection of the Nodes

In the first step, the tracks of interest were selected according to the position of the tide gauges. The points along track at which the tidal constants were computed (also called nodes) are placed on the reference points belonging to the CTOH Topex/Poseidon nominal path (see acknowledgements), with a distance of circa 7 km between two nodes. Each node represents the center of a circular area of influence with radius (ψ_{max}) 15 km. All the SLA observations located within this area are selected for the tidal analysis. In order to account for the different behavior of SLAs, every observation i is weighted with a Gaussian function inversely proportional to its distance from the node ψ_i [3]:

$$w_i = e^{-\beta\psi_i^2} \tag{2}$$

where w_i is the value of the weight, ψ is the distance between the observation and the node, and β is defined as:

$$\beta = \frac{\ln 2}{\tau^2} \tag{3}$$

with $\tau = 0.4\psi_{max}$. The quantity τ is called half-weight width and determines the steepness of the Gaussian function. Namely, it defines the distance from the node for which the weight has value 0.5; in this case the value of τ is 6 km. In Figure 2, the node configuration together with the weighting representation is shown.

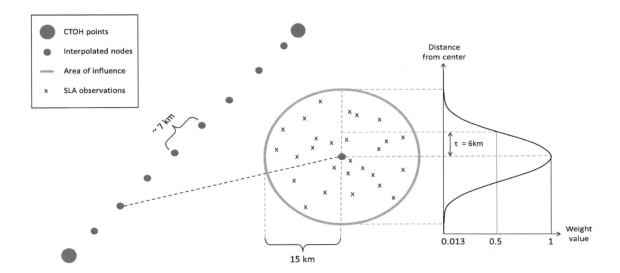

Figure 2. Scheme of the nodes and weighting process. From left to right: the nodes (pale blue) are located over the CTOH track points (red) with a distance of 7 km. For each node a circular area of interest is defined (green), and all the SLA observations (crosses) within this area are selected. The observations are weighted with a Gaussian function dependent on the distance from the node.

3.2. Computation of Tidal Constituents

After collecting the SLA observations, the components of the residual tide signal are computed for the main constituents: M_2, N_2, S_2, K_2, K_1, O_1, P_1, and Q_1. The solution is estimated by a weighted least-squares approach. The unweighted fitting equation for an observation i can be written as [28]:

$$SLA_i = m_j + a \cdot t_i + \sum_{k=1}^{n} \left(A_k \cos P_k \cdot f_k \cos \left(\theta_k + u_k\right) + A_k \sin P_k \cdot f_k \sin \left(\theta_k + u_k\right) \right) \qquad (4)$$

Together with the tidal elements, also the j-th mission offset m_j and the slope a—coming from the SLA time series at the node—are calculated [29]. The summation represents the sum of the n tidal constituents which are defined by the amplitude A_k, the phase P_k, and the given astronomical arguments summarized by the symbol θ_k, dependent on the time of the observation t_i. The nodal corrections f_k and u_k are also given, and can be obtained according to [30]. $A_k \cos P_k$ and $A_k \sin P_k$ are respectively the unknown coefficients of the in-phase and quadrature components and characterize the residual signal of constituent k. All the known right-hand-side elements of Equation (4) are used to form the design matrix of the least-squares approach. In this study, the weight matrix is diagonal and is filled with the weights of the SLAs computed with Equation (2). In combination with the least-squares estimation, a Variance Component Estimation (VCE) is also applied [3]. The VCE is used to weight the contribution of the different missions following the iterative procedure described in e.g., [31]. Finally, the residuals are added to the FES2014 constituents in order to obtain a full tidal signal, from which the amplitude and phase are derived.

4. Evaluation Methods

The different performances between the derived tide models are compared against the GESLA in situ data by computing the Root-Mean-Square difference (RMS) of their harmonic constants. For a k-th constituent, the RMS is computed as:

$$RMS_k = \sqrt{\frac{(A_M \cos P_M - A_T \cos P_T)^2 + (A_M \sin P_M - A_T \sin P_T)^2}{2}} \qquad (5)$$

where A_M and P_M are the amplitude and phase of one solution (*SGDR* or *ALES*) for that constituent, and A_T and P_T the ones given by the tide gauge observations. The absolute *RMS* difference between the *SGDR* and the *ALES* solution is written as $\Delta RMS_k = RMS_{k,SGDR} - RMS_{k,ALES}$ and is measured in cm. The relative *RMS* difference is also shown, which is described as:

$$\Delta RMS_k[\%] = \frac{\Delta RMS_k}{RMS_{k,SGDR}} \cdot 100 \tag{6}$$

This difference is expressed in percentage and indicates the relative improvement or worsening of the ALES solution with respect to the SGDR results. For an overall performance, the Root-Sum Squared (*RSS*) of the available n constituents is also calculated:

$$RSS = \sqrt{\sum_{k=1}^{n} RMS_k{}^2} \tag{7}$$

Both ΔRMS_k and RSS quantities are estimated for all the nodes along-track. In particular, results for the closest node to the tide gauge of interest (henceforth CNTG) are also considered, as they would represent the accuracies with respect to the coastal true values. To highlight the discrepancies among the ALES and SGDR solutions, it was chosen to express the results in terms of absolute differences, such as:

$$\Delta RSS = RSS_{SGDR} - RSS_{ALES} \tag{8}$$

A positive ΔRSS corresponds to higher RSS for the SGDR solutions, and therefore an improvement of ALES solutions with respect to a model using an ordinary retracker. Finally, the internal quality of the models is compared using σ_0, the unit-weight variance of the least-squares fit, and is inversely proportional to the number of observations [32]. The larger is σ_0, the higher is the uncertainty of the fitting.

5. Results and Discussion

5.1. Number of Observations

One of the most advantageous features of ALES retracker is the large amount of valid coastal measurements available along track. In this work this benefit is shown in terms of observations available for each node. In Figure 3 the difference between the number of observations of ALES and the ones retrieved with SGDR are displayed. This difference is expressed as: Δ_{obs}, i.e., observations of ALES minus observations of SGDR. Each dot represents a node along the tracks, plotted against the distance to the coast. The red markers highlight the positive values, that is, the nodes for which ALES provides a larger amount of data with respect to SGDR. The blue dots are used for the negative values. An interesting, yet expected behavior is observed for values below 20 km from the coast: far more observations are available with ALES while approaching the coast, with some exceptions for few points.

5.2. Fitting Uncertainty

An analogous comparison is shown for the variable σ_0, that represents the quality of the least squares fit. In Figure 4a the difference at each node between the σ_0 computed for the SGDR solutions and σ_0 obtained from ALES is shown. A positive value on the Y the change. -axis (red dots) corresponds to a larger fitting error for the SGDR solutions, and negative values (blue dots) for the contrary. From the plot it is clear that in most cases an improvement for σ_0 is achieved with ALES, with exception for few coastal points. The dependence of σ_0 on the number of observations may explain the smaller errors for ALES. However, from Figure 4b one can notice that large improvements in σ_0 are reached also for a lower amount of data. On the other hand, the few cases with larger internal errors may be found at nodes with more data availability. These special cases, which accounts for only the 1.5% of the cases, may be justified by residual erroneous estimations in the ALES data, which were not identified by the outliers analysis.

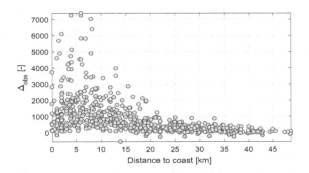

Figure 3. Difference in the number of observations between ALES and SGDR at each node against the distance to coast. The blue dots show the cases for which less observations are available for ALES, while the red dots correspond to a larger amount of data for ALES.

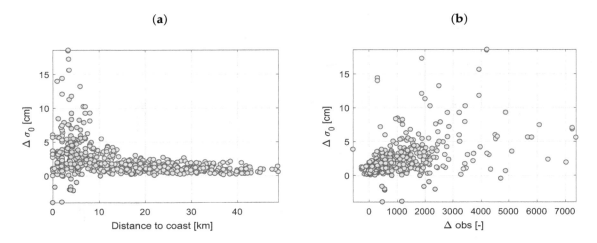

Figure 4. Difference of σ_0 values between SGDR minus ALES at each node against the distance to coast (**a**), and against difference in number of observations between ALES and SGDR (**b**).

5.3. Comparison Against In Situ Data

In this Section the results are compared in terms of RMS and RSS differences against in situ data. The first paragraph gives an overview of the results; the second paragraph discusses the dependency of the results on distance to coast, distance to tide gauge, track direction, and SWH.

5.3.1. General Results

In Figure 5, the spatial distribution of the absolute differences ΔRSS are shown. The differences are computed for the CNTG, for all the 85 tracks. In general, improvements are found for 66 tracks, with an average of 0.4 cm and a maximum value of 1.9 cm. The red dots indicate the highest improvements for the ALES solutions, which are located unevenly between Europe and the American continent.These higher values may be due to improvements to only few single constituents. This can be observed in Figure 6, where the absolute ΔRMS of the closest nodes to the tide gauge of interest are plotted against the longitude of the in situ site. The plot is divided in three rows for an easier visualization, and the ΔRMS of each tidal constant is color-coded according to the legend. An example of large improvements for single constituents can be seen at Prince Rupert, western Canada, or Ringhals, Sweden (respectively longitudes: $-130.32°$ and $12.11°$), where the values of ΔRMS for M_2 and S_2 for Prince Rupert, or Q_1 and O_1 are larger than 2.5 cm. In contrast, there are locations such as La Union, El Salvador, and Swansea, UK (longitudes $-87.82°$ and $-3.98°$) where the ALES solution shows a loss in performance—again differences larger than 2.5 cm—for constituent M_2. The RMS differences for all the 85 tracks are summarized in Table 2. The average values were computed using the single RMS values obtained at each site, at the CNTG. A mean improvement of 2 mm can be measured for the ALES solutions with respect to SGDR. It is important to stress that in the global average, results based on ALES are superior to results based on SGDR for every constituent. For K2, O1 and P1 the improvement is over 10%. For Q1, the improvement is over 25%. For larger RMS (such as M_2 and S_2) a minor effect of ALES is observed for the relative differences.

Table 2. Average of RMS for major constituents for the closest points to the tide gauges. The values are expressed in cm. The last column shows the relative difference between the two solutions.

Constituents	RMS_{ALES} (cm)	RMS_{SGDR} (cm)	ΔRMS (%)
M2	8.0	8.2	2.4
N2	2.1	2.3	8.7
S2	3.5	3.7	5.4
K2	1.4	1.6	12.5
K1	2.1	2.2	4.5
O1	1.4	1.6	12.5
Q1	0.8	1.1	27.3
P1	1.2	1.4	14.3

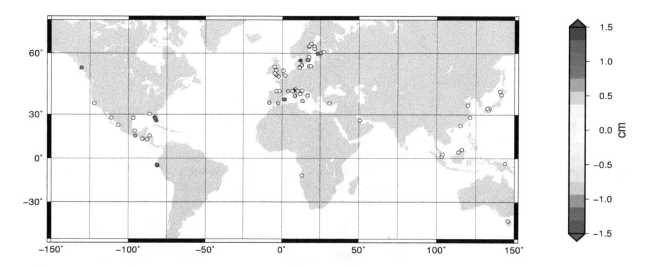

Figure 5. Geographical distribution of the ΔRSS (in cm) for the closest nodes to the tide gauge of interest.

5.3.2. Study of the Dependencies

Also in this section, the CNTGs are used to study the performances of the two retracker solutions. It must be pointed out that the CNTGs may not coincide with the closest points to the coast, as they depend on the position of the track with respect to land. For this reason, it was chosen to analyze the ΔRSS values against the distance to coast (Figure 7a) as well as against the distance to the tide gauge of interest (Figure 7b). The first plot shows not only that the nodes are mostly concentrated within 10 km to the coast, but also that improvements with ALES larger than 0.5 cm occur for nodes closer than 5 km. On the other hand, no visible dependency is observed between the values of ΔRSS and their distance from the tide gauge: in fact, the same improvements over 0.5 cm appear also for distances above 20 km. The dependency on the distance to the coast is also shown for the ΔRMS of the single constituents, Figure 8. Within 10 km from the coast, improvements below 2 cm can be found for all constituents. Larger variability is observed for the major constituents, and single values can reach e.g., ± 5 cm for M_2 and ± 3 cm for S_2.

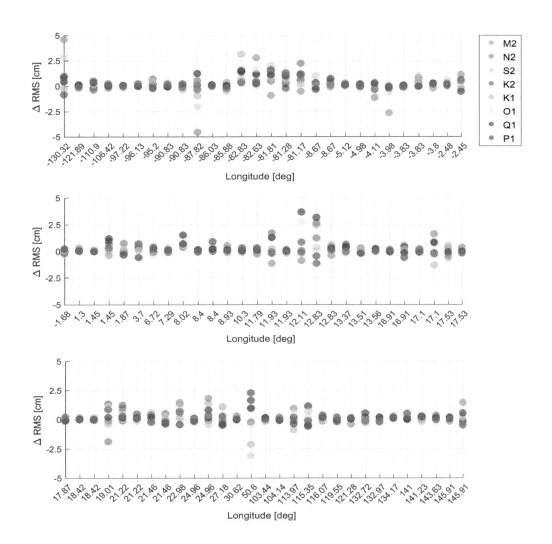

Figure 6. Difference of RMS for major constituents at the CNTG. The values are plotted against the longitude of their location.

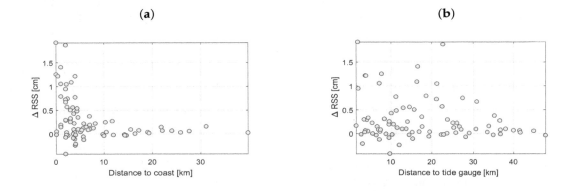

Figure 7. Difference of RSS against distance to coast (**a**) and tide gauge (**b**) in km. The values are shown for the closest nodes to the tide gauge of interest.

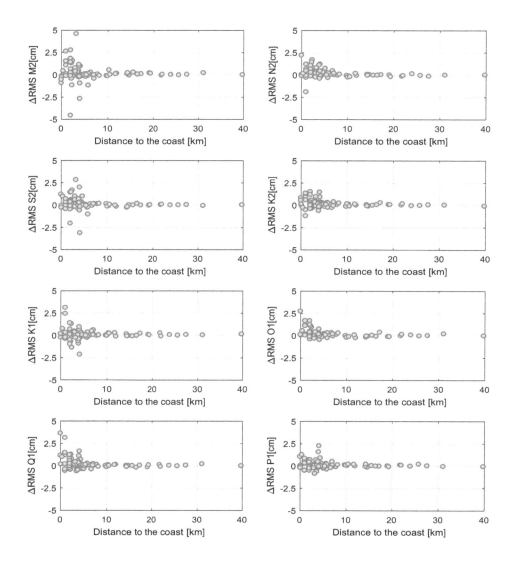

Figure 8. Absolute ΔRMS of major constituents at the CNTG. The values are plotted against the distance to the coast (in km).

Another aspect analyzed is the influence of the track direction on the results, because the performance of a retracker may change whether the satellite approaches land from ocean or flies from ocean to land, as well as if there is a bay (case: land-ocean-land) or the coast is parallel to the track (case: parallel to land). These results are shown in Table 3. The four main headers indicate the track position and the number of tracks used for the RMS average. The RSS computed from the RMS averages for each case are also displayed in the last row of the table. In general, lower values are found for the ALES solution, with exception of few constituents. An interesting result regards the transition land/ocean (i.e., the first four columns): both SGDR and ALES solutions show a higher performance for all constituents for the case ocean-to-land, against the land-to-ocean results, reaching differences larger than 1 cm for single constituents. This can be seen also from the averaged RSS, which show discrepancies of 26 mm between the ALES solutions and 21 mm for SGDR. This situation may be justified by a different behavior of the on-board tracker according to the flight direction, which may consequently influence the performance of the retrackers [7]. A clear example is presented in Figure 9, where the RMS of constituent M_2, computed for the ALES solution, is plotted for the nodes of tracks 111 and 92. Track 111 is ascending, and it goes from ocean to land, while track 92 is descending, going from land to ocean. It can be observed, that even though few nodes at track 92 are closer to the tide gauge, they still show a larger RMS with respect to nodes belonging to track

111. Moreover, larger discrepancies are found between ALES and SGDR for the case ocean-to-land, for which RSS values differ of 6 mm against 1 mm in the land-to-ocean case. Unfortunately, for the cases land-ocean-land and parallel-to-land only few tracks were available. However, from both the single RMS and the RSS values similar performance is found between ALES and SGDR solutions.

Table 3. Average of RMS computed for major constituents at the closest points to the tide gauges. The averages are computed after dividing the tracks according to their position with respect to the coast. The values are in cm.

Constituents	Land to Ocean: 30		Ocean to Land: 34		Land-Ocean-Land: 15		Parallel to Land: 6	
	RMS_{ALES}	RMS_{SGDR}	RMS_{ALES}	RMS_{SGDR}	RMS_{ALES}	RMS_{SGDR}	RMS_{ALES}	RMS_{SGDR}
M2	6.6	6.9	4.8	5.0	19.3	19.2	4.6	4.7
N2	1.7	1.8	1.3	1.6	4.8	5.2	1.4	1.4
S2	3.1	3.2	2.1	2.4	7.7	7.8	2.6	2.5
K2	1.2	1.3	1.0	1.3	2.8	2.9	1.7	1.7
K1	1.9	1.9	1.4	1.5	3.8	4.2	2.2	2.2
O1	1.2	1.3	1.0	1.3	2.5	2.7	1.6	1.6
Q1	0.8	0.9	0.7	1.0	1.3	1.8	0.9	1.0
P1	1.5	1.7	0.7	0.9	1.9	1.9	1.1	1.2
RSS	8.4	8.5	5.8	6.4	22.1	22.8	6.5	6.6

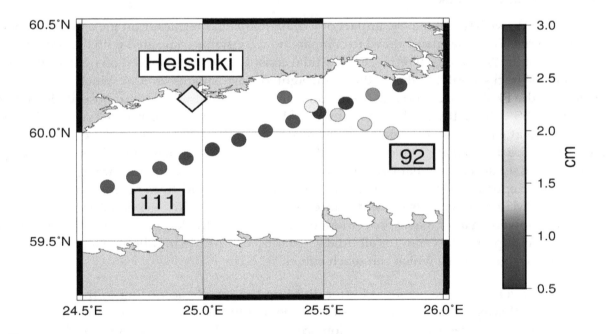

Figure 9. RMS values for M_2 constituent computed with ALES solutions for tracks 111 (**ascending**) and 92 (**descending**). The tracks face the tide gauge station of Helsinki (**diamond-shape marker**). The nodes of each track are represented by the round markers and the color shows the value of the RMS with respect to the tide gauge, in cm.

Finally, the sea state dependency is shown for the absolute ΔRSS. It was chosen to represent the sea state as the average of the SWH at each node, plus its standard deviation. The SWH values are taken from the ALES product. While the improvement of the ALES data for calm sea states (<2.5 m) is expected [33], the available literature concerning data quality in comparison with SGDR for wavy seas is still scarce. Indeed, from Figure 10, relevant improvements (>0.5 cm) are observed for sea states within 2.5 m, while only few examples are available for high states. However, ΔRMS > 1 cm are found above 3 m, showing no sensitive relation between the sea state and the data analyzed.

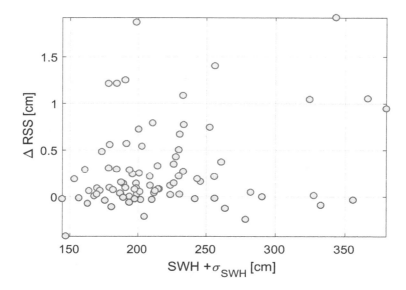

Figure 10. ΔRSS absolute values for the closest node to the tide gauge against the sea state, represented as the averaged SWH computed at each node plus its standard deviation.

6. Conclusions and Outlook

In this work, we have tested the impact of a coast-dedicated retracker on the estimation of ocean tidal constituents. The experiment aimed to compare tidal constants computed with the standard SGDR product against the ALES coastal retracker. The tidal constituents were derived on nodes defined along Jason satellites' tracks, applying the method of weighted least-squares on SLA which were previously corrected for the FES2014 tide model. The results were compared with in situ observations in terms of RMS and RSS values. It was shown that with ALES an increased number of sea level retrievals were available at each node, especially at distances closer than 20 km from the coast. The largest improvements are detected at distances within 10 km to the coast, independently from the geographical location and the sea state. A similar behavior was detected for the least-squares fitting error, which also show no clear dependency on the number of observations. In addition, no evident dependency is found for the RMS improvements to the distance to the tide gauges. The general ΔRSS results over the track nodes showed an average improvement of 0.4 cm for 66 tracks. However, the averaged RMS suggest a mean impact of few mm for all tidal constants. The ΔRMS highlighted a positive impact of ALES for single constituents, which can reach values >2.5 cm.

The RMS averages were presented after dividing the results according to the satellite's flight direction. From this experiment it was possible to see that the performance of both solutions change according to the track direction. The transition ocean-to-land shows smaller RMS for both ALES and SGDR solutions, and for all constituents. Single differences may exceed 1 cm when compared with land-to-ocean solutions.

In conclusion, improvements independent from the node position, together with a lower fitting error and a large data availability, make ALES a favorable choice for coastal tidal analysis. Indeed, the retracker will be exploited within the the new version of the EOT model. However, future research should be extended to minor tidal constituents, as well as dedicated regional analyses.

The improvements shown in this study were found despite the ALES retracking strategy was only applied in the residual analysis of the EOT procedure, while the original FES2014 model, which corrects for most of the tidal variability, is still based on SGDR data. We expect therefore that the use of ALES data could bring a decisive improvement in coastal tide modeling if used as a data source to estimate the full tidal component of the sea level variability. Finally, we recommend further investigations aimed to quantify the impact of additional altimetric corrections on the tidal estimation at the coast and to promote more oceanographic applications of coastal altimetry data.

Author Contributions: G.P. and D.D. conceptualized the research work; G.P. performed the experiments and analyzed the data with D.D. and M.P.; The manuscript was written by G.P. and M.P. contributed in writing the ALES fundamentals. C.S. provided special OpenADB tools used in the processing algorithm and insights on data structure; all the authors revised the manuscript and gave constructive comments.

Acknowledgments: The authors thank NOVELTIS, LEGOS, CLS Space Oceanography Division, CNES and AVISO for providing the FES2014 model. The TOPEX/Poseidon nominal tracks are available at: http://ctoh.legos.obs-mip.fr/products/altimetry/topex-poseidon-nominal-path/view distributed by the CTOH observational service. The tracks were selected by using the AVISO pass locator, accessible on: https://www.aviso.altimetry.fr/en/data/tools/pass-locator.html. The tide-gauge data used to compare our results were taken from the GESLA public dataset, available at: http://gesla.org/.

References

1. Bonnefond, P.; Haines, B.J.; Watson, C. In situ absolute calibration and validation. In *Coastal Altimetry*; Vignudelli, S., Kostianoy, A.G., Cipollini, P., Benveniste, J., Eds.; Springer: Berlin/Heidelberg, Germany, 2011; pp. 259–296, ISBN 978-3-642-12795-3.

2. Shum, C.K.; Woodworth, P.L.; Andersen, O.B.; Egbert, G.D.; Francis, O.; King, C.; Klosko, S.M.; Provost, C.L.; Li, X.; Molines, J.-M.; et al. Accuracy assessment of recent ocean tide models. *J. Geophys. Res.* **1997**, *102*, 25173–25194. [CrossRef]

3. Savcenko, R.; Bosch, W. *EOT11a—Empirical Ocean Tide Model From Multi-Mission Satellite Altimetry*; DGFI Report No. 89; Deutsches Geodätisches Forschungsinstitut: Munich, Germany, 2012. [CrossRef]

4. Andersen, O.B. Shallow water tides in the Northwest European shelf region from TOPEX/POSEIDON altimetry. *J. Geophys. Res.* **1999**, *104*, 7729–7741. [CrossRef]

5. Stammer, D.; Ray, R.D.; Andersen, O.B.; Arbic, B.K.; Bosch, W.; Carrère, L.; Cheng, Y.; Chinn, D.S.; Dushaw, B.D.; Egbert, G.D.; et al. Accuracy assessment of global barotropic ocean tide models. *Rev. Geophys.* **2014**, *52*, 243–282. [CrossRef]

6. Gommenginger, C.; Thibaut, P.; Fenoglio-Marc, L.; Quartly, G.; Deng, X; Gómez-Enri, J.; Challenor, P.; Gao, Y. Retracking altimeter waveforms near the coasts. In *Coastal Altimetry*; Vignudelli, S., Kostianoy, A.G., Cipollini, P., Benveniste, J., Eds.; Springer: Berlin/Heidelberg, Germany, 2011; pp. 61–101, ISBN 978-3-642-12795-3.

7. Passaro, M.; Cipollini, P.; Vignudelli, S.; Quartly, G.D.; Snaith, H.M. ALES: A multi-mission adaptive subwaveform retracker for coastal and open ocean altimetry. *Remote Sens. Environ.* **2014**, *145*, 173–189. [CrossRef]

8. Andersen, O.B.; Piccioni, G. Recent Arctic sea level variations from satellites. *Front. Mar. Sci.* **2016**, *3*, 76. [CrossRef]

9. Cipollini, P.; Benveniste, J.; Birol, F.; Fernandes, M.J.; Passaro, M.; Vignudelli, S. Satellite altimetry in coastal regions. In *Satellite Altimetry over Oceans and Land Surfaces*; Stammer, D., Cazenave, A., Eds.; CRC Press: Boca Raton, FL, USA, 2017; pp. 343–380.

10. Liu, Y.; Weisberg, R.H.; Vignudelli, S.; Roblou, L.; Merz, C.R. Comparison of the X-TRACK altimetry estimated currents with moored ADCP and HF radar observations on the West Florida Shelf. *Adv. Space Res.* **2012**, *50*, 1085–1098. [CrossRef]

11. Passaro, M.; Fenoglio-Marc, L.; Cipollini, P. Validation of Significant Wave Height from improved satellite altimetry in the German Bight. *IEEE Trans. Geosci. Remote Sens.* **2015**, *53*, 4. [CrossRef]

12. Passaro, M.; Dinardo, S.; Quartly, G.D.; Snaith, H.M.; Benveniste, J.; Cipollini, P.; Lucas, B. Cross-calibrating ALES Envisat and CryoSat-2 Delay-Doppler: A coastal altimetry study in the Indonesian Seas. *Adv. Space Res.* **2016**, *58*, 289–303. [CrossRef]

13. Lago, L.S.; Saraceno, M.; Ruiz-Etcheverry, L.A.; Passaro, M.; Oreiro, F.A.; D'Onofrio, E.E.; González, R. Improved sea surface height from satellite altimetry in coastal zones: A case study in Southern Patagonia. *IEEE J. Sel. Top. Appl. Earth Obs. Remote Sens.* **2017**, *10*, 3493–3503. [CrossRef]

14. Andersen, O.B.; Scharroo, R. Range and geophysical corrections in coastal regions—And implications for mean sea surface determination. In *Coastal Altimetry*; Vignudelli, S., Kostianoy, A.G., Cipollini, P., Benveniste, J., Eds.; Springer: Berlin/Heidelberg, Germany, 2011; pp. 103–145, ISBN 978-3-642-12795-3.

15. Picot, N.; Case, K.; Desai, S.; Vincent, P.; Bronner, E. *AVISO and PODAAC User Handbook. IGDR and GDR Jason Products*; SALP-MU-M5-OP-13184-CN (AVISO), JPL D-21352 (PODAAC); Physical Oceanography Distributed Active Archive Center: Pasadena, CA, USA, 2012.

16.   Brown, G. The average impulse response of a rough surface and its applications. *IEEE Trans. Antennas Propag.* **1977**, *25*, 67–74. [CrossRef]

17.   Hayne, G.S. Radar altimeter mean return waveforms from near-normal- incidence ocean surface scattering. *IEEE Trans. Antennas Propag.* **1980**, *28*, 687–692. [CrossRef]

18.   Tran, N.; Thibaut, P.; Poisson, J.-C.; Philipps, S.; Bronner, E.; Picot, N. Impact of Jason-2 wind speed calibration on the sea state bias correction. *Mar. Geod.* **2011**, *34*, 3–4. [CrossRef]

19.   Gómez-Enri, J.; Cipollini, P.; Passaro, M.; Vignudelli, S.; Tejedor, B.; Coca, J. Coastal altimetry products in the strait of Gibraltar. *IEEE Trans. Geosci. Remote Sens.* **2016**, *54*, 99. [CrossRef]

20.   Andersen, O.B.; Stenseng, L.; Piccioni, G.; Knudsen, P. The DTU15 MSS (Mean Sea Surface) and DTU15LAT (Lowest Astronomical Tide) reference surface. In Proceedings of the ESA Living Planet Symposium 2016, Prague, Czech Republic, 9–13 May 2016.

21.   Carrère, L.; Faugère, Y.; Bronner, E.; Benveniste, J. Improving the dynamic atmospheric correction for mean sea level and operational applications of atimetry. In Proceedings of the Ocean Surface Topography Science Team (OSTST) Meeting, San Diego, CA, USA, 19–21 October 2011.

22.   Persson, A. *User Guide to ECMWF Forecast Products*; ECMWF: Reading, UK, 2015.

23.   Scharroo, R.; Smith, W.H.F. A global positioning system-based climatology for the total electron content in the ionosphere. *J. Geophys. Res.* **2010**, *115*, 16. [CrossRef]

24.   Carrère, L.; Lyard, F.; Cancet, M.; Guillot, A.; Picot, N. FES 2014, a new tidal model—Validation results and perspectives for improvements. In Proceedings of the ESA Living Planet Symposium 2016, Prague, Czech Republic, 9–13 May 2016.

25.   McCarthy, D.D.; Petit, G.; (Eds.) *IERS Conventions (2003)*; IERS Technical Note 32; Verlag des Bundesamts für Kartographie und Geodäsie: Frankfurt, Germany, 2004.

26.   Passaro, M.; Kildegaard Rose, S.; Andersen, O.B.; Boergens, E.; Calafat, F.M.; Dettmering, D.; Benveniste, J. ALES+: Adapting a homogenous ocean retracker for satellite altimetry to sea ice leads, coastal and inland waters. *Remote Sens. Environ.* **2018**, in press. [CrossRef]

27.   Woodworth, P.L.; Hunter, J.R.; Marcos, M.; Caldwell, P.; Menendez, M.; Haigh, I. Towards a global higher-frequency sea level data set. *Geosci. Data J.* **2017**, *3*, 50–59. [CrossRef]

28.   Savcenko, R.; Bosch, W. Residual tide analysis in shallow water-contribution of ENVISAT and ERS altimetry. In *ESA-SP636 (CD-ROM), Proceedings of the Envisat Symposium 2007, Montreux, Switzerland, 23–27 April 2007*; Lacoste, H., Ouwehand, L., Eds.; ESA: Noordwijk, The Netherlands, 2007.

29.   Bosch, W.; Savcenko, R.; Flechtner, F.; Dahle, C.; Mayer-Gürr, T.; Stammer, D.; Taguchi, E.; Ilk, K.H. Residual ocean tide signals from satellite altimetry, GRACE gravity fields, and hydrodynamic modelling. *Geophys. J. Int.* **2009**, *178*, 1185–1192. [CrossRef]

30.   Doodson, A.T.; Warburg, H.D. *Admiralty Manual of Tides*; H. M. Stationery Off.: London, UK, 1941.

31.   Teunissen, P.; Amiri-Simkooei, A. Variance component estimation by the method of least-squares. In *VI Hotine-Marussi Symposium on Theoretical and Computational Geodesy, Proceedings of the IAG Symposium, Wuhan, China, 29 May–2 June 2006*. Xu, P, Liu, J., Dermanis, A., Eds.; Springer:Berlin/Heidelberg, 2008; ISBN 978-3-540-74583-9.

32.   Koch, K.-R. *Parameter Estimation and Hypothesis Testing in Linear Models*, 2nd ed.; Springer: Berlin/Heidelberg; Bonn, Germany, 1999; ISBN 978-3-540-65257-1.

33.   Passaro, M.; Cipollini, P.; Benveniste, J. Annual sea level variability of the coastal ocean: The Baltic Sea-North Sea transition zone. *J. Geophys. Res. Oceans* **2015**, *120*, 3061–3078. [CrossRef]

5

The Benefits of the Ka-Band as Evidenced from the SARAL/AltiKa Altimetric Mission: Quality Assessment and Unique Characteristics of AltiKa Data

Pascal Bonnefond [1,*], Jacques Verron [2], Jérémie Aublanc [3], K. N. Babu [4], Muriel Bergé-Nguyen [5], Mathilde Cancet [6], Aditya Chaudhary [4], Jean-François Crétaux [5], Frédéric Frappart [5,7], Bruce J. Haines [8], Olivier Laurain [9], Annabelle Ollivier [3], Jean-Christophe Poisson [3], Pierre Prandi [3], Rashmi Sharma [4], Pierre Thibaut [3] and Christopher Watson [10]

[1] SYRTE, Observatoire de Paris, PSL Research University, CNRS, Sorbonne Universités, UPMC Univ. Paris 06, LNE, 75014 Paris, France

[2] Institut des Géosciences de l'Environnement (IGE)/CNRS, 38041 Grenoble, France; Jacques.Verron@univ-grenoble-alpes.fr

[3] Collecte Localisation Satellites (CLS), 31520 Ramonville Saint-Agne, France; jaublanc@cls.fr (J.A.); aollivier@cls.fr (A.O.); jpoisson@cls.fr (J.-C.P.); pprandi@cls.fr (P.P.); pthibaut@cls.fr (P.T.)

[4] Space Applications Centre (ISRO), Ahmedabad 380015, India; kn_babu@sac.isro.gov.in (K.N.B.); aditya.osd@sac.isro.gov.in (A.C.); rashmi@sac.isro.gov.in (R.S.)

[5] Laboratoire d'Etudes en Géophysique et Océanographie Spatiales (LEGOS), 31400 Toulouse, France; Muriel.Berge-Nguyen@cnes.fr (M.B.-N.); Jean-Francois.Cretaux@legos.obs-mip.fr (J.-F.C.); Frederic.Frappart@legos.obs-mip.fr (F.F.)

[6] NOVELTIS, 31670 Labège, France; Mathilde.Cancet@noveltis.fr

[7] Géosciences Environnement Toulouse (GET), 31400 Toulouse, France

[8] Jet Propulsion Laboratory, California Institute of Technology, Pasadena, CA 91109, USA; Bruce.J.Haines@jpl.nasa.gov

[9] Géoazur—Observatoire de la Côte d'Azur, 06905 Sophia-Antipolis, France; Olivier.Laurain@oca.eu

[10] Surveying and Spatial Science Group, School of Geography and Environmental Studies, University of Tasmania, Hobart 7001, Australia; Christopher.Watson@utas.edu.au

* Correspondence: Pascal.Bonnefond@obspm.fr

Abstract: The India-France SARAL/AltiKa mission is the first Ka-band altimetric mission dedicated to oceanography. The mission objectives are primarily the observation of the oceanic mesoscales but also include coastal oceanography, global and regional sea level monitoring, data assimilation, and operational oceanography. The mission ended its nominal phase after 3 years in orbit and began a new phase (drifting orbit) in July 2016. The objective of this paper is to provide a state of the art of the achievements of the SARAL/AltiKa mission in terms of quality assessment and unique characteristics of AltiKa data. It shows that the AltiKa data have similar accuracy at the centimeter level in term of absolute water level whatever the method (from local to global) and the type of water surfaces (ocean and lakes). It shows also that beyond the fact that AltiKa data quality meets the expectations and initial mission requirements, the unique characteristics of the altimeter and the Ka-band offer unique contributions in fields that were previously not fully foreseen.

Keywords: altimetry; Ka-band; data processing; calibration; validation

1. Introduction

Launched in February 2013, the SARAL/AltiKa mission carries the first Ka-band altimeter ever flown [1]. The primary mission objective is the observation of the oceanic mesoscales, as well as coastal oceanography, global and regional sea level monitoring, data assimilation and operational oceanography. Secondary objectives include ice sheet, sea ice and inland waters monitoring. In order to continue the time series and to benefit from the existing mean sea surface, SARAL flew on the same orbit as Envisat. The SARAL/AltiKa mission was considered as a "gap filler" between Envisat (lost in April 2012) and Sentinel-3A (launched in February 2016) and has permitted to continue the sea level measurements on the historical ground track of the ERS-1&2 and Envisat missions. The nominal phase of the mission was planned to end three years after launch with an objective of five years. Since July 2016, SARAL/AltiKa has left its nominal orbit and entered in a Drifting Phase until its end of life. This phase, seamless in term of accuracy, provides a larger data coverage (Figure 1). The new orbit (altitude increased by 1 km) allows to improve the mean sea surface resolution but has been chosen to maintain also a good temporal resolution (15–20 days) for mesoscale studies.

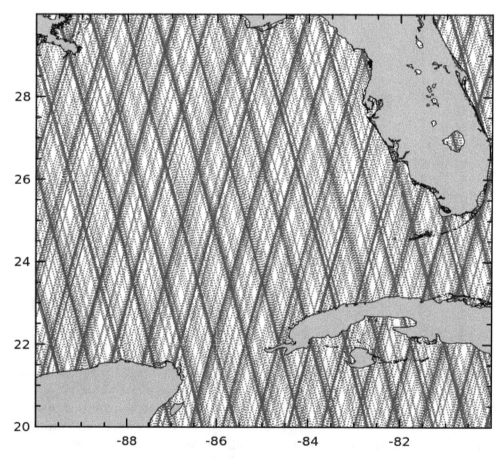

Figure 1. Blue: Repetitive orbit ground track. Red: Drifting orbit ground track.

SARAL/AltiKa's main scientific objective is to provide data products to the oceanographic research user community for studies to improve our knowledge of the ocean mesoscale variability, thanks to the improvement in spatial and vertical resolution brought by SARAL/AltiKa. This main scientific objective is divided into subthemes of mesoscale ocean dynamics: observations, theoretical analyses, modelling, data assimilation, and so forth. This leads to an improvement of our understanding of the climate system through its key ocean component and especially the role and impact of mesoscale features on the climate variability at large spatial and temporal scales. It also contributes to the study of coastal dynamics, which is important for many downstream applications including operational oceanography, which seeks large amounts of in-situ and space-based data.

SARAL/AltiKa's secondary objectives include the monitoring of the main continental water levels (i.e., lakes, rivers, and enclosed seas), the monitoring of large-scale sea level variations, the observation of polar oceans (thanks to the high inclination of its orbit), the analysis and forecast of wave and wind fields, the study of continental ice (thanks to the lower penetration in Ka-band) and sea ice, the access to low rains climatology (enabled in counterpart to the sensitivity of Ka-band to clouds and low rains) and the marine biogeochemistry (notably through the role of the meso- and submeso-scale physics).

The success of SARAL/AltiKa has been made possible by a very fruitful cooperation and efficient joint work between ISRO and CNES and thanks to the highly valuable exchanges between scientists of both countries. The objective of this paper is to highlights salient results in terms of quality assessment and unique characteristics of AltiKa data made since the special issue published in 2015 (Marine Geodesy, 38(S1):1, 2015, [2]). This paper is preceded by a companion paper focusing on the scientific applications where more details about the SARAL/AltiKa mission and its payload are also given [3]. In the first section of this paper, we will focus on the quality assessment of water level through Calibration and Validation and the second section will illustrate some unique contributions of Ka-band information in AltiKa data.

2. Quality Assessment of Water Level through Calibration and Validation

The observation of rising mean sea levels over the Earth's oceans using satellite altimetry has helped inform and shape the debate on global climate change that has emerged over the last two decades [4]. The ongoing monitoring of secular changes in global mean sea level with accuracy tolerances of better than 1 mm/year remains a fundamental science goal within satellite altimetry and represents one of the most challenging objectives in space geodesy. Central to this objective has been the recognition that calibration and validation (cal/val) are vital components of the altimeter measurement system technique. Cal/val defines a multi-disciplinary problem in and of itself, that pushes the limits of available terrestrial, oceanographic and space based observational techniques. Current estimates of regional and global change in mean sea level are only possible with careful and ongoing calibration of altimeter missions. Cross calibration of past, current and future altimeter missions will remain essential for continued sea-level studies.

During the last years, complementary altimetric missions operating concurrently have provided the unprecedented ability to compare measurement systems by undertaking relative calibrations. Studies have demonstrated through global statistics the power of such a technique [5,6]. Given the technical challenges of operating spacecraft for many years in orbit, however, the ability to cross-calibrate multiple missions in this manner cannot be assured. It reinforces the need for a range of complementary calibration methodologies—including a geographically diverse array of in situ absolute calibration sites that can assess changes to instrument behavior in near real time: such facilities, as described in [7], provide in situ sea level measurements in the same reference frame than the satellite altimetry ones and correct them from solid earth vertical motions.

The cal/val activities are focused not only on the important continuity between past, present and future missions but also on the reliability between offshore, coastal and inland altimetric measurement. The ability to sample varied geographically correlated errors and characterize them in an absolute sense are significant benefits of a well-distributed set of calibration sites. There is no doubt, however, that the "calibration task" requires a multifaceted approach, including both in situ calibration sites and regional/global studies.

This section will provide metrics of the SARAL/AltiKa accuracy and precision through absolute calibration sites (Section 2.1), global approach (Section 2.3) and an in between regional approach (Section 2.2). The main achievement of the data quality assessment provided in this section is illustrated in Figure 2 showing that AltiKa data have similar accuracy at the centimeter level in terms of absolute water level whatever the method (from local to global) and the type of water surfaces (ocean and lakes). This illustrates the improvement in altimetry in the last decade thanks to better processing and geophysical corrections but more importantly thanks to new technology (Ka-band, SAR) that paved the way of new frontiers in altimetry. Remaining differences from the various determinations shown in Figure 2 illustrate the characteristics of the studied water level (open-ocean, coastal, inland) but also the remaining errors coming from the methods used and the geodetic datum of each sites. Another important result is that cal/val activities over lakes are now at the same level of accuracy than the ones performed over decades at historical calibration sites (Bass Strait, Corsica, Crete, and Harvest).

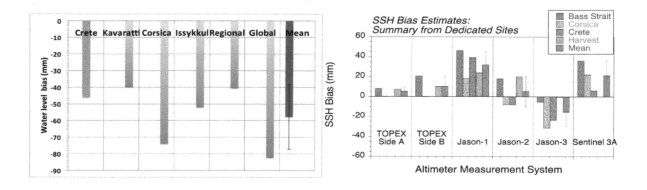

Figure 2. **Left** (this study): water level bias from Crete [8], Kavaratti (Section 2.1.1), Corsica (Section 2.1.2, Lake Issykul (Section 2.1.3), the regional (Section 2.2) and global (Section 2.3) approaches. The mean value of all the determinations is −54 mm with a standard deviation of 15 mm. **Right** [9]: TOPEX, Jason and Sentinel-3A Sea Surface Height (SSH) biases from in situ calibration sites.

2.1. Water Level Calibration from Open-Ocean, Coastal and Inland Waters In Situ Absolute Calibration Sites

2.1.1. Kavaratti, Altimeters' Calibration Site

The Kavaratti site gives an opportunity to do absolute calibration of altimeters over the tropical region. This calibration site do not show altimeter land contamination since it is in the open ocean, which gives it an added advantage. Kavaratti Island, is located southwest of Indian peninsula, ~450 km from the coast, as a platform of scientific observations, it has historic measurements from many Indian scientific institutions (e.g., Survey of India specific to water level). Figure 3 represents the part of Arabian Sea with Lakshadweep Islands at north and Maldives Islands at south. This permanent facility in Kavaratti is situated near the ground tracks of TOPEX/Jason and SARAL/AltiKa missions. The dedicated calibration facility at Kavaratti includes tide gauges, meteorological and oceanographic instruments, and a bottom pressure recorder [10]. This site has long-term sea level measurement at the tide gauge station TG1, which allows for the precise estimation of drift in sea surface height. The site characterization and calibration experiment and its results can be found in [10].

The absolute calibration of SARAL/AtiKa is carried out for its 23 cycles at this site (Figure 4). The determined absolute bias is −39.8 mm with standard deviation of 23.6 mm. In this AltiKa calibration exercise we have not used any local high resolution tide models, also the effects of wind and atmospheric pressures, and vertical land motion for the precise local geoid estimation are disregarded in determining the absolute Sea Surface Height (SSH) bias. These results and the good agreement with other in situ calibration sites shows that the location is robust and encourages us to continue observations to carry out the absolute calibration of altimetric SSH. The errors in estimation of absolute SARAL/AltiKa bias over this site will be improved through future field experiments.

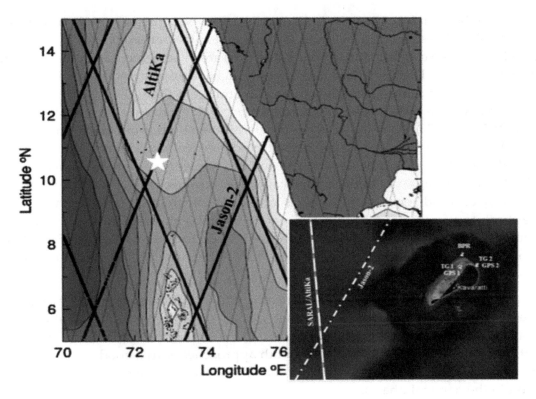

Figure 3. The large map shows the location of Kavaratti island (star) as well as the Jason-2 and SARAL/AltiKa ground tracks. The insert is a zoom of Kavaratti: ISRO's altimeter calibration site, has dual radar tide gauge station and a bottom pressure recorder.

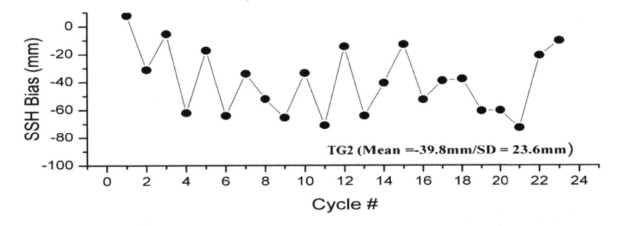

Figure 4. Absolute sea surface height bias of SARAL/AltiKa from cycle 1 to 23 using GDR-T products.

2.1.2. SARAL/AltiKa Absolute Calibration over Corsica during the Nominal and Drifting Phases

The geodetic Corsica site was set up in 1998 in order to perform altimeter calibration of the TOPEX/Poseidon (T/P) mission and subsequently, Jason-1, OSTM/Jason-2 and Jason-3. The scope of the site was widened in 2005 in order to undertake the calibration of the Envisat mission and most recently of SARAL/AltiKa. In [11], we have presented the first results from the latter mission using both indirect and direct calibration/validation approaches. The indirect approach utilizes a coastal tide gauge and, as a consequence, the altimeter derived Sea Surface Height (SSH) needs to be corrected for the geoid slope. The direct approach utilizes a novel GPS-based system deployed offshore under the satellite ground track that permits a direct comparison with the altimeter derived SSH. This latter approach has the advantage being able to be used without any specific infrastructure and have permitted us to make measurements even during the period when the difficulties with the satellite reaction wheels did not permit the ground track to be accurately maintained (from March 2015 until the Drifting Phase in July 2016): this is illustrated in the bottom part of Figure 5 and the SSH biases derived from those configurations (bold red circles) are in agreement with the whole time series. Since the beginning of the Drifting Phase, SARAL/AltiKa have overflown the Corsica calibration sites (Ajaccio and Senetosa) in different configurations illustrated in the right part of Figure 5. The SSH biases have then been computed using Senetosa or Ajaccio tide gauges or even both (cycle 100/pass 735 and cycle 104/pass 677) when the satellite have overflown both geoid areas (shaded in purple on Figure 5 maps). The results for these configurations (bold blue crosses in Figure 5) are also in very good agreement with the whole time series, even in a configuration very close to the coast (~3–4 km, cycle 100/pass 735). The overflights of both Senetosa and Ajaccio sites have permitted us to compare the SSH bias from Ajaccio and Senetosa tide gauges and we found an offset of 30 mm which is due to an issue in the absolute reference of Ajaccio tide gauge. This problem was identified in [11] and has been solved recently. This is discussed in [12] in which we have performed the Sentinel-3A calibration over both Senetosa and Ajaccio sites in a configuration close to the one encountered with SARAL/AltiKa for cycle 104 and pass 677 (see Figure 5).

In conclusion, in this analysis, after correcting the 30 mm offset in the Ajaccio tide gauge measurements, we demonstrate that using either indirect or direct calibration/validation approach gives similar SSH bias, respectively −74 ± 4 mm and −69 ± 11 mm. Even during the Drifting Phase, we can continue to monitor the SSH bias using both approaches with a good agreement compared to the whole time series and a period close to the initial ones (46 days in average compared to the 35-day repeat period in the nominal phase).

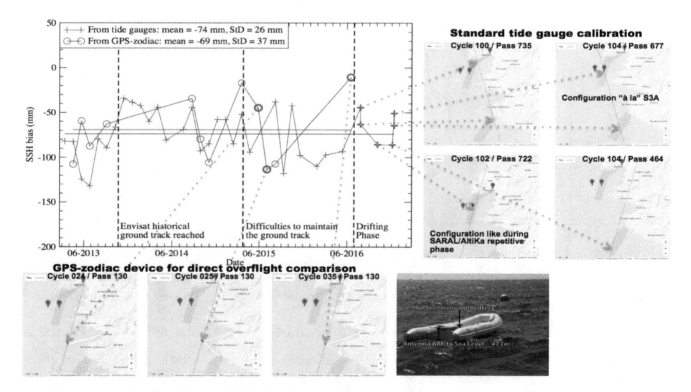

Figure 5. Sea Surface Height (SSH) bias time series for SARAL/AltiKa over Corsica for both direct (GPS-zodiac, red circles) and indirect (tide gauges, blue crosses) approaches. The bold symbols correspond to situations where the satellite was not overflying the calibration sites in a nominal configuration.

2.1.3. Absolute Calibration, Validation of SARAL/AltiKa over the Lake Issykkul

SARAL/AltiKa is used over the continents to measure the water height changes of lakes and rivers [1]. In order to determine the performances of altimeters over inland waters, a dedicated site of calibration/validation has been setup in 2004 over the lake Issykkul. Lake Issykkul (42–43°N, 76–79°E) is a large lake (6000 km^2 of area extent) located in Kyrgyzstan in Central Asia. Initially used as experimental site, thanks to promising first results obtained with TOPEX/Poseidon [13], the Lake Issykkul became permanent cal/val site for several altimeters in orbit. It has served for Jason-1, Jason-2, and Envisat with results coherent with other cal/val sites over the oceans [14,15]. The interest of having cal/val sites over a lake is to benefit from specific conditions of inland water (no tides, no inverse barometer) and to quantify the accuracy of altimeters to measure their water level changes.

Meanwhile, some corrections like the wet or dry tropospheric delay are not measured as accurately as over the ocean. Using ground instrumentation over the lake Issykkul therefore helps quantifying the error budget of altimeters over lakes in general. Since 2004, not less than 15 field campaigns have been done over the Lake Issykkul. The full experimental design is given in [16] and is not described here in details. In few words, it simply consists in measuring the water level changes along the track of the satellite, due to the geoid's gradient (which over the Lake Issykkul reaches several meters over long distance) using a GPS kinematic survey on a boat following the satellite tracks. It allows us to calculate the absolute bias of the altimeter.

First of all, the water level changes of the Lake Issykkul from 2013 to 2016 are calculated every 35 days using the altimetry data from the SARAL/AltiKa instrument. It is compared to the water level changes measured at an historical floating tide gauge located on the north coast of the lake (providing daily water level). It is also compared to hourly water level changes of the Lake measured by a radar installed on the south east lake shoreline. These comparisons allow us to quantify the accuracy of SARAL/AltiKa over the lake Issykkul in particular, which gives an order of magnitude of accuracy of this instrument for large lakes in general. It confirms results obtained in [17] that SARAL/AltiKa allows us to measure water level changes at very high accuracy (3.5 cm for lake Issykkul: Figure 6).

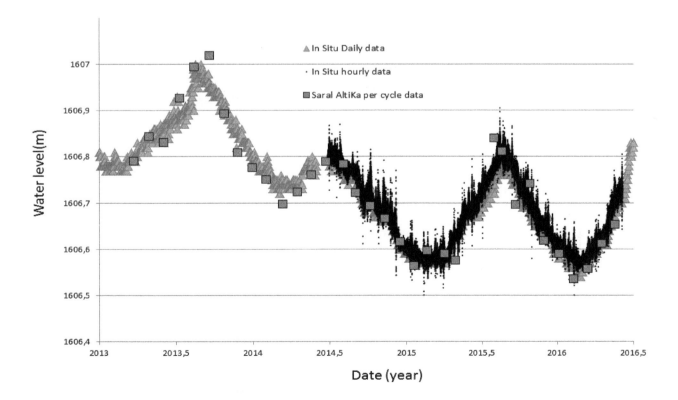

Figure 6. Water level of the lake Issykkul measured by SARAL/AltiKa, a radar and a tide gauge located on respectively east and north coast of the lake, delivering hourly and daily data. RMS of differences between SARAL/AltiKa and the in situ daily data is 3.5 cm.

For specific calculation of altimeter bias of AltiKa, a GPS survey of the track number 554 has been performed during a campaign done on 1 July 2014. The GPS vertical profile along this track is compared to individual measurements of the altimeters over the first 15 cycles using in situ measurements for all corrections:

-   Wet and dry tropospheric delay using a permanent GPS and weather stations
-   The GPS height antenna using a radar installed just below the GPS on the boat: see [16] for details.
-   The in situ daily level changes for the correction of the hydrological signal from cycle 1 to cycle 15. 1 July 2014 is when the GPS survey was done and corresponds to cycle 14.
-   The GINS [18] software was used for the calculation of the GPS 3D coordinates in PPP (Precise Point Positioning) mode.

We then compare the GPS vertical coordinates along the track 554 with the individual measurements of the lake altitude above the ellipsoid using AltiKa measurements. We thus perform the calculation of the absolute bias using the ocean and the ice1-OCOG (Offset Centre Of Gravity) retrackers:

-   With the ocean retracker, the water level bias, which is averaged along the track 554 and using the first 15 cycles of the satellite is given in Figure 7. The absolute bias obtained is: -52 ± 24 mm.
-   With ice1-OCOG retracker, the water level bias, which is averaged along the track 554 and using the first 15 cycles of the satellite, is given in Figure 8. The absolute bias obtained is: $+34 \pm 19$ mm.

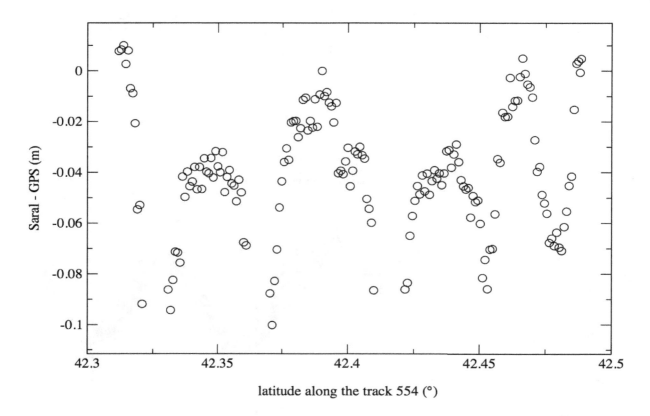

Figure 7. Differences between the GPS profiles and SARAL/AltiKa water altitude above the ellipsoid with ocean retracker at reference points corresponding to the GPS points along the SARAL/AltiKa track 554.

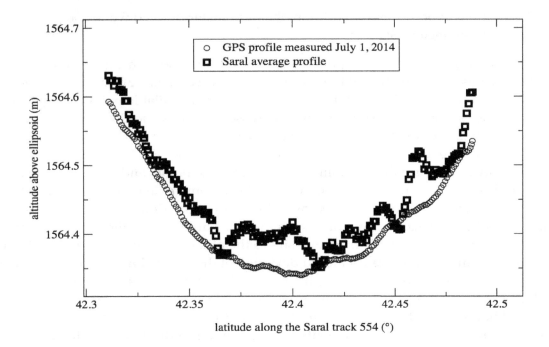

Figure 8. Vertical profiles along the track 554. The altimetry profile is an average over 15 first cycles at reference point corresponding to the GPS points of measurements. Ice1 retracker is represented.

2.2. Regional Calibration of SARAL/AltiKa Sea Surface Height in Corsica, at Harvest and at Bass Strait

In situ calibration of altimeter SSH is usually done at the vertical of a specific cal/val site by direct comparison of the altimeter data with the in situ data. Given that most of the in situ calibration sites (Harvest, Senetosa, Bass Strait and Gavdos) were specifically designed to be located under the TOPEX/Jason orbit, the classical absolute calibration technique can only be used for satellites using this orbit. The regional cal/val technique developed by Noveltis [19,20] aims to assess the altimeter range bias both on satellite passes flying over the calibration site and on satellite passes located several hundreds of kilometres away (Figure 9). In particular, it enables the monitoring of altimetry missions that do not fly directly over the calibration sites, such as SARAL/AltiKa. In principle this technique extends the single site approach to a wider regional scale, thus reinforcing the link between the local and the global cal/val analyses.

The regional method is used to compute the SARAL/AltiKa altimeter biases at three historical TOPEX/Jason-suite calibration sites: the Corsican calibration sites of Senetosa and Ajaccio, the Californian site of Harvest and the Australian site of Bass Strait. These calibration sites are characterized by very different ocean variabilities, which leads to various strategies in terms of corrections for the altimeter SSH.

In Corsica, the ocean variability (from few days to seasons) is rather low (about 50 cm on average), mainly governed by the wind and the atmospheric pressure variations, with low tidal amplitudes (about 20 cm). However, the presence of small islands close to the sites of Ajaccio and Senetosa can have an impact on the quality of the altimeter SSH and the associated corrections, in particular the radiometer wet tropospheric correction. The wet tropospheric correction derived from the ECMWF (European Centre for Medium-Range Weather Forecasts) model is thus used to correct the SARAL/AltiKa SSH in Corsica. In addition, the altimeter and the in situ tide gauge observations are corrected using respectively the COMAPI regional tidal model (Coastal Modeling for Altimetry Product Improvement) [21] for the tides and a global hydrodynamic simulation provided by LEGOS for the DAC (Dynamic Atmospheric Correction) effects [22]. In Harvest and Bass Strait, the ocean variability is much higher than in Corsica (about 2.5 m), with large tidal amplitudes (about 1.5 m). The FES2004 global tidal model [23] and the global DAC simulation is used to correct the altimeter and the in situ observations at both sites. In addition, the radiometer-derived wet tropospheric correction is used to correct the altimeter SSH.

Figure 10 shows the crossover points considered for the computation of the regional SARAL/AltiKa bias estimates at each calibration site. In Harvest and Bass Strait, the number of selected crossover points is rather small in comparison to the number of available points. Large variability is observed in the bias estimates at those surrounding crossover points and they are discarded from the computation (see [19,20] for details). Some previous work on the Envisat mission in Bass Strait [24] showed that using a more recent tidal model with higher resolution (FES2014 global tidal model, [25]) enables us to drastically reduce the variability at some of those points and to reintegrate them in the computation in this specific region. The mean regional bias estimates at each calibration site are given in Table 1. The standard deviation computed over the mean bias estimates at each crossover point is given in the last column. This information shows that the variability of the bias estimates over the considered crossover points is stable from one site to the other, in the order of 1 cm. However, the mean bias estimates and the averaged standard deviation of the bias estimates strongly vary from one site to the other, with mean variability of about 60 mm in Harvest and Bass Strait, twice larger than in Corsica, which is directly linked to local conditions (open ocean with rough seas in Harvest and Bass Strait, sheltered harbours with low ocean variabilities in Corsica). Even after correcting the 30 mm offset in the Ajaccio tide gauge measurements (see Section 2.1.2) the 47 mm difference observed in the bias estimates between Ajaccio and Senetosa, is for the moment unexplained and need further investigation. However, the averaged SSH bias from Ajaccio and Senetosa is 61 mm and is close to the values found from the absolute calibration study (see Section 2.1.2). Further investigation is also needed to understand the ~40 mm difference observed between the bias estimates in Corsica and the estimates in Harvest and Bass Strait. It is more than probable that high precision regional tidal models and dynamical atmospheric correction solutions would improve the computation of accurate altimeter and in situ sea surface heights in these regions.

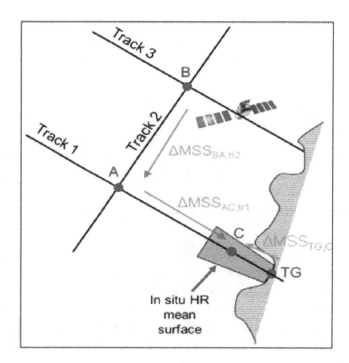

Figure 9. Generic diagram of the regional in situ calibration method. The points A and B represent the crossover points between, respectively, the satellite altimeter tracks 1 and 2, and the tracks 2 and 3. The tracks can belong to different altimeter missions. The in situ high resolution mean surface is used to link the tide gauge (TG) measurements to the altimeter data, and the comparison is done at the point C, located on this surface (adapted from [20]).

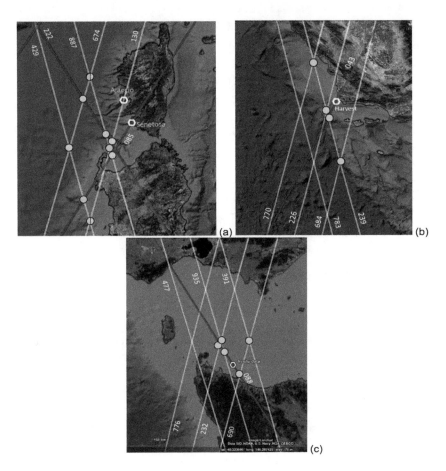

Figure 10. Altimeter crossover points used for the SARAL/AltiKa regional bias estimate in Corsica (**a**: Ajaccio in blue, Senetosa in orange), Harvest (**b**: green dots) and Bass Strait (**c**: green dots). The SARAL/AltiKa and Jason-2 grounds tracks are shown respectively in yellow and red.

Table 1. SARAL/AltiKa regional bias estimates at each site. In Bass Strait, the in situ data available for this study only covers the period until cycle 25.

SARAL/Altika Regional Bias Estimates (mm)	Mean (mm)	Std (mm)	Nb of Cycles	Nb of Xover Points	Std over the Xover Points (mm)
Senetosa (cycle 1–35)	-84 ± 5	32	34	7	5
Ajaccio (cycle 1–35)	-37 ± 6	36	33	6	13
Harvest (cycle 1–35)	-19 ± 12	64	30	4	12
Bass Strait (cycle 1–25)	-22 ± 11	56	24	5	13

2.3. Global Validation of SARAL/AltiKa Sea Surface Height over Ocean

Since the beginning, SARAL/AltiKa data shows a high quality and provides new information with respect to the previous missions [5]. At high latitude, thanks to its orbit reaching 82°, the Arctic Ocean and ice caps are better covered. Near the coast, measurements are more numerous and reliable [26] thanks to its high rate of 40 Hz (one point every 125 m along-track) and to its robust tracking mode (DIODE/Median, except for cycles 10 to 17 in autonomous DIODE mode). Furthermore, high frequency structures are better addressed, thanks to its Ka-band technology [27].

Its data availability is largely over the specification requirements, reaching 99.6% over oceans, Safe Hold Mode periods included, compared to 99.3% on Jason-2. Over other surfaces, the coverage is also very homogeneous as shown on Figure 11. Even the sensitivity of the Ka-band frequency to rain has less impact than expected. Indeed, only 5% of measurements may be not achieved due to rain rates > 1.5 mm/h according to geographic areas [28].

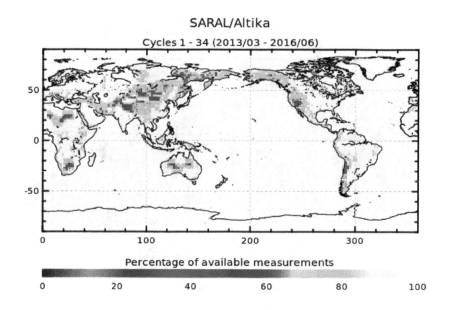

Figure 11. Map of the percentage of available measurements over land on SARAL/AltiKa's cycles 1 to 34.

The relative SSH bias compared to a zero reference of the TOPEX GMSL in 1993 (Global Mean Sea Level, CMEMS 2018 standards (Copernicus Marine Environment Monitoring Service)) is −82.3 mm for SARAL/AltiKa, whereas Sentinel-3A stands for −3.9 mm, Jason-2 for −18.5 mm and Jason-3 for −47.3 mm. Relative SSH bias and its evolution can also be derived from crossovers with a reference mission (such as Jason ones); even if not based exactly on the same length of the time series, results provided in [5] agree at the centimeter level with the ones presented here.

Its long-term stability as well, exceeds the mission requirements, featuring a very consistent mean sea level with respect to the reference record, derived from a combination of TOPEX/Jason-1/Jason-2 time series [29]. SARAL/AltiKa GMSL's shows a 4.8 mm/year evolution on this time period, thus keeping pace with Jason-2 (4.4 mm/year). It is a very satisfying statistic knowing that the time series for SARAL/AltiKa is still under 5 years which is the minimum period required to have a significant trend, see Figure 12.

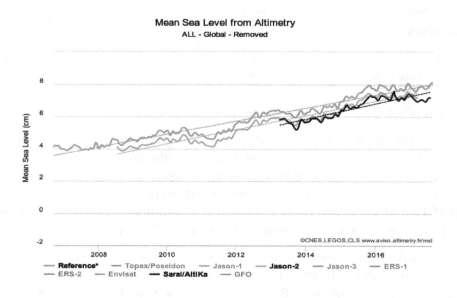

Figure 12. SARAL/AltiKa's Global Mean Sea Level (red) compared to Jason-2 (blue) and the reference one (green).

As for crossovers analysis (see details in [5]), it demonstrates excellent performances, with 5.3 cm of standard deviation compared to 5.4 cm for Jason-2, using the radiometric correction [30]. This illustrates SARAL/AltiKa's very good skills to resolve mesoscale signatures.

Finally, SARAL/AltiKa's Ka-band frequency enables the resolution of small scales of ocean dynamics (mesoscale). Thanks to its higher rate of 40 Hz and a smaller and statistically more homogeneous footprint, its noise level is lower than Jason-2/3 (Figure 13). Provided the application of a fine editing (for 20 Hz measurements: based on the Sea Level Anomaly coherence of consecutives measurements) and a correction [31] dedicated to reduce correlated noise between altimeter range and Significant Wave Height (SWH), it does even better than delay-doppler altimetry missions such as Sentinel-3A. Smaller ocean scales are therefore observable notably around 35 km where the spectral "bump" [32] is clearly reduced compared to "standard" use of the data (see Figure 1 in [3] for comparison).

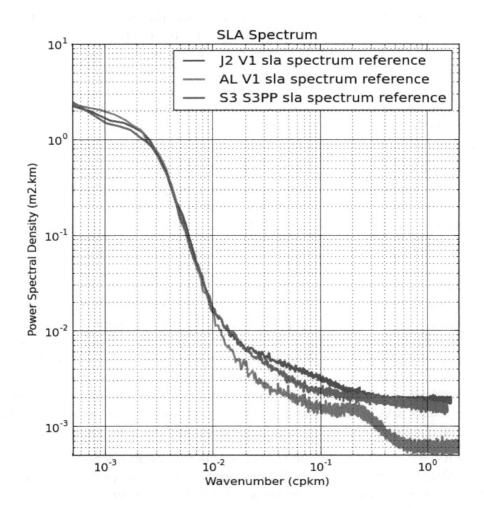

Figure 13. SARAL/AltiKa's SLA power spectrum (green) compared to Jason-2 (blue) and Sentinel-3A (red), with adapted editing and [31] correction for the three missions.

In conclusion, the SARAL/AltiKa mission performance remains excellent, compared to Jason-2, Jason-3 and Sentinel-3A, from fine scale ocean dynamics to long-term stability. And its quality will keep on improving, notably, thanks to several updates of geophysical corrections which are getting ready for the future reprocessing (GDR-E (Geophysical Data Record)) planned for 2018. The benefits of SARAL/AltiKa data to access the fine scale ocean dynamics are highlighted in our companion paper [3].

3. Unique Contributions of Ka-Band Information in AltiKa Data

The key feature of the altimetric payload has been the selection of Ka-band [27]. Using Ka-band avoids the need for a second frequency to correct for the ionosphere delay and eases the sharing of the antenna by the altimeter and the radiometer. The use of the Ka-band also allows the improvement of the range measurement accuracy in a ratio close to 2 (30 cm compared to 47 cm for Jason Ku-band) due to the use of a wider bandwidth and to a better pulse to pulse echo decorrelation. The higher pulse repetition frequency (4 KHz compared with 2 KHz on Jason-2) also permits a better along-track sampling of the surface. Moreover, the enhanced bandwidth (480 MHz compared with 320 MHz on Jason-2) enables a better resolution of the range. Finally, Ka-band antenna aperture is reduced, which limits the pollution within useful ground footprint. The known effect of rain on the Ka-band was such to put some uncertainty on the full availability of data in some regions. In fact, the larger sensitivity to small rates of rain was found to be less constraining than expected and in counterpart provides access to low rains climatology [28].

Apart from the improved accuracy and resolution for the liquid water surfaces (ocean and inland waters), the fact that AltiKa altimeter operates in Ka-band, which is higher than the previous frequencies, offer new paths of investigation. The penetration depth in snow is theoretically reduced from around 10 m in Ku-band to less than 1 m in Ka-band, such that the volume echo originates from the near subsurface. Second, the sharper antenna aperture leads to a narrower leading edge that reduces the impact of the ratio between surface and volume echoes of the height retrieval. Moreover, the volume echo in the Ka-band results from the near subsurface layer and is mostly controlled by ice grain size, unlike the Ku-band [33].

This section gives some illustrations of the unique characteristics of AltiKa instrument that offer unique contributions in fields that where not fully foreseen. Section 3.1 will give insights of the increase of resolution and data availability close to the coasts. Section 3.2 will focus on the Ka-band contribution for altimetry over snow and ice. Section 3.3 will study the rain sensitivity and finally Section 3.4 will give an original use of Ka-band measurements for studying the soil moisture.

3.1. Power Spectrum Density Analysis of SSH Signal from SARAL/AltiKa and Jason-2 (1 Hz) over Bay of Bengal

Indian coasts are vulnerable to high waves and rough sea conditions particularly during the pre- and post-monsoon seasons due to the presence of extreme atmospheric events such as cyclones. Unfortunately, wave models are still not mature enough to predict these waves due to the lack of accurate initial conditions and correct physical parameterizations. With the availability of altimeter data, assimilation of these data in numerical models is often found beneficial in improving the accuracies of the model predictions. One very significant promising development, particularly for the coastal regions, is the SARAL/AltiKa mission carrying onboard altimeter in the Ka-band and thus offering high spatial resolution suitable for coastal studies and assimilation in coastal wave models.

Tracks in the full Bay of Bengal region for the concurrent period of SARAL/AltiKa and Jason-2 (March 2013–July 2016) were used and 1 Hz Sea Surface Height (SSH) spectrum of both altimeters was studied. The Figure 14 shows the SSH spectrums of both the altimeters. Difference is noteworthy at wave length less than 50 km. The higher slopes in the wave lengths below 50 km shows the spectrum for SARAL/AltiKa have lesser noise than Jason-2 in this region. The power spectrum in the mesoscale region (50 km–250 km) is also fitted with power law $a*k^b$ (k is wavenumber). The values of b for AltiKa is -1.96 as compared to b $= -1.79$ for Jason-2. Above 50 km wave length, both SARAL/AltiKa and Jason are comparable to each other. Also the approachability towards the coast has been studied using the SARAL/AltiKa based ISRO Space Applications Center (SAC) coastal product and the Coastal and Hydrology Altimetry product (PISTACH) product from Jason-2. Figure 15a shows the 20-Hz SWH from Jason-2 towards the coast and Figure 15b shows 40-Hz SWH from SARAL/AltiKa. Clearly the SARAL/AltiKa based coastal product provides more valid data in proximity to the coast that will help to improve the coastal models.

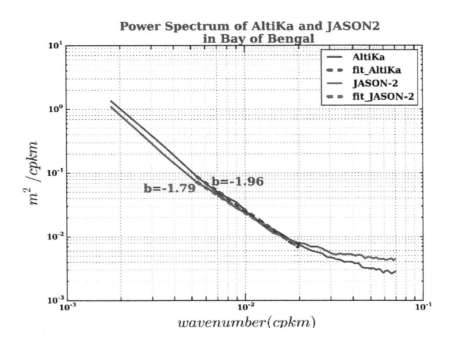

Figure 14. The power spectra of sea surface heights from SARAL/AltiKa and Jason-2 in Bay of Bengal. Green line represents (1/50 km).

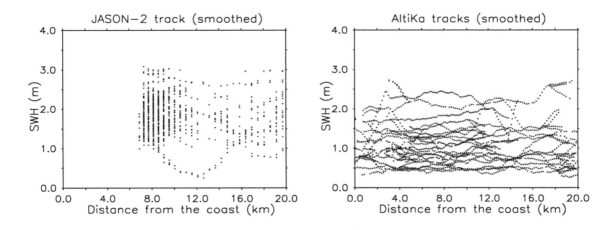

Figure 15. SWH from Jason-2 PISTACH product (**left**) and SAC coastal product (**right**) based on AltiKa data.

3.2. Complementarity between Ku and Ka Measurements over Ice Sheet Surfaces

Exploiting Ka frequency properties over land ice or sea ice is of particular interest to better assess the level of surfaces covered by ice or snow. Comparing measurements in Ku and Ka frequencies allows us to understand and quantify the penetration properties of the radar waves. Over water surfaces, Ku and Ka signals reflect at the air/water interface. It is commonly understood that in presence of snow or ice, the radar wave penetrates the ice pack if its wavelength is larger than the snow grain size (typically 0.5 mm but grain size depends on the temperature (snow metamorphism) and snow accumulation). According to [34], higher temperatures cause the snow grains to grow quickly, whereas higher accumulation rate causes them to slowly grow. Ku- and Ka-band wavelengths are respectively 2.21 cm and 0.84 cm. The factors between signal wavelength and grain size are respectively 44 and 17 in Ku- and Ka-bands. For a given wavelength, according to Mie theory [35], the scattering coefficient is conversely proportional to the radar wavelength at a power of 4. From Ku- to Ka-band, the scattering coefficient consequently increases by a factor 55. This leads to a penetration depth over snow surface between 0.1 m and 0.3 m in Ka-band [36]. Analysis performed on the altimeter signal acquired in Ku- and Ka-bands over the Antarctic ice sheet are consistent with this assumption.

To illustrate this theoretical introduction, we represent in Figure 16, the different penetration effects in Ku- and Ka-bands by comparing mean waveforms computed over ocean (SWH = 1 m ± 20 cm and similar epochs) and at a cross-over over Lake Vostok [37]. This lake in Antarctic is known to be quite perfectly flat, avoiding all effects induced by surface slopes (that can be different in Ku- and Ka-bands due to the different antenna gain patterns of the two missions). Over ocean, for SARAL/AltiKa (Ka-band) and Sentinel-3A (Ku-band in Pseudo-Low Resolution Mode), 1498 and 377 individual waveforms have been aggregated. Over Lake Vostok, for SARAL/AltiKa and Sentinel-3A, 1763 and 359 individual waveforms have been aggregated. We must note that waveforms have been artificially aligned to be compared.

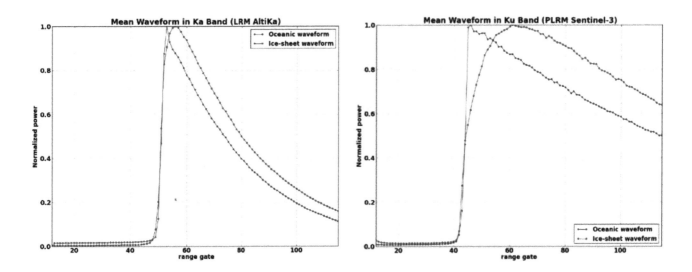

Figure 16. Mean SARAL/AltiKa (**left**) and Sentinel-3A (**right**) waveforms over ocean (blue) and Lake Vostok (red).

It appears on the left plot of Figure 16 that ocean and land ice leading edges are quite identical except for the very last points of the leading edge. Additional energy appears on the trailing edge samples for ice returns. This clearly indicates that the radar signal has been backscattered in the snow pack. However, considering that the distribution of energy in the leading edge is similar over ocean and over land ice, we can assume that an unique reflective surface is responsible for the main reflection (as for ocean) and that a retracking processing will provide the range corresponding to the distance between the satellite and this reflective surface. This surface can be assumed to be at the air/snow interface.

The plot on the right of Figure 16 shows that in Ku-band, returns are completely different over ocean and land ice. Clearly, the red echo can be considered as the summation of different contributions coming from different layers inside the snow pack, each contribution being attenuated depending on the absorption characteristics of each layer. In these cases, determining a range indicating the position of the main reflection (its distance from the radar) is very hard. If a threshold empirical retracker is used, the value of the threshold arbitrarily determines the reflective surface inside the snowpack.

For a better comparison of Ku and Ka measurements over Lake Vostok, waveforms in both bands are superimposed in Figure 17. The waveform acquired in Synthetic Aperture Radar (SAR) mode (Sentinel-3A, Ku-band) has been added (in red). As for SARAL/AltiKa, Sentinel-3A SAR mode mean echo over Lake Vostok presents a leading edge not impacted by penetration effects.

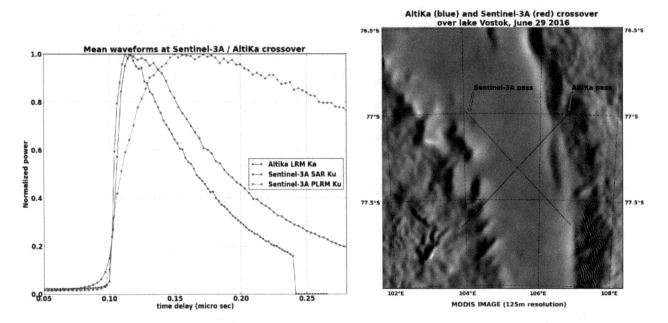

Figure 17. Left: Mean SARAL/AltiKa (blue) and Sentinel-3A (SAR in red, P-LRM in green) waveforms at a cross-over between SARAL/AltiKa and Sentinel-3A over Lake Vostok. **Right**: MODIS image of Lake Vostok with SARAL/AltiKa and Sentinel-3A tracks locations.

This analysis illustrates the complementarity between Ku and Ka measurements over ice sheet surfaces and the potential relevant information that could be retrieved by combining them. Of particular interest is the snow thickness that could be derived from the exploitation of the Ku and Ka waveform shapes, in particular their trailing edges.

Analysis of SARAL/AltiKa measurements over ice sheet have not only brought a new insight of the radar measurements of the snowpack but also encourage the development of new algorithms able to derive new parameters such as snow or ice pack properties. Scientific applications based on Ka-band unique characteristics and complementarity with Ku-band over ice sheet, icebergs and sea ice are developed in [3].

3.3. Impact of Rain on SARAL/AltiKa Measurements, a New Opportunity to Observe and Study the Rainfall Climatology

Since the first developments of the SARAL/AltiKa mission, one of the major concerns has been the Ka-band sensitivity to atmospheric conditions, and especially to atmospheric liquid water (both rain and clouds). Hence, the instrument power and gain have been optimized in order to maximize the Signal to Noise Ratio (SNR) and to minimize atmospheric attenuation issues [27]. As a result, the data loss over ocean is lower than anticipated (<0.1%) and the SNR is higher than the preflight expected value (14 dB) thanks to margins taken in the altimeter link budget (see details in [27]). An example of an AltiKa SNR map over ocean is provided in Figure 18 for cycle number 110, showing values varying from 18 to 32 dB. The altimeter link budget is thus excellent and allows a full data coverage over ocean.

Despite the very low number of lost data, the high sensitivity of the Ka-band to atmospheric liquid water (ten times larger than in Ku-band) impacts the altimeter waveform shape. AltiKa echoes are attenuated and distorted by rain events of heavy clouds [38–40], as shown in Figure 19a. The waveform amplitude is not only strongly attenuated but its shape is distorted, especially on the trailing edge part of the waveform. The example shown in Figure 19a is collocated to WindSat and SSMIS-F16/F17 rainfall rate measurements indicating a rain rate of about 4.5 mm/h during this event.

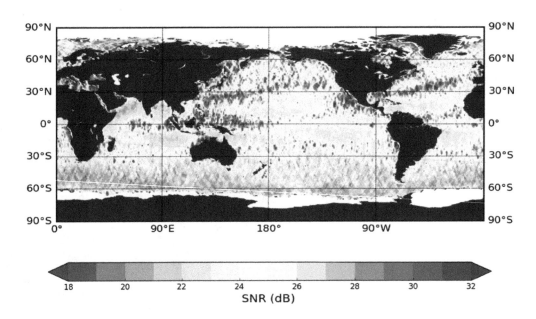

Figure 18. Map of Signal to Noise Ratio computed from standard MLE-4 estimates over ocean for cycle number 110 (from 2017/06/19 to 2017/07/24).

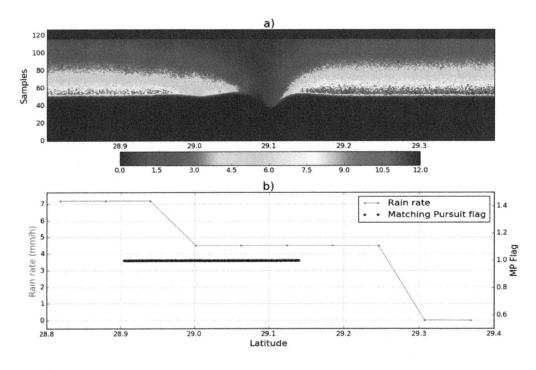

Figure 19. (a) AltiKa waveforms corrected from Automatic Gain Control during a rain event, pass 121 cycle 5; (b) collocated rainfall rate from SSMI and WindSat measurement in grey and Matching Pursuit flag in black [41].

Even if measurements are not lost during a rain event, the resulting waveform corruption may degrade the retracker performances and the geophysical estimates [42]. Indeed, operational ocean retrackers implemented in the ground segment are not designed to correctly process rain waveforms: these retrackers consider that the altimeter waveform footprint contains homogeneous backscattering properties. This is clearly not the case during rain events. As a result, the noise on each geophysical estimate increases when SARAL/AltiKa is overflying a rain cell. Impact of rain pollution on the range noise is illustrated in Figure 20. The collocation between SARAL/AltiKa and the WindSat/SSMIS measurements has been performed accounting for a maximum time delay of 1 hour. A clear rain rate dependency of the altimeter range noise (about 0.35 cm/mm/h) can be observed. The other geophysical estimates (significant wave height and backscattering coefficient) are impacted as well. Recent developments (in particular the "Adaptive" retracker) have shown very promising results to deal with such corrupted echoes [43].

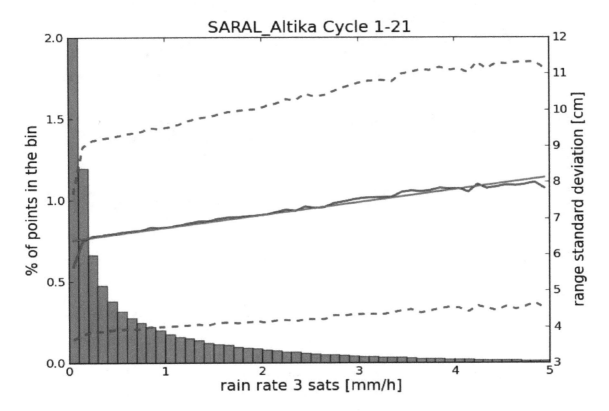

Figure 20. Variations of range std against rain rate. Please note that the first bin corresponding to dry measurements is truncated on the plot as it contains more than 70% of the measurement points. The plain blue line is the mean of the range standard deviation and the dashed blue lines represent the standard deviation for this mean. The green line is a linear regression of the mean (plain blue line).

The editing of altimeter data impacted by rain has been largely addressed in previous studies [44–46] and different methods exist to identify rain measurements. In the SARAL/AltiKa ground processing, the Matching Pursuit algorithm has been selected to flag rain data. This algorithm is based on the analysis of the short scale variations of the off-nadir angle estimate to identify rain cells in the SARAL/AltiKa measurements [41]. The performances of this algorithm have been assessed in [28] and show that the Matching Pursuit algorithm combined with the radiometer Integrated Liquid Water Content (ILWC) > 0.1 kg/m^2 provide a good rain flag. A map of 1 Hz flagged points is shown in Figure 21 illustrating the geographical distribution of the near 10% flagged measurements. As expected, most of the flagged points over ocean are located in the Intertropical Convergence Zone which reflects the global distribution of precipitations [47]. The comparison of the rain flag performances with WindSat and SSMIS-F16/F17 rainfall rates has shown acceptable performances but studies have been performed to improve and simplify rain flagging based on continuous wavelet transform. Results have been presented in [48].

Flagging SARAL/AltiKa measurements corrupted by rain presents a new opportunity to observe and study the rainfall climatology, seasonal and inter-annual variability. Moreover, the link between the signal attenuation and the rainfall rate can be exploited to extract valuable information on rain events. On-going studies seek to exploit Ka-band measurements to derive rain rate information, coupling them with the brightness temperature measurements provided by the radiometers for climatological monitoring.

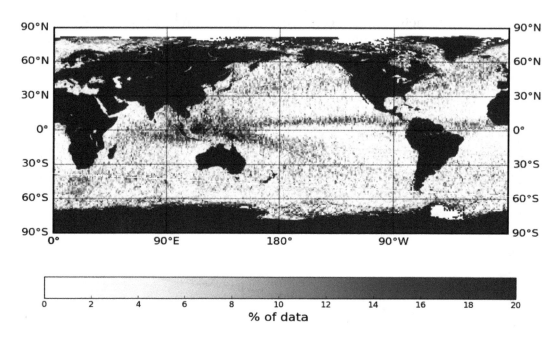

Figure 21. Percentage of 1 Hz flagged samples using the operational MP algorithm for which the ILWC >0.1 kg/m^2 for 1 year (cycles 100 to 110).

3.4. Relationship between Surface Soil Moisture and Radar Altimetry Backscattering Coefficients

Spatial and temporal variations of radar altimetry backscattering coefficients were related to the dynamics of surface properties. In semi-arid areas, soil moisture drives several surface processes including soil organic matter mineralization [49], vegetation productivity [50], land surface fluxes [51] and land surface atmosphere interactions [52]. A first inversion of the surface soil moisture (SSM) from the Ice-1 derived backscattering coefficients at Ku-band from Envisat RA-2 data was performed over Sahelian savannahs in the Gourma region of Mali [53]. Due to the lower sensitivity of nadir-looking altimeters to vegetation cover compared with side-looking SAR and scatterometers [54], higher correlation was found between in-situ SSM measurements and altimetry (R = 0.88, [53]) than with SAR (R = 0.85, [55]) and scatterometer (R = 0.63, [56]) backscattering coefficients over the sandy sites of the same semi-arid study area. Using the 16 first available cycles of the SARAL/AltiKa mission, the correlation is 0.88 between radar altimetry Ice-1 derived backscattering coefficients at Ka-band and level-3 SSM products derived from the Soil Moisture and Ocean Salinity satellite (SMOS) passive microwave observations over Sahelian savannahs [57]. The results obtained were generally better than the ones obtained using 3.5 times more numerous backscattering coefficients at Ku and C-bands from Jason-2 during the same observation period or the ones obtained using backscattering coefficients at Ku and S-bands from Envisat during its whole observation period (and comparing them to level-3 SSM products derived from AMSR-E) or passive microwave observations [57]. Time series of altimetry backscattering coefficients (obtained using Ice-1 retracking algorithm) at Ka-band from SARAL/AltiKa and Ku and C bands from Jason-2 are presented in Figure 22 from February 2013 to May 2016 over Sudano-Sahelian savanahs. This period corresponds to the 35 cycles of SARAL/AltiKa on its nominal orbit (and to cycles 169 to 291 of Jason-2). The backscattering level during the dry season decreases as the frequency increases—around 22.5 dB at C-band, 17 dB at Ku-band, and between 5 dB and 6 dB at Ka-band—whereas the amplitude of variations increases with the frequency—up to 10 dB at C-band, 20 dB at Ku-band and 30 dB at Ka-band. The larger dynamic of the backscattering coefficient at Ka-band during the rainy season is a strong advantage to accurately monitor the time-variations of SSM in semi-arid areas. Correlation coefficients between backscattering coefficients and SSM from SMOS (derived using the approach from [58]) were computed on the common period of availability of the two datasets, from February 2013 to May 2015. Due to the rapid response of SSM to rainfall—strong increase after a rainfall event followed by a fast decrease due to evaporation—in semi-arid environments, correlation were re-estimated applying a three-day smoothing window to limit the effect of the difference of acquisition time between altimeters and SMOS. Very similar results are obtained for both ascending and descending tracks:

(i)   R = 0.80 and 0.79 at Ka-band for SARAL/AltiKa tracks 0001 and 0846 respectively,
(ii) R = 0.58 and 0.54 at Ku-band for Jason-2 tracks 046 and 161 respectively,
(iii) R = 0.64 and 0.65 at C-band for Jason-2 tracks 046 and 161 respectively.

As the backscattering coefficient is linearly related to SSM over sand, better estimates of SSM can be expected from the inversion of altimetry backscattering coefficient at Ka-band than at other frequency bands commonly used in altimetry. In spite of the low temporal resolution of the SARAL/AltiKa mission with its repeat-period of 35 days, altimetry-derived SSM at Ka-band would be useful for the monitoring of the seasonal variations of this key variable along the altimeter tracks at high spatial resolution ($\sim$165 m using the high-frequency 40 Hz data) and could be used for the calibration and validation of land-surface models and low resolution active (scatterometry) and passive (radiometry) microwave-based products. The characteristics of the backscattering coefficient at Ka-band illustrated in this study is also used to study the snow and ice state and important scientific results are given in [3].

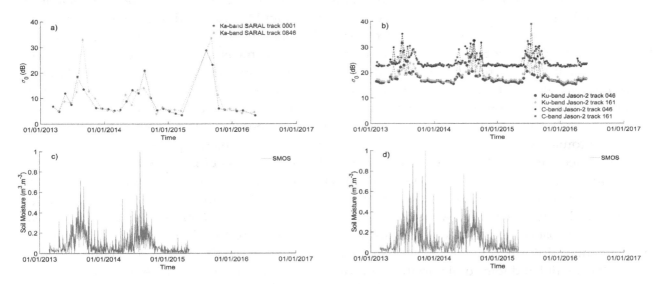

Figure 22. Time-series of backscattering coefficient (dB) for (**a**) SARAL/AltiKa (Ka-band) 0001 (blue) and 0846 (green) tracks and (**b**) Jason-2 (Ku and C-bands) 046 (blue and black) and 161 (light green and red), of volumetric soil moisture (m3.m-3), from SMOS (blue) at SARAL/AltiKa (**c**) and Jason-2 (**d**) sites.

4. Conclusions

Altimetry has been dominated by the Ku-band since the beginning and strong fears about a possible degradation by using Ka-band had limited the studies about launching a mission embarking on such technology. Vincent et al. [36] demonstrated the feasibility of such an altimeter but also importantly the need of higher resolution altimetry to address one of the most problematic features of ocean circulation, the mesoscale eddies and fronts. These features are essential to understanding the dynamics of ocean circulation on all space and time scales and theoretical studies of the cascade of energy over scales from 100–1000 km have always suffered from the lack of high-resolution observations. From [36] a Ka-band altimeter technology looked to be able to provide higher resolution to help answering these key issues. Selecting the Ka-band allows a larger bandwidth (480 MHz), which provides a vertical resolution of 0.3 m instead of 0.5 m in Ku-band. With such a resolution, the altimeter is close to a beam limited one: there is no 'plateau' in the echo as in Ku-band, since it strongly attenuates shortly after the leading edge, due to the small antenna aperture. This greatly reduces the pollution of 'land gates' into 'ocean gates', when considering land-sea transition areas. Moreover, due to the smaller antenna beamwidth, the Brown echo has a sharper shape in Ka-band than what is obtained with conventional Ku-band altimeters. Finally, the shorter decorrelation time of sea echoes at Ka-band also enables to double the number of independent echoes per second compared with Ku-band altimeters, allowing high frequency measurements (40 Hz, 20 Hz for Ku-band).

SARAL/AltiKa mission is the realization of the [36] study and in this paper we show that the quality of its measurements meets and generally exceeds the expectation of the visionary designers of this mission. Firstly, one of the main result is probably the fact that the AltiKa data quality and accuracy are met at a similar level whatever the water surfaces and whatever the scale (Section 2). This is demonstrated using global, regional or local methods and based on comparisons with various external data (in situ, other altimetric missions). The precision is at least at the same level or even better than for Ku-band for the open-oceans but improves substantially the along-track resolution (e.g., Figure 13 or Figure 14). These results will continue to improve with the reprocessing in 2018 with upgraded standards (GDR-E). Moreover, the characteristics of the AltiKa instrument allows to observe smaller water surfaces with equivalent precision and are of very great interest for coastal and inland waters studies [3]. The benefits of this increase of precision and resolution are highlighted through Science Applications of SARAL/Altika data [3]. Apart of these great achievements in term of precision and resolution, the characteristics of the altimeter and the Ka-band offer unique contributions in fields that where not fully foreseen. Before launch, concerns were raised about the sensitivity of the Ka-band to rain events, leading to missing and invalid measurements. However, in practice, the SARAL/AltiKa data return is remarkably high with few missing data (e.g., Figures 11 and 18). In fact, the sensitivity of Ka-band measurements to rain may also lead to a way of estimating very light rainfall over the oceans, for which we dramatically lack of information and which could lead to a great improvement of our knowledge of the oceanic rain climatology. Another specificity of the Ka-band is that electromagnetic wave penetration effects are minimized, which is particularly important over continental ice surfaces. From Ku- to Ka- bands, the scattering coefficient is increased by a factor 55: volume scattering is then being clearly dominant over surface scattering and gives information on snow or ice pack properties. Inversely, the radar wave extinction is also increased, leading to a penetration depth over the snow surface between 0.1 m and 0.3 m (versus 2 to 10 m in Ku-band). The altimetric observation and height restitution thus correspond to a thin subsurface layer and give complementary information than the one given historically by the Ku-band (see Section 3.2 and [3]). Finally, the spatial and temporal variations of Ka-band backscattering coefficients that are related to the dynamics of surface properties, show valuable information for studying snow and ice state but also surface soil moisture (e.g., Section 3.4). This last field of studies is at his infancy but results presented in this paper are clearly encouraging.

With the upcoming Surface Water & Ocean Topography (SWOT) mission embarking on a Ku-band nadir altimeter and a Ka-band interferometer (KaRIN) and the phase 0 of a new cryosphere satellite mission (bi-frequency altimeter in Ku- and Ka-band) that is currently studied by CNES and ESA, it is more than likely that SARAL/AltiKa Ka-band altimetric mission will continue to help preparing these missions but also help to understand Ku-band better.

Acknowledgments: The SARAL/AltiKa mission is an achievement of a very fruitful cooperation between ISRO and CNES. The contribution of EUMETSAT to data distribution is also strongly appreciated. We acknowledge the support of all the Investigators, PIs and Co-Is, of the SARAL/AltiKa Mission. Most of the studies performed by the French authors have been conducted and financed thanks to Centre National d'Etudes Spatiales (CNES), Centre National de la Recherche Scientifique (CNRS), and French Ministry of Research. The altimetry data used in Section 3.4 were made available by "Centre de Topographie des Océans et de l'Hydrosphère" (CTOH)—https://ctoh.legos.obs-mip.fr/. The SMOS data were obtained from the "Centre Aval de Traitement des Données SMOS" (CATDS), operated for the "Centre National d'Etudes Spatiales" (CNES, France) by IFREMER (Brest, France)—https://www.catds.fr/sipad/.

Author Contributions: Pascal Bonnefond and Jacques Verron coordinated the whole paper. K. N. Babu designed the study, analyzed the data and wrote the Section 2.1.1. Pascal Bonnefond and Olivier Laurain designed the study, analyzed the data and wrote the Section 2.1.2. Jean-François Crétaux and Muriel Bergé-Nguyen designed the study, analyzed the data and wrote the Section 2.1.3. Mathilde Cancet designed the study, analyzed the data and wrote the Section 2.2 while Pascal Bonnefond, Bruce J. Haines, Olivier Laurain and Christopher Watson provided the data and contributed to the writing and analysis. Annabelle Ollivier and Pierre Prandi designed the study, analyzed the data and wrote the Section 2.3. Aditya Chaudhary and Rashmi Sharma designed the study, analyzed the data and wrote the Section 3.1. Pierre Thibaut and Jérémie Aublanc designed the study, analyzed the data and wrote the Section 3.2. Jean-Christophe Poisson, Pierre Prandi and Pierre Thibaut designed the study, analyzed the data and wrote the Section 3.3. Frédéric Frappart designed the study, analyzed the data and wrote the Section 3.4. All the other co-authors helped in the analysis and paper writing.

References

1. Verron, J.; Sengenes, P.; Lambin, J.; Noubel, J.; Steunou, N.; Guillot, A.; Picot, N.; Coutin-Faye, S.; Sharma, R.; Gairola, R.M.; et al. The SARAL/AltiKa altimetry satellite mission. *Mar. Geod.* **2015**, *38*, 2–21, doi:10.1080/01490419.2014.1000471.
2. Verron, J.; Picot, N. Preface. *Mar. Geod.* **2015**, *38*, 1, doi:10.1080/01490419.2015.1052711.
3. Verron, J.; Bonnefond, P.; Aouf, L.; Birol, F.; Bhowmick, S.A.; Calmant, S.; Carret, A.; Conchy, T.; Crétaux, J.F.; Dibarboure, G.; et al. The benefits of the Ka-band as evidenced from the SARAL/AltiKa altimetric mission: Scientific applications. *Remote Sens.* **2017**, under review.
4. Church, J.; Clark, P.; Cazenave, A.; Gregory, J.; Jevrejeva, S.; Levermann, A.; Merrifield, M.; Milne, G.; Nerem, R.; Nunn, P.; et al. Climate change 2013: The physical science basis. Contribution of Working Group I to the fifth assessment report of the intergovernmental panel on climate change. In *Sea Level Change*; Cambridge University Press: Cambridge, UK; New York, NY, USA, 2013; Chapter 13.
5. Prandi, P.; Philipps, S.; Pignot, V.; Picot, N. SARAL/AltiKa global statistical assessment and cross-calibration with Jason-2. *Mar. Geod.* **2015**, *38*, 297–312, doi:10.1080/01490419.2014.995840.
6. Ablain, M.; Philipps, S.; Picot, N.; Bronner, E. Jason-2 global statistical assessment and cross-calibration with Jason-1. *Mar. Geod.* **2010**, *33*, 162–185, doi:10.1080/01490419.2010.487805.
7. Bonnefond, P.; Haines, B.; Watson, C. Coastal altimetry. In *In Situ Calibration and Validation: A Link from Coastal to Open—Ocean Altimetry*; Springer: Berlin/Heidelberg, Germany, 2011; Chapter 11, pp. 259–296, ISBN 978-3-642-12795-3.
8. Mertikas, S.P.; Daskalakis, A.; Tziavos, I.N.; Vergos, G.; Fratzis, X.; Tripolitsiotis, A. First calibration results for the SARAL/AltiKa altimetric mission using the gavdos permanent facilities. *Mar. Geod.* **2015**, *38*, 249–259, doi:10.1080/01490419.2015.1030052.
9. Bonnefond, P.; Desai, S.; Haines, B.; Leuliette, E.; Picot, N. Regional and Global CAL/VAL for Assembling a Climate Data Record Summary. In Proceedings of the Ocean Surface Topography Science Team Meeting, Miami, Florida, 2017. Available online: https://tinyurl.com/ycu32zeo (accessed on 9 January 2018).
10. Babu, K.N.; Shukla, A.K.; Suchandra, A.B.; Kumar, S.V.V.A.; Bonnefond, P.; Testut, L.; Mehra, P.; Laurain, O. Absolute calibration of SARAL/AltiKa in Kavaratti during its initial Calibration-validation phase. *Mar. Geod.* **2015**, *38*, 156–170, doi:10.1080/01490419.2015.1045639.
11. Bonnefond, P.; Exertier, P.; Laurain, O.; Guillot, A.; Picot, N.; Cancet, M.; Lyard, F. SARAL/AltiKa absolute calibration from the multi-mission Corsica facilities. *Mar. Geod.* **2015**, *38*, 171–192, doi:10.1080/01490419.2015.1029656.
12. Bonnefond, P.; Laurain, O.; Exertier, P.; Boy, F.; Guinle, T.; Picot, N.; Labroue, S.; Raynal, M.; Donlon, C.; Féménias, P.; et al. Calibrating SAR SSH of sentinel-3A and CryoSat-2 over the Corsica facilities. *Remote Sens.* **2018**, *10*, 92, doi:10.3390/rs10010092.
13. Crétaux, J.F.; Calmant, S.; Romanovski, V.; Shabunin, A.; Lyard, F.; Bergé-Nguyen, M.; Cazenave, A.; Hernandez, F.; Perosanz, F. An absolute calibration site for radar altimeters in the continental domain: Lake Issykkul in Central Asia. *J. Geod.* **2009**, *83*, 723–735, doi:10.1007/s00190-008-0289-7.
14. Crétaux, J.F.; Calmant, S.; Romanovski, V.; Perosanz, F.; Tashbaeva, S.; Bonnefond, P.; Moreira, D.; Shum, C.K.; Nino, F.; Bergé-Nguyen, M.; et al. Absolute Calibration of Jason Radar Altimeters from GPS Kinematic Campaigns over Lake Issykkul. *Mar. Geod.* **2011**, *34*, 291–318, doi:10.1080/01490419.2011.585110.
15. Crétaux, J.F.; Bergé-Nguyen, M.; Calmant, S.; Romanovski, V.; Meyssignac, B.; Perosanz, F.; Tashbaeva, S.; Arsen, A.; Fund, F.; Martignago, N.; et al. Calibration of Envisat radar altimeter over Lake Issykkul. *Adv. Space Res.* **2013**, *51*, 1523–1541, doi:10.1016/j.asr.2012.06.039.

16. Cretaux, J.F.; Bergé-Nguyen, M.; Calmant, S.; Djamangulova, N.; Satylkanov, R.; Lyard, F.; Perosanz, F.; Verron, J.; Samine Montazem, A.; Le Guilcher, G.; et al. Absolute calibration / validation of the altimeters on sentinel-3A and Jason-3 over the lake Issykkul. *Remote Sens.* **2017**, under review.

17. Arsen, A.; Crétaux, J.F.; del Rio, R.A. Use of SARAL/AltiKa over Mountainous Lakes, Intercomparison with Envisat Mission. *Mar. Geod.* **2015**, *38*, 534–548, doi:10.1080/01490419.2014.1002590.

18. Marty, J.C.; Loyer, S.; Perosanz, F.; Mercier, F.; Bracher, G.; Legresy, B.; Portier, L.; Capdeville, H.; Fund, F.; Lemoine, J.M.; et al. GINS: The CNES/GRGS GNSS scientific software. In Proceedings of the 3rd International Colloquium Scientific and Fundamental Aspects of the Galileo Programme, ESA Proceedings WPP326, Copenhagen, Denmark, 31 August–2 September 2011; Volume 31.

19. Jan, G.; Ménard, Y.; Faillot, M.; Lyard, F.; Jeansou, E.; Bonnefond, P. Offshore Absolute Calibration of space-borne radar altimeters. *Mar. Geod.* **2004**, *27*, 615–629, doi:10.1080/01490410490883469.

20. Cancet, M.; Bijac, S.; Chimot, J.; Bonnefond, P.; Jeansou, E.; Laurain, O.; Lyard, F.; Bronner, E.; Féménias, P. Regional in situ validation of satellite altimeters: Calibration and cross-calibration results at the Corsican sites. *Adv. Space Res.* **2013**, *51*, 1400–1417, doi:10.1016/j.asr.2012.06.017.

21. Cancet, M.; Lux, M.; Pénard, C. COMAPI: New Regional Tide Atlases and High Frequency Dynamical Atmospheric Correction. In Proceedings of the Ocean Surface Topography Science Team Meeting, Lisbon, Portugal, 2010. Available online: https://tinyurl.com/y98xedyq (accessed on 9 January 2018).

22. Carrère, L.; Lyard, F. Modeling the barotropic response of the global ocean to atmospheric wind and pressure forcing—Comparisons with observations. *Geophys. Res. Lett.* **2003**, *30*, 1275.

23. Lyard, F.; Lefevre, F.; Letellier, T.; Francis, O. Modelling the global ocean tides: Modern insights from FES2004. *Ocean Dyn.* **2006**, *56*, 394–415, doi:10.1007/s10236-006-0086-x.

24. Cancet, M.; Watson, C.; Haines, B.; Bonnefond, P.; Lyard, F.; Laurain, O.; Guinle, T. Regional CALVAL of Jason-2 and SARAL/AltiKa at Three Calibration Sites. In Proceedings of the Ocean Surface Topography Science Team Meeting, Reston, USA, 2015. Available online: https://tinyurl.com/ybm9huug (accessed on 9 January 2018).

25. Carrère, L.; Lyard, F.; Cancet, M.; Guillot, A.; Picot, N.; Dupuy, S. FES2014: A New Global Tidal Model. In Proceedings of the Ocean Surface Topography Science Team Meeting, Reston, USA, 2015. Available online: https://tinyurl.com/y9j9p4mf (accessed on 9 January 2018).

26. Valladeau, G.; Thibaut, P.; Picard, B.; Poisson, J.C.; Tran, N.; Picot, N.; Guillot, A. Using SARAL/AltiKa to improve Ka-band Altimeter measurements for Coastal Zones, Hydrology and ice: The PEACHI Prototype. *Mar. Geod.* **2015**, *38*, 124–142, doi:10.1080/01490419.2015.1020176.

27. Steunou, N.; Desjonquères, J.D.; Picot, N.; Sengenes, P.; Noubel, J.; Poisson, J.C. AltiKa Altimeter: Instrument description and in flight performance. *Mar. Geod.* **2015**, *38*, 22–42, doi:10.1080/01490419.2014.988835.

28. Tournadre, J.; Poisson, J.C.; Steunou, N.; Picard, B. Validation of AltiKa matching pursuit rain flag. *Mar. Geod.* **2015**, *38*, 107–123, doi:10.1080/01490419.2014.1001048.

29. Meyssignac, B.; Cazenave, A. Sea level: A review of present-day and recent-past changes and variability. *J. Geodyn.* **2012**, *58*, 96–109, doi:10.1016/j.jog.2012.03.005.

30. Picard, B.; Frery, M.L.; Obligis, E.; Eymard, L.; Steunou, N.; Picot, N. SARAL/AltiKa wet tropospheric correction: In-flight Calibration, retrieval strategies and performances. *Mar. Geod.* **2015**, *38*, 277–296, doi:10.1080/01490419.2015.1040903.

31. Zaron, E.D.; de Carvalho, R. Identification and reduction of retracker-related noise in altimeter-derived sea surface height measurements. *J. Atmos. Ocean. Technol.* **2016**, *33*, 201–210, doi:10.1175/JTECH-D-15-0164.1.

32. Dibarboure, G.; Boy, F.; Desjonqueres, J.D.; Labroue, S.; Lasne, Y.; Picot, N.; Poisson, J.C.; Thibaut, P. Investigating short-wavelength correlated errors on low-resolution mode altimetry. *J. Atmos. Ocean. Technol.* **2014**, *31*, 1337–1362, doi:10.1175/JTECH-D-13-00081.1.

33. Rémy, F.; Flament, T.; Michel, A.; Blumstein, D. Envisat and SARAL/AltiKa observations of the Antarctic Ice Sheet: A comparison between the Ku-band and Ka-band. *Mar. Geod.* **2015**, *38*, 510–521, doi:10.1080/01490419.2014.985347.

34. Legrésy, B.; Rémy, F. Using the temporal variability of the radar altimetric observations to map surface properties of the Antarctic ice sheet. *J. Glaciol.* **1998**, *44*, 197–206.

35. Wriedt, T. Mie Theory: A Review. In *The Mie Theory: Basics and Applications*; Hergert, W., Wriedt, T., Eds.; Springer: Berlin/Heidelberg, Germany, 2012; pp. 53–71.

36. Vincent, P.; Steunou, N.; Caubet, E.; Phalippou, L.; Rey, L.; Thouvenot, E.; Verron, J. AltiKa: A Ka-band Altimetry Payload and System for Operational Altimetry during the GMES period. *Sensors* **2006**, *6*, 208–234.

37. Aublanc, J.; Thibaut, P.; Lacrouts, C.; Boy, F.; Guillot, A.; Picot, N.; Rémy, F.; Blumstein, D. Altimetry Mission Performances over the Polar Ice Sheets: Cryosat-2, AltiKa and Sentinel-3A. In Proceedings of the Ocean Surface Topography Science Team Meeting, Miami, Florida, 2017. Available online: https://tinyurl.com/y7e7bckx (accessed on 9 January 2018).

38. Quartly, G.; Guymer, T.; Srokosz, A. The effects of rain on Topex radar altimeter data. *J. Atmos. Ocean. Technol.* **1996**, *13*, 1209–1229.

39. Tournadre, J.; Morland, J.C. The effects of rain on TOPEX/Poseidon altimeter data. *IEEE Trans. Geosci. Remote Sens.* **1997**, *35*, 1117–1135, doi:10.1109/36.628780.

40. Tournadre, J.; Lambin-Artru, J.; Steunou, N. Cloud and rain effects on AltiKa/SARAL Ka-band radar altimeter—Part I: Modeling and mean annual data availability. *IEEE Trans. Geosci. Remote Sens.* **2009**, *47*, 1806–1817, doi:10.1109/TGRS.2008.2010130.

41. Tournadre, J.; Lambin-Artru, J.; Steunou, N. Cloud and rain effects on AltiKa/SARAL Ka-band radar altimeter—Part II: Definition of a rain/cloud flag. *IEEE Trans. Geosci. Remote Sens.* **2009**, *47*, 1818–1826, doi:10.1109/TGRS.2008.2010127.

42. Prandi, P.; Debout, V.; Ablain, M.; Labroue, S. *SARAL/Altika Validation and Cross Calibration Activities: Annual Report 2016*; Technical Report; CLS: Ramonville Saint-Agne, France, 2016.

43. Thibaut, P.; Piras, F.; Poisson, J.C.; Moreau, T.; Aublanc, J.; Amarouche, L.; Picot, N. Convergent Solutions for Retracking Conventional and Delay Doppler Altimeter Echoes. In Proceedings of the Ocean Surface Topography Science Team Meeting, Miami, Florida, 2017. Available online: https://tinyurl.com/ybppndx6 (accessed on 9 January 2018).

44. Tournadre, J. Validation of Jason and envisat altimeter dual frequency rain flags. *Mar. Geod.* **2004**, *27*, 153–169, doi:10.1080/01490410490465616.

45. Tran, N.; Obligis, E.; Ferreira, F. Comparison of two Jason-1 altimeter precipitation detection algorithms with rain estimates from the TRMM Microwave Imager. *J. Atmos. Ocean. Technol.* **2005**, *22*, 782–794.

46. Tran, N.; Tournadre, J.; Femenias, P. Validation of envisat rain detection and rain rate estimates by comparing with TRMM data. *IEEE Geosci. Remote Sens. Lett.* **2008**, *5*, 658–662, doi:10.1109/LGRS.2008.2002043.

47. Adler, R.F.; Huffman, G.J.; Chang, A.; Ferraro, R.; Xie, P.P.; Janowiak, J.; Rudolf, B.; Schneider, U.; Curtis, S.; Bolvin, D.; et al. The Version-2 Global Precipitation Climatology Project (GPCP) monthly precipitation analysis (1979–Present). *J. Hydrometeorol.* **2003**, *4*, 1147–1167, doi:10.1175/1525-7541(2003)004<1147:TVGPCP>2.0.CO;2.

48. Poisson, J.; Thibaut, P.; Hoang, D.; Boy, F.; A. Guillot, A.N.P. Wavelet analysis of AltiKa measurements. In Proceedings of the Ocean Surface Topography Science Team, Constance, Germany, 28–31 October 2014.

49. Zech, W.; Senesi, N.; Guggenberger, G.; Kaiser, K.; Lehmann, J.; Miano, T.; Miltner, A.; Schroth, G. Factors controlling humification and mineralization of soil organic matter in the tropics. *Geoderma* **1997**, *79*, 117–161.

50. Hiernaux, P.; Mougin, E.; Diarra, L.; Soumaguel, N.; Lavenu, F.; Tracol, Y.; Diawara, M.; Jarlan, L. Rangeland response to rainfall and grazing pressure over two decades: Herbaceous growth pattern, production and species composition in the Gourma, Mali. *J. Hydrol.* **2009**, *375*, 114–127.

51. Brümmer, C.; Falk, U.; Papen, H.; Szarzynski, J.; Wassmann, R.; Brüggemann, N. Diurnal, seasonal, and interannual variation in carbon dioxide and energy exchange in shrub savanna in Burkina Faso (West Africa). *J. Geophys. Res. Biogeosci.* **2008**, *113*, doi:10.1029/2007JG000583.

52. Taylor, C.M.; Harris, P.P.; Parker, D.J. Impact of soil moisture on the development of a Sahelian mesoscale convective system: A case-study from the AMMA Special Observing Period. *Q. J. R. Meteorol. Soc.* **2010**, *136*, 456–470, doi:10.1002/qj.465.

53. Fatras, C.; Frappart, F.; Mougin, E.; Grippa, M.; Hiernaux, P. Estimating surface soil moisture over Sahel using ENVISAT radar altimetry. *Remote Sens. Environ.* **2012**, *123*, 496–507, doi:10.1016/j.rse.2012.04.013.

54. Fatras, C.; Frappart, F.; Mougin, E.; Frison, P.L.; Faye, G.; Borderies, P.; Jarlan, L. Spaceborne altimetry and scatterometry backscattering signatures at C- and Ku-bands over West Africa. *Remote Sens. Environ.* **2015**, *159*, 117–133, doi:10.1016/j.rse.2014.12.005.

55. Baup, F.; Mougin, E.; de Rosnay, P.; Hiernaux, P.; Frappart, F.; Frison, P.L.; Zribi, M.; Viarre, J. Mapping surface soil moisture over the Gourma mesoscale site (Mali) by using ENVISAT ASAR data. *Hydrol. Earth Syst. Sci.* **2011**, *15*, 603–616, doi:10.5194/hess-15-603-2011.

56. Gruhier, C.; de Rosnay, P.; Hasenauer, S.; Holmes, T.; de Jeu, R.; Kerr, Y.; Mougin, E.; Njoku, E.; Timouk, F.; Wagner, W.; et al. Soil moisture active and passive microwave products: Intercomparison and evaluation over a Sahelian site. *Hydrol. Earth Syst. Sci.* **2010**, *14*, 141–156, doi:10.5194/hess-15-603-2011.

57. Frappart, F.; Fatras, C.; Mougin, E.; Marieu, V.; Diepkilé, A.; Blarel, F.; Borderies, P. Radar altimetry backscattering signatures at Ka, Ku, C, and S bands over West Africa. *Phys. Chem. Earth Parts A/B/C* **2015**, *83*, 96–110, doi:10.1016/j.pce.2015.05.001.

58. Kerr, Y.H.; Waldteufel, P.; Richaume, P.; Wigneron, J.P.; Ferrazzoli, P.; Mahmoodi, A.; Al Bitar, A.; Cabot, F.; Gruhier, C.; Juglea, S.E.; et al. The SMOS soil moisture retrieval algorithm. *IEEE Trans. Geosci. Remote Sens.* **2012**, *50*, 1384–1403, doi:10.1109/TGRS.2012.2184548.

6

Evaluation of Coastal Sea Level Offshore Hong Kong from Jason-2 Altimetry

Xi-Yu Xu [1,2,3,*], Florence Birol [2] and Anny Cazenave [2,4]

[1] The CAS Key Laboratory of Microwave Remote Sensing, National Space Science Center, Chinese Academy of Sciences, Beijing 100190, China
[2] Laboratoire d'Etudes en Géophysique et Océanographie Spatiales (LEGOS), Observatoire Midi-Pyrénées, 31400 Toulouse, France; florence.birol@legos.obs-mip.fr (F.B.); anny.cazenave@legos.obs-mip.fr (A.C.)
[3] State Key Laboratory of Remote Sensing Science, Institute of Remote Sensing and Digital Earth, Chinese Academy of Sciences, Beijing 100094, China
[4] International Space Science Institute, 3102 Bern, Switzerland
* Correspondence: xuxiyu@mirslab.cn

Abstract: As altimeter satellites approach coastal areas, the number of valid sea surface height measurements decrease dramatically because of land contamination. In recent years, different methodologies have been developed to recover data within 10–20 km from the coast. These include computation of geophysical corrections adapted to the coastal zone and retracking of raw radar echoes. In this paper, we combine for the first time coastal geophysical corrections and retracking along a Jason-2 satellite pass that crosses the coast near the Hong-Kong tide gauge. Six years and a half of data are analyzed, from July 2008 to December 2014 (orbital cycles 1–238). Different retrackers are considered, including the ALES retracker and the different retrackers of the PISTACH products. For each retracker, we evaluate the quality of the recovered sea surface height by comparing with data from the Hong Kong tide gauge (located 10 km away). We analyze the impact of the different geophysical corrections available on the result. We also compute sea surface height bias and noise over both open ocean (>10 km away from coast) and coastal zone (within 10 km or 5 km coast-ward). The study shows that, in the Hong Kong area, after outlier removal, the ALES retracker performs better in the coastal zone than the other retrackers, both in terms of noise level and trend uncertainty. It also shows that the choice of the ocean tide solution has a great impact on the results, while the wet troposphere correction has little influence. By comparing short-term trends computed over the 2008.5–2014 time span, both in the coastal zone and in the open ocean (using the Climate Change Initiative sea level data as a reference), we find that the coastal sea level trend is about twice the one observed further offshore. It suggests that in the Hong Kong region, the short-term sea level trend significantly increases when approaching the coast.

Keywords: Jason-2; Hong Kong coast; retracking; X-TRACK; ALES; PISTACH

1. Introduction

Sea level rise is one of the most threatening consequences of present-day global warming. About 10% of the world population currently lives in the world's coastal zones and this number will increase in the future. Therefore, it is crucial to monitor and understand sea level variations along coastlines [1]. Although the tide gauge network has expanded in recent years, some highly populated areas like western Africa remain devoid of any station. For 25 years, satellite altimetry routinely monitored sea level changes over the global open ocean, but was largely unexploited in the coastal areas. Indeed, satellite altimetry was originally designed to precisely measure sea level in the open ocean, where the shape of the pulse-limited radar altimeter echo (i.e., after reflection on the sea surface; called waveform), is well described by the classical mathematical Brown model [2], based on the assumption of a homogeneous rough sea surface within the radar footprint. In that case, the sea level parameters (range between satellite and sea surface, significant wave height and backscatter coefficient) are extracted from the model via a Maximum Likelihood Estimation (MLE) approach. In a coastal band of a few kilometers wide (corresponding to the footprint size of the altimeter antenna), radar echoes integrate reflections from nearby land, leading to complex waveforms that significantly depart from the standard Brown model (e.g., [3–5] and references therein). Besides, in some shallow shelf areas, the sea surface might be so calm that the waveform can display several peaks due to specular reflection. Figure 1 shows two examples of waveforms, one over the open ocean (a) and the other in a coastal zone (b).

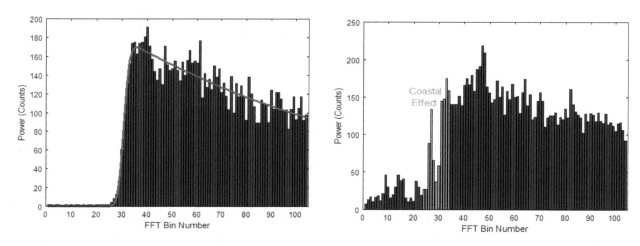

Figure 1. Examples of typical open ocean waveform (**left**; the red line corresponds to the fitted Brown model) and coastal ocean waveform (**right**).

Another difficulty arises from the geophysical corrections applied to the altimeter measurements that usually suffer large uncertainties in the coastal zone. The most limiting corrections are wet tropospheric delay, ocean tides correction, dynamic atmospheric correction (DAC), and sea state bias. Near the coast, these corrections either have large spatio-temporal variability poorly reproduced by numerical models (e.g., tides and DAC corrections), or are less precisely estimated by onboard sensors than over the open ocean (e.g., the radiometer-based wet tropospheric correction that suffers severe land contamination). As a consequence, most altimeter data in a band of 10–30 km from land are declared invalid and discarded from the standard products. For about a decade, significant efforts have been realized by the altimetry community and space agencies to overcome this difficulty and retrieve as much as possible historical altimeter measurements near the coast. Corresponding coastal altimetry products are based on improved coastal geophysical corrections [6,7] and/or dedicated analysis approaches to extract the sea level parameters (range, significant wave height and backscatter coefficient) from non-standard waveforms (a process called 'retracking') [3–5].

In this study, we investigate the relative performances of the experimental coastal altimetry products available for the Jason-2 mission. Jason-2 is chosen here because it was consistently operating for nearly one decade and, above all, because it has been reprocessed by most of the coastal altimetry groups. Satellite pass 153 is considered. We used the following coastal altimetry products: (1) X-TRACK [6] developed by LEGOS (Laboratoire d'Etudes en Géophysique et Océanographie Spatiales, France); (2) PISTACH (Prototype Innovant de Système de Traitement pour l'Altimétrie Côtière et l'Hydrologie, [8]) developed by CLS (Collecte Localisation Satellites, France); and (3) ALES (Adaptive Leading Edge Sub-waveform, [3]) developed by NOC (National Oceanography Centre, UK). The products propose either improved geophysical corrections or waveform retracking algorithms, thus have different contents. X-TRACK products provide sea level time series on a nominal mean track and are available for all altimeter missions except HY-2A and Sentinel-3A. ALES products provide products for Jason-2 and Envisat missions. The successor of the PISTACH product, PEACHI (Prototype for Expertise on Altimetry for Coastal, Hydrology, and Ice), is dedicated to the SARAL/AltiKa mission and provides sea level data along original, cycle-by-cycle tracks [9]. Because of these important differences between products, very few attempts to inter-compare coastal sea level data have been carried out so far, despite the importance of this type of exercise for defining the best processing strategy to derive coastal altimetry data.

In this paper we focus on the South China Sea (Hong-Kong area), not only because of a variety of climate processes impacting sea level (ENSO–El Niño Southern Oscillation-, eddies, storm surges, monsoon, etc.), but also for its importance in the economy and security of Southeastern Asia and the Western Pacific regions. An important consideration when selecting the study area was also the availability of external tide gauge-based sea level data for validation. A dozen tide gauges exist along the China coast, but all stations located along the Chinese mainland ceased to provide data after 1997. Fortunately, a tide gauge remained in operation during the Jason-2 mission at Quarry Bay, Hong-Kong. Corresponding hourly data are available from the University of Hawaii Sea Level Center (UHSLC) [10]. In this study, we considered the closest Jason-2 satellite track (which is ascending and numbered 153) of the HK tide gauge (10.05 km away).

The paper is organized as follows: The study area and the data set are presented in Sections 2 and 3, respectively. Section 4 describes the methodology while Section 5 shows the results. Some elements of discussion are proposed in Section 6, followed by a conclusion (Section 7).

2. Study Area

Hong-Kong (HK) is located just south of the Tropic of Cancer. The climate is predominantly subtropical and displays clear seasonal variations. The southwesterly/northeasterly monsoon give rise to warm wet summers and cool dry winters. HK is also frequently impacted by typhoons. On the western side of the HK island flows the Zhujiang River (Pearl River), which brings abundant freshwater (~3.5 × 10^{11} m^3 per year [11]), resulting in a high salinity gradient. All these factors increase the complexity of the HK environment and have an impact on the regional sea level variations at different spatio-temporal scales.

The HK coast has also an extremely complex geomorphology. As shown on Figure 2, tiny islands lie within the radar footprint of the Jason-2 track chosen for this study. As a consequence, the corresponding altimeter and radiometer measurements are expected to be severely impacted by land effects. This makes this area particularly relevant for analyzing the performances of coastal altimetry data.

The definition of the "coastal zone" in altimetry is somewhat arbitrary. In some studies the criteria of 50 km from land is used (for example, ALES data are provided only in the 50-km coastal band), while some others focus on the first 10 km or even 5 km off the coastline. In this paper we have chosen to base the definition of the area on the "rad_surf_type" parameter provided in the Geophysical Data Record products (GDRs). This surface classification flag is derived from the radiometer measurements and indicates the type of observed surface. For the Jason-2 pass #153, the coastal area based on this flag corresponds to the area where the satellite flies less than 70 km from the closest land (including small islands).

The HK coastal topography is extremely irregular. Figure 3 shows a bathymetric profile along the Jason-2 pass, from southwest to northeast (blue dashed line; from 21.8°N to 22.3°N latitude). Despite a narrow band between 21.8°N and 22°N, where the depth is steeply falling down to ~−60 m, the study area corresponds to very shallow waters. We can thus expect complex local tides and currents influencing sea level variations. Figure 3 also shows the corresponding along track distance to land (mainland or island). The "distance to land" profile is rather complicated. Coastward, it first decreases to ~8 km, slightly increases to ~10 km, and then fluctuates between 10 and 0 km. It even reaches ~0 km when the satellite flies over a small island called Wailingding Island (see Figure 2). Considering both distance to coast and water depth, we define three cases for our comparison exercise, corresponding to three different overlapping segments along the Jason-2 pass. Segments 1, 2, and 3, correspond to cases 1, 2, and 3. In Figure 2, these can be identified by the orange, green, and dashed blue colors, respectively. They cover distances of 50 km, 30 km, and 10 km, and reflect increasing coastal conditions (thus increasing difficulty to retrieve a coherent physical signal from altimetry data). For each case, each satellite cycle, and each product analyzed, we spatially averaged all available 20-Hz (~0.3 km resolution) along-track sea level data along the corresponding pass segment (up to the last valid measurement at HK coast). We finally obtained three mean sea level values for each product and each date. The corresponding altimetry-based sea level time series were then compared with the tide gauge data for validation.

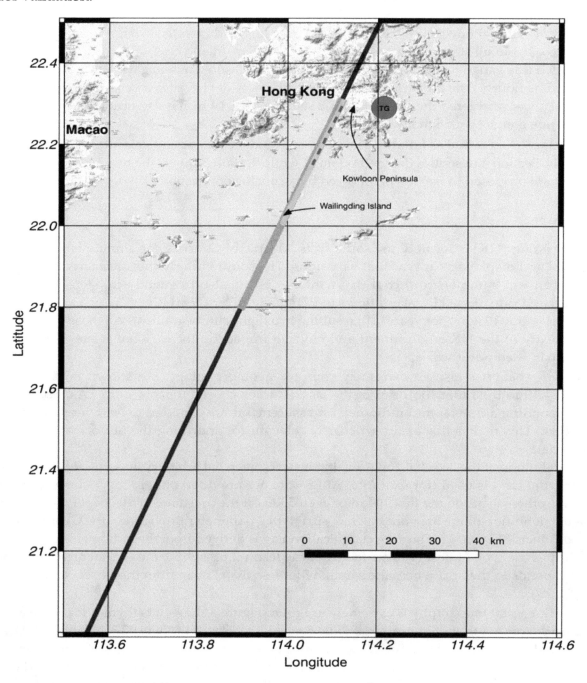

Figure 2. Map showing the study area, the selected Jason-2 pass 153 (black and colored line) and the Quarry Bay tide gauge (red circle). The latter is located ~10 km away from the Jason-2 pass. The along-track sections corresponding to study cases 1, 2, and 3 are also indicated (orange, green, and blue-green dashed lines, respectively). The background map is from Google Earth.

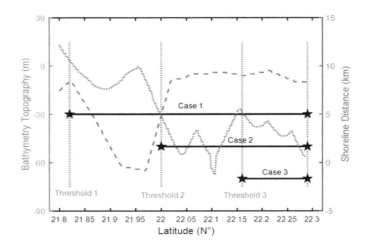

Figure 3. Bathymetry and shoreline distance at the Hong-Kong (HK) coast along 3 segments of the Jason-2 pass identified as cases 1, 2, and 3 (see location in Figure 2).

3. Data Sets

The Jason-2 mission started operating in July 2008, while the high-quality tide gauge data were accessible until 2014 only. Besides, the ALES data were available for cycle 1 to 252 (July 2008 to May 2015). These constraints define the study time period: from July 2008 to December 2014, corresponding to cycles 1-238 of Jason-2 satellite, i.e., a time span of 6.5 years.

3.1. Coastal Altimetry Data Sets

3.1.1. General Description

The experimental coastal Jason-2 products analyzed in this study are X-TRACK (CTOH/LEGOS, [6]), ALES (PODAAC, [3]) and PISTACH (CNES, [8]). These data sets are available from the following websites: ftp.legos.obs-mip.fr; ftp://podaac.jpl.nasa.gov/allData/coastal_alt/L2/ALES and ftp://ftpsedr.cls.fr/pub/oceano/pistach.

We have to keep in mind the differences between the products considered in this study. X-TRACK is a level 3 (L3) product: using the GDR data and state-of-the-art altimetry corrections, along-track sea level time series projected onto reference tracks (points at same locations for every cycle) are computed at 1-Hz (~6 km along-track resolution). Both a raw and a spatially filtered version (40 km cutoff frequency) of the product are distributed. It is simple to use, and is based on improved geophysical corrections near the coast (see [6] for details) but its current version only includes the standard MLE4 retracker adapted to open ocean conditions. ALES and PISTACH are level 2 (L2, i.e., cycle-by-cycle) products: the measurements are provided at 20-Hz (~0.3 km along-track resolution) and are not projected onto reference tracks, which means that their location varies from one cycle to another (this processing step needs to be done by the user). The ALES and PISTACH products do not include improved geophysical corrections, but provide altimeter range estimates obtained using different retracking algorithms (Section 3.1.2). PISTACH also includes a classification of the altimeter waveforms which can be used to study the type of waveform analyzed. It also indicates if the waveform shape is consistent to the retracker used.

3.1.2. Waveform Retrackers

Waveform retrackers can be classified into two categories: physically-based retrackers and empirical retrackers [12]. The model-free retracker is purely based on the statistics of the waveforms and does not require any echo model. Among the multiple model-free retrackers available, the improved threshold retracker [13] is usually considered as the best one. It combines the advantages of the OCOG (Offset Center of Gravity, [14]) and simple threshold [15] algorithms.

Over the last decade, a new approach called "sub-waveform" has been developed. It consists of fitting only the portion of the waveform that contains the leading edge but excludes the trailing edge where most artifacts appear (e.g., [3,8,16–18]).

The retrackers that are available in the different L2 products are summarized in Table 1. The standard GDRs include two solutions: MLE3 and MLE4. The rationale of these two retrackers is similar: fitting the waveform to a Brown model based on the MLE (essentially non-linear least squares) techniques. The main difference between them is the number of parameters included in the model. MLE3 estimates three parameters: epoch (i.e., altimetric range), Significant Wave Height (SWH) and amplitude (i.e., backscatter coefficient-sigma-0), while MLE4 also retrieves the square of off-nadir angle.

Table 1. Overview of different retrackers applied in different altimetry products.

Retracker	Product	Idea	Sub-Waveform	Comments
MLE4	SGDR [1]	Brown model	No	Official standard retracker.
MLE3	SGDR [1]	Brown model	No	
OCE3	PISTACH	Brown model	No	Same as MLE3
RED3	PISTACH	Brown model	Fixed: bins: $t_0 + [-10:20]$	Simplified version of ALES
ALES	ALES	Brown model	Adaptive to the SWH	Two-pass retracker
ICE1	PISTACH	Modified threshold	No	
ICE3	PISTACH	Modified threshold	Fixed: bins: $t_0 + [-10:20]$	

[1]: SGDR contains all GDR parameters, plus some supplementary items such as the retracker outputs and waveforms.

The PISTACH products provide four retrackers: OCE3, RED3, ICE1, and ICE3 [8]. OCE3 is essentially the same as the MLE3 parameters in the GDRs. ICE 1 is a modified threshold retracker. RED3 and ICE3 are the counterparts of OCE3 and ICE1 respectively, where the retracker is executed in a sub-waveform version instead of on the entire waveform.

ALES is based on an advanced retracker called sub-waveform retracker [3]. It is an improved version of RED3, and the difference between them is the estimation of the sub-waveform. In RED3, the sub-waveform has always 31 bins ($t_0 + [-10:20]$, where t_0 is the 32nd bin of the entire waveform), while in ALES, the sub-waveform length can vary from 39 bins (for SWH = 1 m) to 104 bins (i.e., the entire waveform, for SWH $\geq$17 m). In practice, the content of the ALES product is the same as the standard SGDR product (Sensor GDR) plus seven additional parameters specific to to the ALES retracker.

3.1.3. Geophysical Corrections

In PISTACH and X-TRACK, state-of-the-art geophysical corrections other than those of the official GDR are provided. For X-TRACK, only the ocean tide solution and the DAC are provided individually, while in PISTACH, two to three values are given for each correction. Different sets of correction terms obviously lead to different coastal sea level estimates.

3.2. Tide Gauge Data

The Quarry Bay tide gauge is a float-type instrument. It provides sea level data with an accuracy of 1 cm for a single measurement and is regularly calibrated every other year [19]. The tide gauge is located at 114.22°E, 22.28°N, near the northern coast of the HK Island, separated from the Kowloon Peninsula by the Victoria Harbor (see Figure 2). Note that ~95% of the Victoria Harbor shoreline is shaped by human activity [11]. Thus sea level on this area is likely influenced by anthropogenic local-scale factors, in addition to more regional and global ocean variations.

A harmonic analysis was first applied to the tide gauge data in order to compute and remove the tidal signals from the sea level time series. A time-averagedsea level value was also removed from the time series in order to be consistent with the altimetry sea level data. Finally, the hourly tide gauge data were interpolated to the time of the Jason-2 observations. Note that the dynamic atmospheric correction was not removed from the tide gauge data. This will be discussed in Section 5.5. The tide gauge-based sea level time series interpolated to the closest Jason-2 observations is shown in Figure 4. We observe a large seasonal cycle, due to the monsoon, modulated by important high-frequency variations which can reach several tens of cm. In particular, a peak is observed at cycle 228. It is caused by a storm surge associated with the violent Typhoon Kalmaegi that sideswiped the HK coast before dawn on 16 September 2014. The Jason-2 altimeter flew over the HK area at 3.45 am (local time) on 16 September 2014, and the peak in the tide gauge sea level series is coincident with the typhoon event. Therefore, this peak was eliminated in our analysis as an outlier.

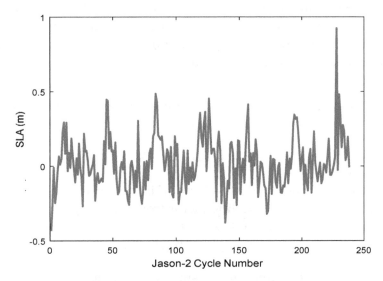

Figure 4. HK tide gauge sea level time series (in meters) interpolated to Jason-2 observations.

3.3. ESA Climate Change Initiative (CCI) Sea Level Data

Global altimetry sea level data sets are regularly produced by different groups worldwide. Five of them are solely based on the TOPEX-Poseidon/Jason-1/2/3 satellites (reference missions): AVISO (Archivage, Validation et Interprétation des données des Satellites Océanographiques) in France, NOAA (National Oceanic and Atmospheric Administration), GSFC (Goddard Space Flight Center) and CU (University of Colorado) in the United States, and CSIRO (Commonwealth Scientific and Industrial Research Organization) in Australia. The Climate Change Initiative (CCI) sea level (sl_CCI) project of the European Space Agency (ESA) has been developed in the recent years to realize the full potential of multi-mission altimetry as a significant contribution to climate research [20–22]). In the context of the CCI, altimetry missions have been reprocessed using improved geophysical corrections and improved links between missions, providing reduced errors of sea level products. In this study, we use the sl_CCI (v 2.0) data from July 2008 to December 2014 to estimate sea level trends in the open ocean away from the HK coast. The sl_CCI v 2.0 data set is a gridded product at 0.25° × 0.25° resolution. The Sl_CCI project was not dedicated to estimate sea level near the coast. However, it can be used as a good independent reference to link open ocean sea level variations to coastal altimetry sea level, thus providing a regional context to interpret the coastal results obtained in this study.

4. Methodology

4.1. Altimetry Data Processing

As indicated above, current coastal altimetry products differ in terms of content. Thus a first step consists of processing the different data sets to obtain homogeneous variables for further comparison. Because there is no waveform data in PISTACH, we used the waveforms provided in ALES, and merged PISTACH and ALES using the measurement time common to all products. In Section 5.4, we also projected all along-track, cycle-by-cycle L2 data onto the X-TRACK 1-Hz reference grids to benefit from the XTRACK improved geophysical corrections.

4.1.1. Sea Level Anomaly (SLA) Estimation

From the altimeter range, once all the propagation and geophysical corrections mentioned below are removed, we can deduce the sea surface height (SSH, i.e., sea level referred to a reference ellipsoid). If we further remove a mean sea surface in order to overcome the problem of geoid estimation, we obtain sea level anomaly (SLA) data. In this study we use SLA data, computed as follows:

$$SLA = H - R - \Delta R_{iono} - \Delta R_{dry} - \Delta R_{wet} - \Delta R_{ssb} - \Delta R_{tide} - \Delta R_{DAC} - MSS \tag{1}$$

In Equation (1), H is the orbital height, R is the Ku-band altimeter range, ΔR_{iono} is the ionospheric correction, ΔR_{dry} and ΔR_{wet} are the dry and wet tropospheric corrections, respectively, ΔR_{ssb} is the sea state bias, ΔR_{tide} is the tide correction (composed of the ocean tide, pole tide and solid Earth tide), ΔR_{DAC} is the dynamic atmospheric correction, and MSS is mean sea surface (in practice we used the MSS_CNES_CLS-2011 model).

In some cases, R is not directly available (for instance, there is no specialized range parameter in the ALES product), so it can be computed as follows:

$$R = T + E \times (c/2) + D + M + 0.180 \tag{2}$$

where T is the onboard tracking range in the Ku band, E is the retracked offset (with time dimension), the factor "$c/2$ (c is light velocity)" is the scaling factor from traveling time to range, D is the Doppler correction in Ku band, M is the instrument bias due to PTR and LPF features ([23,24] and 0.180 is a bias (in meters) due to wrong altimeter antenna reference point [25].

In this study, we computed SLA time series using the altimeter ranges issued from six retrackers: ALES, MLE3, MLE4, RED3, ICE1, and ICE3. For the MLE3, MLE4, and ALES retrackers, we first used Equation (2) to compute R, and then applied Equation (1) to compute the SLA (because R is often flagged in the GDR and ALES products). To validate our calculation method, we compared the MLE4 SLA obtained with the equivalent official "ssha" parameter provided in the GDRs, and found good consistency. For RED3, ICE1, and ICE3, we applied Equation (1) to directly compute the SLA, because there were no valid E values in the PISTACH product.

4.1.2. Choice of the Geophysical Corrections

The geophysical corrections near the coast also need specific considerations. The first error source comes from the wet tropospheric correction because the onboard radiometer suffers from land contamination in the coastal area. A simple but effective approach is to extrapolate a model-based correction (using for example atmospheric reanalyses from the European Center for Medium-Range Weather Forecasts, ECMWF) but the corresponding spatial resolution is relatively low for coastal applications. Other approaches include an improved radiometer-based correction accounting for the land contamination effect [26], or the computation of GNSS-derived Path Delay (GPD, [27]). Since the GPD has not been included in the current versions of the three coastal products analyzed here, we used the decontaminated radiometer solution provided in PISTACH.

Concerning the ionospheric correction, the imperfect coastal altimeter range measurements lead to significant errors, generating outliers in the correction values. We use the MAD (median absolute deviation) technique described in detail in Section 4.1.3 to detect and remove the outliers. The along-track profile of ionospheric corrections is further spatially low-pass filtered using a LOESS (LOcally Estimated Scatterplot Smoothing) method with a cutoff frequency at 100 km.

The coastal ocean tide corrections, provided by global models, are also far from accurate. There are five different ocean tide solutions available in the products, computed by two scientific teams: (1) the Goddard Ocean Tide (GOT) models developed by Ray et al. [28], and the Finite Element Solution (FES) models developed by Lyard et al. [29]. Two ocean tide solutions are provided in the official GDRs:

GOT4.8 and FES2004. PISTACH contains two older versions of GOT: GOT00.2 and GOT4.7 as well as an upgraded version of FES: FES 2012. In Equation (1) we adopt the GOT4.8 ocean tide solution.

Ray [30] compared different tide solutions against 196 shelf-water tide gauges and 56 coastal tide gauges. Their accuracy was characterized by the RSS (root sum square) error of the eight main tidal components (Q1, O1, P1, K1, N2, M2, S2, K2). For the shelf-water gauges, the accuracy of GOT4.8 was 7.04 cm along European coasts and 6.11 cm elsewhere, while the accuracy of FES2012 was 4.82 cm and 4.96 cm respectively for the European coasts and elsewhere. For the coastal tide gauges, the accuracy of GOT4.8 and FES2012 were 8.45 cm and 7.50 cm respectively. In comparison, the accuracy of GOT4.7 and FES2004 in shelf-water were 7.77 cm and 10.15 cm respectively. These results illustrate the significant improvement in coastal ocean tide solution during the last 10 years. In Section 5.5, we also compare three tide solutions, GOT4.8, FES 2004, and FES 2012, to assess their relative performances.

Another important altimetry correction is the sea state bias (SSB). The SSB depends on the retracking algorithm, because it contains the tracker bias. A careful analysis showed that for Jason-2 GDRs, the SLA obtained from MLE3 and MLE4 retrackers has large bias. From a statistical analysis based on cycles 1 to 238 for a couple of altimeter passes over the open ocean, we obtained: $SLA_{MLE3} - SLA_{MLE4}$ = +2.3 cm. Near the coast, this bias appeared to be even larger and even more critical as it was not constant. Figure 5 shows both MLE3 and MLE4 SSB corrections as a function of SWH for an arbitrary pass (cycle 16, pass #153). MLE3 SSB has a clear bias (~+3 cm) relative to MLE4 SSB. Moreover, MLE3 SSB seems to have many outliers, in particular near the coast. We concluded that the bias observed between MLE3 and MLE4 sea level estimates corresponds to a bias in the SSB corrections.

Deeper investigation showed that the MLE3 SSB outliers are often related to large off-nadir angle values (not shown), probably erroneous given the good attitude control of Jason-2. For that reason, we adopted the MLE4 SSB in the computation of all SLAs, resulting in a relative bias <1 cm for all retrackers.

In Equation (1), some parameters are available at 20-Hz (e.g., E), and others at 1-Hz (e.g., D). We interpolated all 1-Hz parameters at 20-Hz and finally computed 20-Hz SLA. The 20-Hz SLA is more useful for the retracker performance analysis, and can also be used for subtle feature detection.

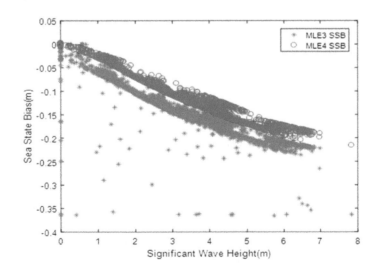

Figure 5. Sea state bias (SSB) difference with respect to Significant Wave Height.

4.1.3. Data Editing Strategy

Near the coast, the number of outliers in data sets is obviously much larger than over the open ocean. Thus the data must be edited before being used. The outliers are detected and removed using the MAD filter. It is based on the median value of the data analyzed, which is more robust than the mean value in the presence of outliers. For a Gaussian random signal, MAD is defined as:

$$MAD = \sqrt{\frac{2}{\pi}} \cdot \sigma \tag{3}$$

where σ is the standard deviation of the data. For the Chauvenet's criterion (0.5% outliers for a Gaussian series, [31]), it will be appropriate to set a threshold of $2.81 \times \sigma$, equivalent to $3.52 \times MAD$. Therefore, values satisfying:

$$|x - Median| > 3.52 \times MAD \tag{4}$$

are defined as outliers. This is very similar to the criterion ($3.5 \times MAD$) adopted by Birol et al. [6] in the ionospheric correction editing process of the X-TRACK product.

Objectively, all outliers cannot be thoroughly detected by the above criterion. In the presence of distorted waveforms, retrackers such as ALES can occasionally produce unrealistic SLA values, up to tens of meters and hence increase the MAD. Some outliers may remain after editing in this case. So we have defined a threshold value before applying the MAD editing. All the SLA values beyond ± 2 m are deleted. Some high amplitude real oceanographic features (such as storm surge) may also be deleted. But for the purpose of the present study, the storm surge events can be removed from the analysis. Finally, almost all outliers are detected after this threshold is applied.

A typical along track profile of 20-Hz SLA before and after editing is shown in Figure 6 for a given cycle (cycle #5 is arbitrarily chosen here). ALES is not as robust as the MLE algorithms for the last few coastal measurements: it has many outliers in almost all cycles (in fact, the mean number of outliers beyond ± 10 m in ALES is ~9 per cycle, much larger than for MLE3 or MLE4). However, after editing, ALES performs better than MLE3 and MLE4, since it displays the lowest noise level.

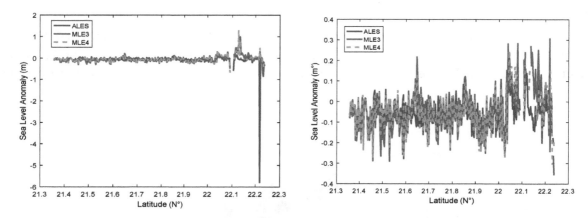

Figure 6. 20-Hz sea level anomaly (SLA) before (**left**) and after (**right**) editing. Jason-2 cycle 5 is considered here.

4.2. Sea Level Data Analysis

After computing the SLA for each cycle and each retracking, the time series can be analyzed to retrieve useful oceanography information. Because of the monsoon, the annual and semi-annual signals are both significant near the HK coast. Therefore, at a first order, SLA variations can be modeled as follows:

$$\begin{aligned} SLA(t) &= a_1 \cos(2\pi t / T_{year}) + a_2 \sin(2\pi t / T_{year}) \\ &+ a_3 \cos(4\pi t / T_{year}) + a_4 \sin(4\pi t / T_{year}) + a_5 t + a_6 + \varepsilon(t) \end{aligned} \tag{5}$$

where T_{year} = 365.2425 days, $\varepsilon(t)$ is the residual SLA, a_1 to a_6 are the regression coefficients to be estimated: a_6 is a bias, a_5 is the sea level trend, the annual/semi-annual amplitude and phase can be deduced from a_1 to a_4:

$$A_{annual} = \sqrt{a_1^2 + a_2^2}; A_{semi-annual} = \sqrt{a_3^2 + a_4^2} \tag{6}$$

$$\Phi_{annual} = \arctan(a_2/a_1); \Phi_{semi-annual} = \arctan(a_4/a_3) \qquad (7)$$

Coefficient uncertainty is defined as the square root of the diagonal elements in the covariance matrix of the coefficient vector.

In practice, we estimated the seasonal terms and trend separately: the seasonal coefficients were removed first from the detrended time series; then the short-term trend was estimated from the residuals (i.e., initial time series corrected for the seasonal terms) after reintroducing the initial trend. In the latter step, a nine-point moving window (corresponding to an about 3-month cutoff frequency) was applied to the residuals to reduce the intrinsic 59-day erroneous signal presented in Jason altimetry missions [32,33].

Note that the computed trends based on only 6.5 years of data essentially represent the inter-annual variability. These short-term trends are expected to be significantly different for long-term trends estimated over the whole altimetry era. In the following the term 'trend' means 'short-term trend'. In this work, it is used as a diagnostic to analyze the quality of near-coastal altimetry data and not the long term sea level trend related to climate change.

5. Results

5.1. A First Analysis Based on Jason-2 Waveforms Observed along the Track

The HK coast considered in this study has a very complex topography, so the waveforms are diverse. In order to go further in our analysis we used the waveform classification provided in the PISTACH product. It classifies the altimeter waveforms into 16 classes, including a "doubt" class. Based on considerations on all waveform shapes, we grouped these classes into five categories (see Table 2). In Table 2, for each category, we provide a general description of the surface observed and propose a specific retracker.

Table 2. Overview of the five categories defined from the 16 PISTACH classes.

Category	Waveform Characteristics	Possible Surface Observed	Retracking Strategy
Brown	The waveform is close to the Brown model.	The land contribution in the altimeter footprint is null or small.	MLE4
Distorted Brown	The waveform is similar to the Brown model, but with distortions (either with an increasing plateau, or with a sharply decreasing plateau, or with a too broad noise floor).	There is land signature at the fringe of the altimeter footprint, with different reflection than over open ocean surface.	ALES
Peak dominated	One (or a few) large peak(s) dominate(s) the waveform; there may be a Brown shape in the waveform, but its maximum power is much lower than the peak(s).	There are one or a few strong bright targets (e.g., extremely calm water surface or effective corner reflector, see [34]) within the altimeter footprint.	MLE based on Gaussian model or improved threshold
Brown + Peak	The waveform is a mixture of Brown shape and one (or a few) peak(s) with comparable power levels. The location of the peak can be in the leading edge or the plateau.	The portions of ocean and land surfaces within the altimeter footprint are equivalent.	MLE based on BAGP (Brown with Asymmetric Gaussian Peak [35])
Others	Unexplained waveform patterns (e.g., very noisy echoes or linear echoes).	Unexplainable surface features (e.g., very composite geomorphology or some extreme events).	Should be rejected.

We use these five categories to classify all Jason-2 waveforms used in our study. The resulting percentages of waveforms as a function of category and case (i.e., distance to the coast) are shown in Figure 7. In all three cases the most frequent category is "Brown", and the second is "Distorted Brown". Not surprisingly, the percentage of Brown-like waveforms decreases when approaching land. In case 2, more than 52% of the waveforms are Brown-like and in case 3 this number decreases to 37%. Concerning the distorted-Brown waveforms, logically, their percentage increases when the distance to the coast decreases. It varies from 17% in case 1 to 27% in case 3. These results suggest that within a shoreline distance of 5 km, a significant number of waveforms are Brown-like and that using a classical open-ocean retracker can still provide good quality sea level data in this very nearshore area. However, the use of ALES is expected to increase the number of retrieved accurate data, which is what we found in the previous section.

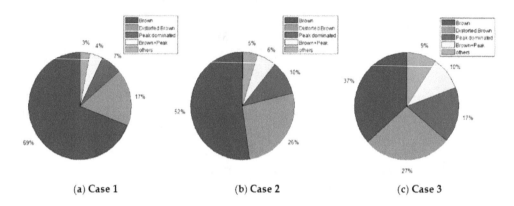

(a) Case 1   (b) Case 2   (c) Case 3

Figure 7. Percentages of waveform categories in the 3 cases.

5.2. Sea Level Trend

5.2.1. A regional View

We first used the gridded CCI sea level data to obtain a regional picture of the sea level trends from July 2008 to December 2014 (Figure 8a). To compare the gridded CCI sea level with the coastal sea level estimated in this study (see below), we interpolated the CCI trend grid along the Jason-2 pass at 1-Hz resolution (~7 km). Corresponding CCI-based along-track sea level trends are shown in Figure 8b. The uncertainty associated with the trend estimation is also provided (blue bars). These error bars are rather large because the time span of analysis is short and the inter-annual variability is large. Trend estimates using the whole CCI altimetry record (i.e., 23-years, from January 1993 to December 2015) display much smaller errors, but a behavior similar to Figure 8b is also observed, i.e., larger errors around 16°N than near the coast.

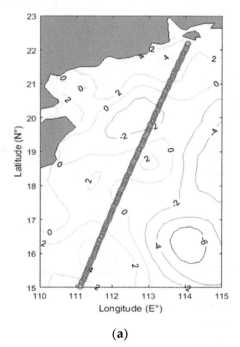

(a)

Figure 8. *Cont.*

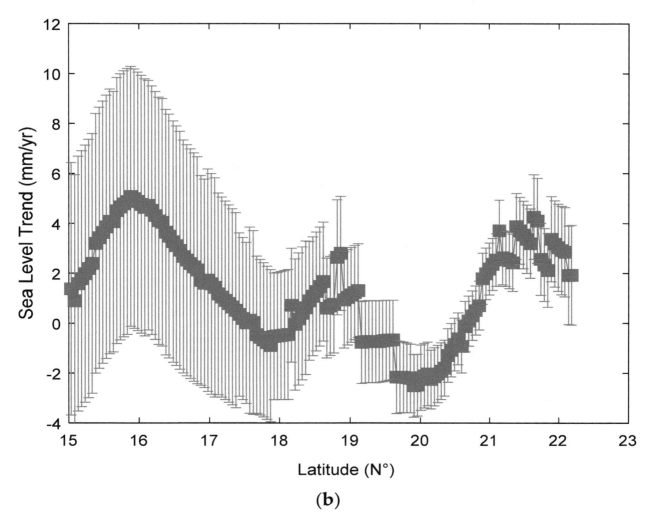

(b)

Figure 8. Regional 6.5 yr-long trends in sea level derived from the CCI sea level data. (**a**) using gridded data and (**b**) interpolated along the Jason-2 pass 153. Vertical blue bars in Figure 8b represent the uncertainty on the trend estimation.

In parallel, in order to see the impact of the different altimetry products used here (thus of the different data processing) on the results, the linear 6.5-year trends were also estimated along the Jason-2 pass from both the raw and filtered versions of the X-TRACK SLA time series. Results obtained from the filtered X-TRACK SLA are shown in Figure 9 as a function of latitude.

From Figures 8 and 9, we observe a significant spatial variability in the regional sea level trends. The CCI map (Figure 8a) displays values of −6 to −4 mm/yr over the open ocean which tend to increase to +2 to +4 mm/yr near the coast. Figure 8b illustrates how Jason 2 pass 153 captures part of this regional trend pattern, with values varying significantly from −2 mm/yr to +4 mm/yr along the track. In X-TRACK (Figure 9), around 20°N, the linear trend is also relatively small (~2–3 mm/yr), in agreement with the CCI data. A change is observed near 19°N, corresponding to the along-track influence of a narrow trench (the bathymetry rapidly decreases to ~−2500 m and then recovers to ~−100 m). As in the CCI data, the X-TRACK trends increase towards the coast but the values obtained in the vicinity of HK are much larger than for the CCI trends: +8 to +9 mm/yr against +2 mm/yr. The sea level trend at the last point of the filtered X-TRACK SLA is: +8.6 ± 2.2 mm/yr. However, gridded altimetry products, such as in the CCI data set, have too low resolution to capture near-shore sea level variations. This also explains why the trends of the gridded CCI data have larger uncertainty than the unsmoothed X-TRACK data.

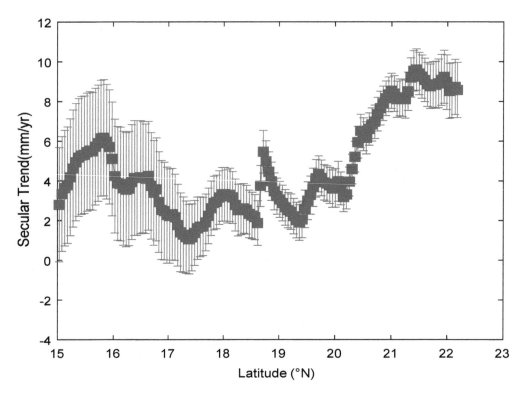

Figure 9. Regional 6.5 yr-long trends in sea level derived from filtered X-TRACK data along the Jason-2 pass 153. Vertical blue bars represent the uncertainty on the trend estimation.

5.2.2. The Tide Gauge Reference

Using Equation (5), we computed the tide gauge trend over the 6.5-yr time span considered in this study. We found a linear sea level trend value of +5.2 ± 2.0 mm/yr (after eliminating the outlier corresponding to the storm-surge event). However for further comparison with the altimetry-based sea level data, the tide gauge time series needs to be corrected for vertical land motion (VLM) [36]. This can be done using GPS precise positioning techniques. Yuan et al. (2008) [37] analyzed data from a GPS station named HKOH, which is close to the tide gauge. VLM at HKOH was estimated to +0.22 ± 0.46 mm/yr. Lau et al. (2010) [38] analyzed GPS data (2006–2009) from HKOH and obtained a VLM estimate of +0.3 mm/yr (uncertainty not provided). The time spans of these studies are different from ours, but because the VLM at HK appeared to be fairly stable over the last decade, we assume a VLM value of +0.3 mm/yr. After correcting for VLM, we find a trend of +5.5 ± 2.0 mm/yr at the tide gauge site.

5.2.3. Solutions Derived from the Different Retrackers

For each of the three cases defined in Section 2 and each retracker, we computed a spatially averaged 20-Hz SLA time series (see Section 4.1.1) as well as the associated 20-Hz noise level (estimated by the standard deviation of the 20-Hz SLA series). The results are shown in Figure 10. In all cases, ALES solution provides the lowest noise level after editing, and MLE4 is slightly less noisy than MLE3. Concerning the three experimental retrackers used in PISTACH, ICE3 has the lowest noise level, and RED3 is slightly less noisy than ICE1.

Note that a problem was detected in some of the PISTACH SLA solutions: around cycle #150, both ICE3 and RED3 have much larger SLA than ICE1, the relative bias being about 0.2 m. Comparison with the tide gauge sea level showed that ICE1 SLA roughly followed the tide gauge data, while ICE3 and RED3 data display large jumps. The latter severely influences the corresponding sea level trend estimates.

Sea level trends obtained from Equation (5) are summarized in Table 3 (except for OCE3 in PISTACH, which is the same as MLE3).

Table 3. Estimated linear trend and associated uncertainty (mm/yr) as a function of sea level data source and case.

Data Source	Case 1	Case 2	Case 3
ALES	+5.9 ± 1.5	+9.7 ± 1.6	+17.3 ± 2.3
MLE3	+5.0 ± 1.6	+8.1 ± 2.2	+2.6 ± 3.1
MLE4	+4.2 ± 1.6	+4.3 ± 2.2	+5.5 ± 3.5
ICE1	−29.1 ± 2.4	−27.5 ± 2.8	−22.9 ± 4.0
ICE3	+57.5 ± 2.3	+60.1 ± 2.5	+52.9 ± 3.0
RED3	+55.3 ± 2.1	+58.0 ± 2.3	+58.5 ± 3.0
XTRACK		+8.6 ± 2.2	
Tide Gauge (in-situ reference)		+5.2 ± 2.0	
Tide Gauge (After VLM correction)		+5.5 ± 2.0	
Regional trend from the CCI data		2.7 ± 2.0	

From Table 3 we note that, using the tide gauge as reference, case 1 provides the closest estimate to the tide gauge trend. Besides, we note a difference of less than 1 mm/yr for ALES and M LE3. This can easily be explained. As shown in Figure 2, case 2 includes a number of small islands while case 1 contains more land-free areas, thus more accurate SLA data. Finally, case 3 has very few valid altimetry measurements (usually less than five valid points), which is not enough to provide robust results.

MLE3 and ALES trends are both close to the tide gauge trend (within 0.5 mm/yr). The trends estimated from MLE4 are slightly lower than for ALES and MLE3 but the difference is within the error bar. The trends deduced from the PISTACH retrackers highly disagree with the tide gauge trend: both ICE3 and RED3 show unrealistic large values (>+5 cm/yr), while ICE1 shows a negative trend of −2 cm/yr. The ICE1 retracker may be inherently not accurate enough to derive trends, but concerning ICE3 and RED3 retrackers, since the processing procedure used in this work is homogeneous, the large errors may not come from the retracker algorithm itself, but more likely from errors in the basic products. For example, in a new PISTACH version, the 18 cm calibration bias used in Equation (2) may no longer be applied in the computation of R. Since the retracked offset E parameter is no more accessible in the current PISTACH version, this hypothesis could not be verified. Anyway, in the remaining part of the study we discard ICE1, ICE3, and RED3 solutions and concentrate on MLE3, MLE4, and ALES which, in the context of our study, appear as the best available retrackers to capture coherent coastal sea level signals from altimetry.

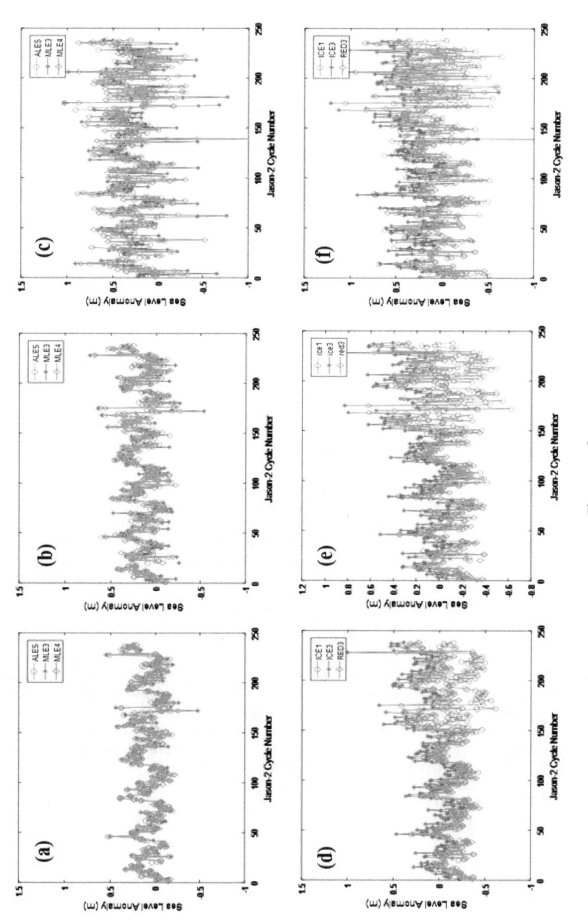

Figure 10. *Cont.*

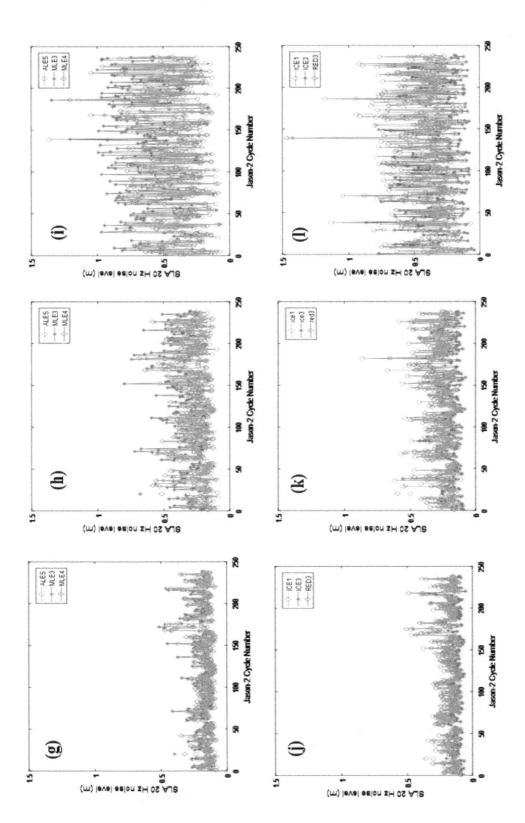

Figure 10. SLA means and noise levels. (**a**) Case 1, mean SLA for ALES, MLE3 and MLE4; (**b**) Case 2, mean SLA for ALES, MLE3 and MLE4; (**c**) Case 3, mean SLA for ALES, MLE3 and MLE4; (**d**) Case 1, mean SLA for ICE1, ICE3 and RED3; (**e**) Case 2, mean SLA for ICE1, ICE3 and RED3; (**f**) Case 3, mean SLA for ICE1, ICE3 and RED3; (**g**) Case 1, noise levels for ALES, MLE3 and MLE4; (**h**) Case 2, noise levels for ALES, MLE3 and MLE4; (**i**) Case 3, noise levels for ALES, MLE3 and MLE4; (**j**) Case 1, noise levels for ICE1, ICE3 and RED3; (**k**) Case 2, noise levels for ICE1, ICE3 and RED3; (**l**) Case 3, noise levels for ICE1, ICE3 and RED3.

5.3. Coastal Seasonal Signal along the Jason-2 Pass

Using Equation (5) we also computed the amplitude and phase of the seasonal signal for all sea level time series. The resulting values are shown in Table 4 (for the annual signal) and Table 5 (for the semi-annual signal). All three cases give rather similar results that agree well with X-TRACK. The annual phases lie between 300°–360° and are significantly larger than the tide gauge-based phase. Amplitudes are also slightly larger. The semiannual phases lie between 220°–270° and are close to the tide gauge-based phase. Amplitudes are also slightly larger. We cannot exclude that some local seasonal signal at the tide gauge site is responsible for the difference observed between altimetry and tide gauge data.

Table 4. Estimated annual amplitude (cm) and phase (degree) as a function of sea level data source and case.

Data Source	Case 1	Case 2	Case 3
ALES	13.05/344	12.49/348	11.19/2
MLE3	13.29/338	12.39/339	13.17/343
MLE4	12.96/339	11.44/340	11.12/348
XTRACK		13.23/338	
Tide Gauge		11.46/311	

Table 5. Estimated semiannual amplitude (cm) and phase (degree) as a function of sea level data source and case.

Data Source	Case 1	Case 2	Case 3
ALES	6.03/235	5.67/239	7.38/260
MLE3	6.17/241	6.76/252	9.24/270
MLE4	6.02/236	6.96/245	11.80/262
XTRACK		6.81/223	
Tide Gauge		7.62/236	

5.4. Relative Performances of MLE4, MLE3, and ALES Near Hong Kong

Here, we examine the relative performance of the MLE3, MLE4, and ALES retrackers in the H-Kcoastal zone. Because ALES and PISTACH are L2 products, we used the L3 X-TRACK product to project ALES and PISTACH MLE3 and MLE4 SLA onto regular 1-Hz reference points along the track, allowing us to obtain SLA time series over the study period. Figure 11a shows the percentage of valid measurements obtained for the different retrackers. MLE3 seems to be the most robust near the coast in the sense that it provides more valid data. Figure 11b shows the time-averaged SLA values for the three retrackers. Over the open ocean, ALES and MLE4 agree well (within 1 cm); MLE3 has a negative bias of 1–2 cm. Near the coast, ALES shows less variations, while MLE3 and MLE4 display large peaks likely due to retracking errors. Figure 11c shows the trend estimates for the different retrackers. ALES displays the smoothest pattern, which seems to indicate a better performance than MLE3 and MLE4.

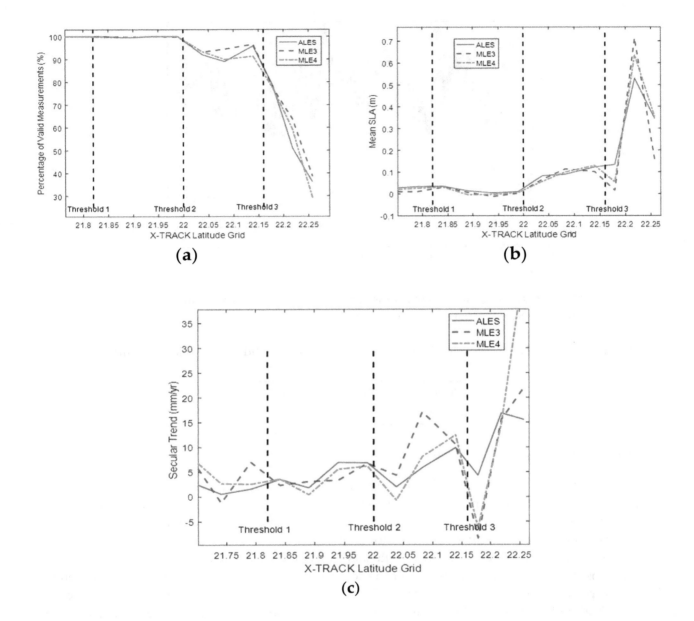

Figure 11. (**a**) Percentage of valid measurements; (**b,c**) mean values and 6.5-yr trend estimates for ALES, MLE3, and MLE4 retrackers. Results are presented as a function of latitude.

The sea level residuals obtained after removing the trend and seasonal signal are shown in Figure 12 for MLE3, MLE4, ALES and the tide gauge data. A 3-month low pass filter was applied to the different SLA time series. For most cycles, the altimetry-based SLA have variations similar to the tide gauge, but the latter occasionally shows larger anomalies. In the first few cycles, the tide gauge residuals have surprising low values, and around cycle 135–160 have a few large negative peaks. Very local small-scale tides, waves, and currents may cause these sea level signals. The standard deviations of the altimetry SLA residuals with respect to the tide gauge residuals, before and after the 3-month smoothing, are given in Table 6. The improvement due to the smoothing is significant, the standard deviations decreasing by more than 50%. The consistency between the altimetry and tide gauge residuals is about 5 cm, which is encouraging given that the study area which is quite complex. ALES SLA has slightly larger standard deviation with respect to tide gauge sea level.

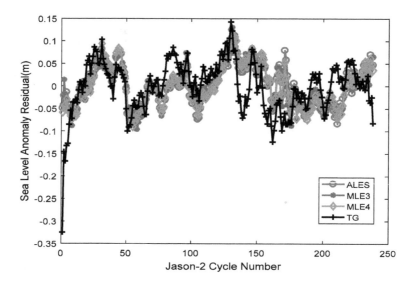

Figure 12. Detrended and deseasoned SLA time series based on ALES, MLE3, and MLE4, with 3-month smoothing (tide gauge SLA—noted TG—is shown as reference).

Table 6. Deseasoned and detrended SLA standard deviation w.r.t. tide gauge sea level (cm) for case 1.

SLA Series	ALES	MLE3	MLE4
Agreement	5.12	4.82	4.88

5.5. Impact of the Geophysical Corrections

As mentioned above, different solutions for some of the geophysical corrections are provided with the different coastal altimetry products. This is the case in particular for the wet tropospheric delay and ocean tides correction. In order to test their relative impact on the quality of the coastal altimetry sea level data, as well as to define the best choice of corrections, we use the different possibilities in Equation (1) to derive different sea level data sets (only one term varies in each case). Concerning the wet tropospheric correction, we tested both the decontaminated solution from the radiometer [26] and the composite correction derived from ECMWF and the radiometer. Concerning the tides, we considered both FES2004 and FES2012, besides GOT4.8. We also computed a version without the DAC. Given the results obtained above, we concentrated on case 1 area. Different diagnostics are considered to evaluate the impact of the correction choice:

(1)   Secular trend and uncertainty (mm/yr) computed for different retrackers (results are shown in Table 7), and
(2)   Standard deviation of the SLA residuals with respect to the tide gauge data (results are shown in Table 8).

Table 7. Secular trend and associated uncertainty (mm/yr) for case 1, with different geophysical corrections choices. DAC = dynamic atmospheric correction.

Choice	ALES	MLE3	MLE4
Standard choice (Table 4)	+5.9 ± 1.5	+5.0 ± 1.6	+4.2 ± 1.6
composite wet delay	+6.1 ± 1.5	+5.1 ± 1.6	+4.3 ± 1.6
FES 2004 ocean tide	+5.7 ± 1.5	+4.6 ± 1.6	+4.7 ± 1.6
FES 2012 ocean tide	+9.1 ± 1.3	+7.8 ± 1.5	+7.1 ± 1.5
No DAC applied	+6.3 ± 1.6	+5.1 ± 1.8	+4.5 ± 1.7

Table 8. Residual SLA, standard deviation w.r.t. tide gauge-based sea level (cm), for case 1, after 3-month smoothing, with different geophysical corrections choices.

Choice	ALES	MLE3	MLE4
Standard choice (Table 4)	5.12	4.82	4.88
composite wet delay	5.02	4.80	5.01
FES 2004 ocean tide	4.87	4.76	4.82
FES 2012 ocean tide	4.89	4.56	4.88
No DAC applied	4.84	4.56	4.76

Inspection of Tables 6 and 7 indicates that the impact of the wet tropospheric correction is rather low. For the trend estimates, the uncertainties are the same but the composite wet delay results are +0.1–0.2 mm/yr larger. Some values of the standard deviations of the differences w.r.t the tide gauge data increase, but some others decrease.

The impact of the ocean tide correction appears much more important. The use of FES tide solutions introduces slightly less residual errors than GOT 4.8. FES2012 decreases the trend uncertainties, but the trend estimate seems too large (large difference with respect to the tide gauge). Correlations for different tide solutions with respect to the tide gauge data are also shown in Table 9. All ocean tide solutions lead to excellent correlations between altimetry-based and tide gauge-based sea level. FES 2012 gives the highest correlation, FES2004 the lowest.

Table 9. Correlations of different tide solutions with respect to the tide gauge tide height.

Ocean Tide Solution	GOT 4.8	FES 2004	FES 2012
Correlation	0.91	0.89	0.95

Remember that we did not correct the HK tide gauge data from the DAC. Assuming that the tide gauge and altimeter roughly capture the same dynamic atmospheric effects, we should obtain better statistics if no DAC correction is applied to altimetry. It corresponds to what is observed in Table 8.

In conclusion, we note that the choice of the retracker and of the altimetry geophysical corrections, especially the tide correction, has a very significant impact on the estimated coastal sea level trend.

6. Discussion

There is a long-standing debate in the coastal altimetry community concerning the choice of the best retracking algorithms and still today, no consensus has emerged. From the results of this study, we consider that regarding the complexity of the corresponding radar echoes associated with the variety of existing coastal land surfaces, the choice of the retracker should be specific to the coastal area considered. We have however to keep in mind that this option does not suit for regional or global applications because we need homogeneous data sets, which would be difficult to obtain if one switches from one retracker to another as a function of the oceanic region considered. This would certainly produce bias in the estimated sea level.

Although the results obtained in this study with ALES in the HK coastal area are already encouraging, we believe that they could be further improved. For example, one of the most useful advances in altimeter waveform processing is the Singular Value Decomposition (SVD) approach proposed by Ollivier [39]. The idea of SVD is to eliminate the waveform components which are relative to the minor singular value, since those components probably arise from system noises rather than real ocean signals. Near the coast, the waveforms can be very different, so only the "Brown-like" waveforms were singled out (according to the "wf_class" parameter in the PISTACH product) to establish the waveform matrix. Results for cycle #100 are shown here as an arbitrary example. For cycle #100, there are 420 waveforms considered as ocean waveforms offshore HK, 342 of which are Brown-like. Therefore it is sufficient to carry out SVD processing. We tested two typical thresholds: 90% and 80%.

The 20-Hz epoch noise characteristics are provided in Table 10 (the 1-Hz noise is roughly similar to the 20-Hz one). The measurements are split into two regions: far from the coast (shoreline distance >10 km) and near the coast (shoreline distance $\leq$10 km). For both regions, the epoch noises are significantly improved when using the SVD. The present study is somewhat preliminary since we have not computed nor analyzed the SLA performances after the SVD processing, but this strategy looks like a serious option to consider in the future for coastal altimetry processing.

Table 10. 20-Hz epoch noise (cm) for different singular value decomposition (SVD) configurations (cycle #100).

Threshold Percentage	No SVD	90% SVD	80% SVD
Far from the coast	8.66	8.05	7.52
Near the coast	9.41	8.90	8.60
The last ocean-like waveform	39.17	36.18	34.68

7. Conclusions

In this paper, we attempted to compare for the first time the three most advanced coastal altimetry products for Jason-2 satellite altimetry: ALES, PISTACH, and XTRACK. The HK coastal zone was selected because it is the only candidate in the South China Sea with available hourly tide gauge data for comparison and validation. Besides, the HK coast presents extremely complex geomorphological and environmental conditions that lead to quite different waveforms. We considered six retrackers: MLE4, MLE3, ALES, ICE1, ICE3, and RED3, and computed Jason-2 sea level for each retracker over a 6.5-year time span (Jason-2 cycles 1–238). We generated a time series (cycle-by-cycle) for each sea level estimate in three cases, each covering different spatial scales. We found that case 1 is a good compromise between coherence and accuracy, and may represent the best sea level estimate at the HK coast. This is probably because, in the study zone, it contains more land-free areas than cases 2 and 3 much contaminated by the presence of small islands. An interesting outcome is that averaging along-track altimetry data over distances larger than 10 km provides better agreement with the tide gauge record than using only altimetry data very close to the coast.

ICE3 and RED3 retrackers show surprising large jumps (~+0.2 m) around cycle #150, that prevents further quantitative analysis. Data from other cycles need to be analyzed to determine whether these jumps result from a software bug (possibly different algorithm versions). A number of studies have shown that improved threshold retrackers sometimes outperform physically-based retrackers, so if any jump is always present, the data set needs to be corrected.

The results indicate that, in spite of the presence of large outliers in ALES (up to tens of meters), after outlier-editing, ALES performs better than MLE4 and MLE3, both in terms of noise level and uncertainty in the sea level trend estimate. We validated the coastal altimetry-based sea level by comparing with data from the HK tide gauge (located ~10 km away). Applying a 3-month smoothing, the standard deviation between the altimetry and tide gauge sea level time series is about 5 cm, which is quite encouraging given the complexity of the study area.

Another interesting result is that the computed sea level trend within 5 km from the coast is about twice the trend estimated at larger distances from the coast. This result, based on merging the CCI open ocean-based sea level with the retracked coastal data, suggests that offshore of the HK region, the short-term trend significantly increases when approaching the coast. This result cannot be generalized for several reasons, in particular because the estimated short-term trends essentially represent the inter-annual variability that may be different from one region to another.

Although the results presented here are encouraging, as shown in Section 6, the HK coastal altimetry sea level data could be further improved by applying an SVD processing before the waveform analysis. There is also certainly room to develop better altimeter waveform retrackers. Geophysical corrections, in particular tide corrections, also appear as an important limiting factor for coastal altimetry applications. In parallel, even if they provide much shorter time series of data than the TOPEX/Poseidon and -Jason missions, we could consider additional altimetry missions (such as HY-2A as well as SARAL/AltiKa and Sentinel-3 based respectively on Ka-band and delay-Doppler technology). In the near future, it will be also highly beneficial to systematically combine the best retrackers and geophysical corrections, to provide a coastal sea level data set with global coverage, usable for climate studies and coastal impact investigations.

Acknowledgments: This study is supported by the National Key R & D Program of China under contract #2017YFB0502800 and #2017YFB0502802, and by the Open Fund of State Key Laboratory of Remote Sensing Science (Grant No. OFSLRSS201705). We are very grateful to Fernando Nino for providing us with Figure 2. Xi-Yu Xu was funded by the China Scholarship Council (Grant No. 201604910329) during a one year visit in LEGOS. Jason-2 SGDR data are archived and maintained by AVISO. X-TRACK, ALES, and PISTACH data are generated and distributed by LEGOS/CTOH, NOC and CLS respectively. Special gratitude is also given to the three peer-reviewers, whose insightful comments have improved the manuscript significantly.

Author Contributions: Xi-Yu Xu collected and processed the coastal altimetry and CCI data and computed all results presented in the paper. Florence Birol processed the tide gauge data and checked the results. All three authors contributed to discuss the results and wrote the manuscript.

References

1. Nicholls, R.J. Planning for the impacts of sea level rise. *Oceanography* **2011**, *24*, 144–157. [CrossRef]
2. Brown, G. The average impulse response of a rough surface and its applications. *IEEE Trans. Antennas Propag.* **1977**, *25*, 67–74. [CrossRef]
3. Passaro, M.; Cipollini, P.; Vignudelli, S.; Quartly, G.D.; Snaith, H.M. ALES: A multimission adaptive subwaveform retracker for coastal and open ocean altimetry. *Remote Sens. Environ.* **2014**, *145*, 173–189. [CrossRef]
4. Cipollini, P.; Calafat, F.M.; Jevrejeva, S.; Melet, A.; Prandi, P. Monitoring sea level in the coastal zone with coastal altimetry and tide gauges. *Surv. Geophys.* **2017**, *38*, 33–57. [CrossRef]
5. Cipollini, P.; Birol, F.; Fernandes, M.J.; Obligis, E.; Passaro, M.; Strub, P.T.; Valladeau, G.; Vignudelli, S.; Wilkin, J. Satellite altimetry in coastal regions. In *Satellite Altimetry over Oceans and Land Surfaces*; Stammer, D., Cazenave, A., Eds.; CRC Press: Boca Raton, FL, USA, 2017; ISBN 978-1-4987-4345.
6. Birol, F.; Fuller, N.; Lyard, F.; Cancet, M.; Niño, F.; Delebecque, C.; Fleury, S.; Toublanc, F.; Melet, A.; Saraceno, M. Coastal applications from nadir altimetry: Example of the X-TRACK regional products. *Adv. Space Res.* **2017**, *59*, 936–953. [CrossRef]
7. Vignudelli, S.; Kostianoy, A.G.; Cipollini, P.; Benveniste, J. (Eds.) *Coastal Altimetry*; Springer: Berlin/Heidelberg, Germany, 2011; 565p. [CrossRef]
8. Mercier, F.; Rosmorduc, V.; Carrere, L.; Thibaut, P. *Coastal and Hydrology Altimetry Product (PISTACH) Handbook*; CLS-DOS-NT-10-246; CNES: Paris, France, 2010.
9. Valladeau, G.; Thibaut, P.; Picard, B.; Poisson, J.C.; Tran, N.; Picot, N.; Guillot, A. Using SARAL/AltiKa to improve Ka-band altimeter measurements for coastal zones, hydrology and ice: The PEACHI prototype. *Mar. Geodesy* **2015**, *38* (Suppl. 1), 124–142. [CrossRef]
10. Caldwell, P.C.; Merrifield, M.A. *Joint Archive for Sea Level Data Report*; JIMAR Contribution No. 15-392, Data Report No. 24; University of Hawaii at Manoa, Joint Institute for Marine and Atmospheric Research: Manoa, HI, USA, 2015.
11. Lai, W.S.; Matthew, J.P.; Ho, K.Y.; Astudillo, J.C.; Yung, M.N.; Russell, B.D.; Williams, G.A.; Leung, M.Y. Hong Kong's marine environments: History, challenges and opportunities. *Reg. Stud. Mar. Sci.* **2016**, *8*, 259–273. [CrossRef]
12. Gommenginger, C.; Thibaut, P.; Fenoglio-Marc, L.; Quartly, G.; Deng, X.; Gómez-Enri, J.; Challenor, P.; Gao, Y. Retracking Altimeter Waveforms Near the Coasts. In *Coastal Altimetry*; Vignudelli, S., Kostianoy, A., Cipollini, P., Benveniste, J., Eds.; Springer: Berlin/Heidelberg, Germany, 2011; pp. 61–101.
13. Hwang, C.; Guo, J.; Deng, X.; Hsu, H.-Y.; Liu, Y. Coastal gravity anomalies from retracked Geosat/GM altimetry: Improvement, limitation and the role of airborne gravity data. *J. Geodesy* **2006**, *80*, 204–216. [CrossRef]
14. Wingham, D.J.; Rapley, C.G.; Griffiths, H. New techniques in satellite tracking systems. In Proceedings of the IGARSS'86 Symposium, Zürich, Switzerland, 8–11 September 1986; pp. 1339–1344.

15. Davis, C.H. Growth of the Greenland ice sheet: A performance assessment of altimeter retracking algorithms. *IEEE Trans. Geosci. Remote Sens.* **1995**, *33*, 1108–1116. [CrossRef]

16. Mercier, F.; Picot, N.; Thibaut, P.; Cazenave, A.; Seyler, F.; Kosuth, P.; Bronner, E. CNES/PISTACH Project: An Innovative Approach to Get Better Measurements over in-Land Water Bodies from Satellite Altimetry: Early Results. In *EGU General Assembly Conference Abstracts, Proceedings of the European Geosciences Union General Assembly 2009, Vienna, Austria, 19–24 April 2009*; Copernicus Publications: Gettingen, Germany, 2009; Volume 11, p. 11674.

17. Bao, L.; Lu, Y.; Wang, Y. Improved retracking algorithm for oceanic altimeter waveforms. *Prog. Nat. Sci.* **2009**, *19*, 195–203. [CrossRef]

18. Yang, L.; Lin, M.; Liu, Q.; Pan, D. A coastal altimetry retracking strategy based on waveform classification and subwaveform extraction. *Int. J. Remote Sens.* **2012**, *33*, 7806–7819. [CrossRef]

19. Chan, Y.W. Tide Reporting and Applications in Hong Kong, China. 2006. Available online: www.gloss-sealevel.org/publications/documents/hong_kong2006.pdf (accessed on 7 February 2018).

20. Ablain, M.; Cazenave, A.; Larnicol, G.; Balmaseda, M.; Cipollini, P.; Faugère, Y.; Fernandes, M.J.; Henry, O.; Johannessen, J.A.; Knudsen, P.; et al. Improved sea level record over the satellite altimetry era (1993–2010) from the Climate Change Initiative project. *Ocean Sci.* **2015**, *11*, 67–82. [CrossRef]

21. Ablain, M.; Legeais, J.F.; Prandi, P.; Marcos, M.; Fenoglio-Marc, L.; Dieng, H.B.; Benveniste, J.; Cazenave, A. Satellite altimetry-based sea level at global and regional scales. *Surv. Geophys.* **2017**, *38*, 7–31. [CrossRef]

22. Legeais, J.F.; Ablain, M.; Zawadzki, L.; Zuo, H.; Johannessen, J.A.; Scharffenberg, M.G.; Fenoglio-Marc, L.; Fernandes, J.; Andersen, O.B.; Rudenko, S.; et al. An Accurate and Homogeneous Altimeter Sea Level Record from the ESA Climate Change Initiative, submitted. *Earth Syst. Sci. Data* **2017**. [CrossRef]

23. Thibaut, P.; Amarouche, L.; Zanife, L.O.Z.; Stunou, N.; Vincent, P.; Raizonville, P. Jason-1 altimeter ground processing look-up correction tables. *Mar. Geodesy* **2004**, *27*, 409–431. [CrossRef]

24. Xu, X.Y.; Xu, K.; Wang, Z.Z.; Liu, H.G.; Wang, L. Compensating the PTR and LPF Features of the HY-2A Satellite Altimeter Utilizing Look-Up Tables. *IEEE J. Sel. Top. Appl. Earth Obs. Remote Sens.* **2015**, *8*, 149–159. [CrossRef]

25. Dumont, J.P.; Rosmorduc, V.; Carrere, L.; Picot, N.; Bronner, E.; Couhert, A.; Desai, S.; Bpnekamp, H.; Scharroo, R.; Lillibridge, J. OSTM/Jason-2 Products Handbook (Issue: 1 rev 11). SALP-MU-M-OP-15815-CN. 2017. Available online: https://www.aviso.altimetry.fr/fileadmin/documents/data/tools/hdbk_j2.pdf (accessed on 7 February 2018).

26. Brown, S. A Novel Near-Land Radiometer Wet Path-Delay Retrieval Algorithm: Application to the Jason-2/OSTM Advanced Microwave Radiometer. *IEEE Trans. Geosci. Remote Sens.* **2010**, *48*, 1986–1992. [CrossRef]

27. Fernandes, M.J.; Lázaro, C.; Ablain, M.; Pires, N. Improved wet path delays for all ESA and reference altimetric missions. *Remote Sens. Environ.* **2015**, *169*, 50–74. [CrossRef]

28. Ray, R.D.; Egbert, G.D.; Erofeeva, S.Y. Tide predictions in shelf and coastal waters: Status and prospects. In *Coastal Altimetry*; Vignudelli, S., Kostianoy, A.G., Cipollini, P., Benveniste, J., Eds.; Springer: Berlin/Heidelberg, Germany, 2011; Chapter 7.

29. Lyard, F.; Lefevre, F.; Letellier, T.; Francis, O. Modelling the global ocean tides: Modern insights from FES2004. *Ocean Dyn.* **2006**, *56*, 394–415. [CrossRef]

30. Ray, R. Status of Modeling Shallow-water Ocean Tides: Report from Stammer international model comparison project. In Proceedings of the NASA/CNES Surface Water and Ocean Topography (SWOT) Science Definition Team (SDT) Meeting, Toulouse, France, 26–28 June 2014.

31. Glover, D.M.; Jenkins, W.J.; Doney, S.C. *Modeling Methods for Marine Science*; Cambridge University Press: Cambridge, UK, 2011; 571p.

32. Masters, D.; Nerem, R.S.; Choe, C.; Leuliette, E.; Beckley, B.; White, N.; Ablain, M. Comparison of global mean sea level time series from TOPEX/Poseidon, Jason-1, and Jason-2. *Mar. Geodesy* **2012**, *35*, 20–41. [CrossRef]

33. Chambers, D.P.; Cazenave, A.; Champollion, N.; Dieng, H.; Llovel, W.; Forsberg, R.; Schuckmann, K.; Wada, Y. Evaluation of the Global Mean Sea Level Budget between 1993 and 2014. *Surv. Geophys.* **2017**, *38*, 309–327. [CrossRef]

34. Xu, X.Y.; Liu, H.G.; Yang, S.B. Echo phase characteristic of interferometric altimeter for case of random surface plus one strong point scatter. In Proceedings of the 2016 IEEE International Geoscience and Remote Sensing Symposium (IGARSS), Beijing, China, 10–15 July 2016; pp. 6456–6459.

35. Halimi, A.; Mailhes, C.; Tourneret, J.-Y.; Thibaut, P.; Boy, F. Parameter estimation for peaky altimetric waveforms. *IEEE Trans. Geosci. Remote Sens.* **2013**, *51*, 1568–1577. [CrossRef]

36. Cazenave, A.; Dominh, K.; Ponchaut, F.; Soudarin, L.; Crétaux, J.F.; Le Provost, C. Sea level changes from Topex-Poseidon altimetry and tide gauges, and vertical crustal motions from DORIS. *Geophys. Res. Lett.* **1999**, *26*, 2077–2080. [CrossRef]

37. Yuan, L.G.; Ding, X.L.; Chen, W.; Guo, Z.H.; Chen, S.B.; Hong, B.S.; Zhou, J.T. Characteristics of daily position time series from the Hong Kong GPS fiducial network. *Chin. J. Geophys.* **2008**, *51*, 1372–1384. [CrossRef]

38. Lau, D.S.; Wong, W.T. *Monitoring Crustal Movement in Hong Kong Using GPS: Preliminary Results*; Hong Kong Observatory: Hong Kong, China, 2010.

39. Ollivier, A. Nouvelle Approche Pour L'extraction de Paramtres Gophysiques Partir Des Mesures en Altimtrie Radar. Ph.D. Thesis, Institut National Polytechnique de Grenoble, Grenoble, France, 2006.

7

Monitoring Water Levels and Discharges using Radar Altimetry in an Ungauged River Basin

Sakaros Bogning [1,2,3,*], Frédéric Frappart [3,4], Fabien Blarel [3], Fernando Niño [3], Gil Mahé [5], Jean-Pierre Bricquet [5], Frédérique Seyler [6], Raphaël Onguéné [2], Jacques Etamé [1], Marie-Claire Paiz [7] and Jean-Jacques Braun [4]

[1] Département de Sciences de la Terre, Université de Douala, BP 24 157 Douala, Cameroun; etame.jacques@yahoo.fr
[2] Jeune Equipe Associée à l'IRD—Réponse du Littoral Camerounais aux Forçages Océaniques Multi-Échelles (JEAI-RELIFOME), Université de Douala, BP 24 157 Douala, Cameroun; ziongra@yahoo.fr
[3] LEGOS, Université de Toulouse, CNES, CNRS, IRD, UPS OMP, 14 Av. E. Belin, 31400 Toulouse, France; frederic.frappart@legos.obs-mip.fr (F.F.); fabien.blarel@legos.obs-mip.fr (F.B.); fernando.nino@ird.fr (F.N.)
[4] GET, Université de Toulouse, CNRS, IRD, UPS OMP, 14 Av. E. Belin, 31400 Toulouse, France; jjbraun1@gmail.com
[5] HydroSciences Montpellier, Université de Montpellier, CNRS, IRD, 300 Av. Pr E. Jeanbrau, 34090 Montpellier, France; gil.mahe@ird.fr (G.M.); jean-pierre.bricquet@ird.fr (J.-P.B.)
[6] ESPACE-DEV, Université de Montpellier, IRD, Université des Antilles, Université de Guyane, Université de La Réunion, Maison de la Télédétection, 500 Rue J-F. Breton, 34093 Montpellier, France; frederique.seyler@ird.fr
[7] The Nature Conservancy Gabon Program Office, Lot 114 Haut de Gué-Gué, 13553 Libreville, Gabon; mcpaiz@tnc.org
* Correspondence: sakaros.bogning@legos.obs-mip.fr

Abstract: Radar altimetry is now commonly used for the monitoring of water levels in large river basins. In this study, an altimetry-based network of virtual stations was defined in the quasi ungauged Ogooué river basin, located in Gabon, Central Africa, using data from seven altimetry missions (Jason-2 and 3, ERS-2, ENVISAT, Cryosat-2, SARAL, Sentinel-3A) from 1995 to 2017. The performance of the five latter altimetry missions to retrieve water stages and discharges was assessed through comparisons against gauge station records. All missions exhibited a good agreement with gauge records, but the most recent missions showed an increase of data availability (only 6 virtual stations (VS) with ERS-2 compared to 16 VS for ENVISAT and SARAL) and accuracy (RMSE lower than 1.05, 0.48 and 0.33 and R^2 higher than 0.55, 0.83 and 0.91 for ERS-2, ENVISAT and SARAL respectively). The concept of VS is extended to the case of drifting orbits using the data from Cryosat-2 in several close locations. Good agreement was also found with the gauge station in Lambaréné (RMSE = 0.25 m and R^2 = 0.96). Very good results were obtained using only one year and a half of Sentinel-3 data (RMSE < 0.41 m and R^2 > 0.89). The combination of data from all the radar altimetry missions near Lamabréné resulted in a long-term (May 1995 to August 2017) and significantly improved water-level time series (R^2 = 0.96 and RMSE = 0.38 m). The increase in data sampling in the river basin leads to a better water level peak to peak characterization and hence to a more accurate annual discharge over the common observation period with only a 1.4 $m^3 \cdot s^{-1}$ difference (i.e., 0.03%) between the altimetry-based and the in situ mean annual discharge.

Keywords: altimetry; water level; discharge

1. Introduction

Inland waters have a crucial role in the Earth's water cycle through complex processes at interfaces with the atmosphere and oceans. They also strongly influence socio-economic practices through their impacts on primary needs, agricultural and industrial activities [1]. Recent and future global changes and increase of population will intensify the stress on water resources [2,3]. However, in many parts of the world, reliable field measurements of water level and water discharge are either completely unavailable or difficult to access for addressing integrated water resource management, use in operational flood forecasting or disaster mitigation [4,5]. In the large rainforest of Central Africa, hosting the Congo river basin and associated small neighboring river basins, the number of hydrological stations has dramatically dropped off because of the irregular maintenance of stations; furthermore, the spatial distribution of the stations often hinder the effectiveness of the network [6].

Spaceborne radar altimetry, although originally designed for the study of ocean topography by continuously measuring the distance between the Earth's surface and the sensor onboard the satellite [7], has proved to be very useful for continental hydrology [8]. In spite of limitations over land, radar altimetry has demonstrated a strong capability to accurately retrieve water levels of large lakes and enclosed seas where the observed surfaces are sufficiently homogeneous [9,10] but also in large river basins where the cross-sections between river and altimetry ground-tracks can reach several kilometers [11,12]. These early results were obtained using Geosat and Topex/Poseidon data processed with the Ocean retracking algorithm. With the launch of ENVISAT in 2002, other retracking algorithms started to be commonly used for processing radar altimetry data. Among them, the Offset Center Of Gravity (OCOG, also known as Ice-1) was found to provide, most of the times, the best results for the determination of river water stages [13]. Combined with availability of land dedicated corrections of the ionosphere and the wet troposphere delays and improvements in the data processing, this allowed the generalization of the use of radar altimetry for the monitoring of inland waters. Currently, all these improvements allow detection of water bodies of a few or below one hundred meters of width (e.g., [14–16]).

With the decrease in number and availability of river discharges around the world, altimetry-based water stages are used to estimate river discharges among several other techniques based on remote sensing [17]. Among them, some commonly used are (i) the application of either power law or a polynomial relationship between stage and discharge (rating curve) to altimetry-based water stages [18–21], (ii) the use of flood routing and hydrodynamics models to derive rating curves under altimetry ground-tracks [22–24], (iii) the calibration of hydrodynamics and hydrological models using altimetry-based water stages [25–29].

This study presents the multi-mission altimetry-based hydrological network setup in the Ogooué River Basin (ORB) to provide a continuous water stage and discharge monitoring in this almost ungauged basin. It aims at answering the following questions:

- what are the performances of the different altimetry missions, from ERS-2 to Sentinel-3A, to retrieve water levels?
- how does the combination of data from several altimetry virtual stations improve the retrieval of the annual discharge in the ORB?

2. Study Area

The Ogooué River is the largest Gabonese river. Its length is about 900 km from its source in the Mounts Ntalé, in Congo, at an altitude close to 840 m.a.s.l. to its mouth in the southern part of the Atlantic coast of Gabon. The Ogooué river flows northwestward in upstream until the confluence with the Ivindo and southwestward from the confluence with the Ivindo river to a 100 km long and 100 km width delta it forms in the south of Port Gentil where it discharges into the Atlantic ocean [30]. The Ogooué river basin (ORB) is located between 9° and 15°E, and 3°S and 2.5°N (Figure 1) stretching on about 80% of the total area of Gabon and It is bounded on the east by the Congo basin, on the south by the Niari and Nyanga basins, on the west and north-west by the coastal river basins [31].

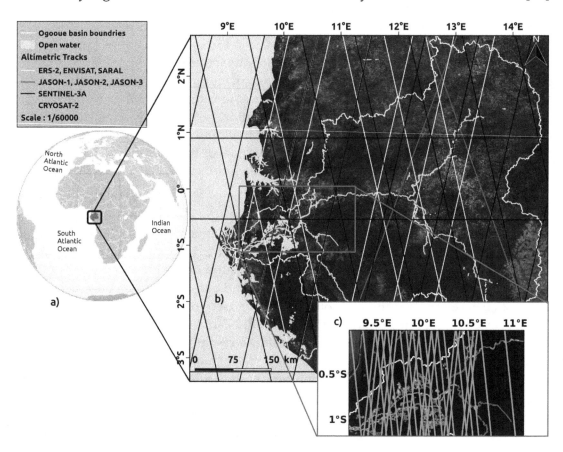

Figure 1. (**a**) Location of the Ogooué River Basin in Gabon in Equatorial Africa. (**b**) In this basin, delineated with a white line, the Ogooué and its major tributaries appear in light blue. Altimetry tracks are represented in red for the missions on a 10-day repeat cycle on their nominal track (Jason-1/2/3), in black for Sentinel-3A on its nominal track (27-day repeat cycle), in yellow for the missions on a 35-day repeat cycle on their nominal track (ERS-2/ENVISAT/SARAL), (**c**) zoom of the downstream of the Ogooué River Basin with altimetric tracks of Cryosat-2 on its nominal track (369-day repeat cycle) in cyan lines.

Due to its location crossing the Equator, the ORB receives the largest annual precipitation in Africa (1600–2200 mm yr^{-1}), making the annual discharge of the river of 4750 m$^3 \cdot$s^{-1}, the third along the African West Coast after the Congo River (40,000 m$^3 \cdot$s^{-1}) and the Niger River (5590 m$^3 \cdot$s^{-1}) [30,32]. The annual variation of the discharge of the Ogooué river passes by two maximum : in spring and autumn corresponding to the rainy seasons [31]. On the ORB at Lambaréné, before the 1970's, river discharge of spring floods were equivalent to those of the autumn floods. After the 1970's, spring floods differ significantly from autumn floods with differences between 2000 and 3000 m$^3 \cdot$s^{-1} [33].

3. Datasets

3.1. Radar Altimetry Data

Radar altimetry data used in this study comes from the measurements on the nominal orbit of the following missions: Jason-2 (06/2008–10/2016), Jason-3 (since 01/2016), ERS-2 (06/1995–07/2003), ENVISAT (06/2002–10/2010), SARAL (02/2013–07/2016), Sentinel-3A (since 02/2016) and Cryosat-2 (since 04/2010—operating in low resolution mode—LRM), but also from the second (drifting) orbit of ENVISAT (10/2010–06/2012). The Jason missions have a 10-day, ERS-2, ENVISAT and SARAL a 35-day, Cryosat-2 and 369-day and Sentinel-3A a 27-day repeat-periods. Jason-2 and Jason-3 data come from Geophysical Data Records (GDRs) D, GDR v2.1 for ENVISAT, GDR T for SARL, GDR C for Cryosat-2 and GDR ESA IPF 06.07 land products for Sentinel-3A delivered by CNES/ESA/NASA processing centers. These data were made available by Centre de Topographie des Océans et de l'Hydrosphère (CTOH) [34]. ERS-2 data were reprocessed by CTOH to ensure the continuity with ENVISAT for land studies [35].

3.2. In Situ Water Levels and Discharges

Long-term datasets of field measured water level and discharge are not available in the ORB since it is was completely ungauged between the 1980s and 2001 [31]. Only the Lambaréné gauge station gathered data from July 2001 to September 2017, sometimes with non operating periods up to a month. These in situ data of water level (collected at Lambaréné by the Société de l'Energie et de l'Eau du Gabon, SEEG) were used to calculate the river discharge, using a historical calibration formula provided by the HydroSciences Montpellier (HSM) laboratory of the University of Montpellier (France). Both are used for validating altimetry-based water level and discharge.

4. Methods

4.1. Altimetry-Based Water Levels

Initially developed to provide accurate measurements of the sea surface topography, radar altimetry is now commonly used for the monitoring of inland water levels (see [8] for a recent review). The variations of the altimeter height from one cycle to the other can be associated to changes in water level.

In this study, we used the Multi-mission altimetry Processing Software (MAPS), frequently used for processing altimetry data over land and ocean (e.g., [36–40]), that allows a refined selection of the valid altimeter data to build time-series of water levels at a so-called virtual station. Data processing is composed of four main steps:

(i)   the rough delineation of the cross-section between the altimeter tracks and the rivers using Google Earth,

(ii)   the loading of the altimetry over the study area and the computation of the altimeter heights from the raw data contained in the GDRs,

(iii) a refined selection of the valid altimetry data through visual inspection,

(iv)   the computation of the water level time-series as the median of the selected water levels every cycle.

A detailed description of the processing of altimetry data using MAPS can be found in [41]. MAPS is made available by CTOH. Previous studies showed that Ice-1-derived altimetry heights are the more suitable for hydrological studies in terms of accuracy of water levels and availability of the data (e.g., [13,37]) among the commonly available retracked data present in the GDRs. In this study, the data used were processed using the Offset Center of Gravity (OCOG) [42] also named Ice-1 or Ice retracking algorithm depending on the mission for all the missions.

Time series of water levels are generally obtained processing data from altimetry missions with a repeat period between 10 and 35 days [8]. It is what was done in this study to build the network of altimetry virtual stations under Jason-1/2/3, ERS-2/ENVISAT/SARAL and Sentinel-3A groundtracks. Considering the life cycles of these different missions and excluding the Jason missions whose crosstrack is too coarse (315 km at the Equator) to allow the definition of a dense network of virtual stations in, most of the river basins, there is a lack of data between the end of the ENVISAT mission in October 2010 on the nominal orbit, or in April 2012 considering its extended orbit, and the launch of the SARAL mission in February 2013. The only option to fill this gap is to consider the data acquired by the Cryosat-2 mission. Due to its long revisit time (369 days), its data are used to retrieve time series of water levels over large lakes [43], but are generally considered useless to define virtual stations over rivers [44,45]. The long repeat cycle of Cryosat-2 is compensated by a small crosstrack of 7.5 km at the Equator leading to a large number of cross-sections between the river and the altimetry tracks in a close vicinity. Under the assumption that the temporality of water stages is not changing on a few tenths of kilometers of distance, it is possible to build VS gathering Cryosat-2 from tracks in a short distance thanks to the high spatial coverage and the sub-cycle period of 30 days of this mission. Changes in river characteristics (width, depth) over distances of a few kilometers are likely to impact the amplitude of the water stage but not its dynamics, a linear regression between altimetry-based water levels and in-situ data was applied to altimetry data to retrieve time series of water levels (see (2) in sub-section 4.2 Conversion into river discharges).

4.2. Discharge Estimates

River discharge is classically estimated from water level measurements through a functional relationship between the two quantities known as stage-discharge rating or rating curve (Rantz et al., 1982). It has the following form:

$$Q(t) = \alpha(h(t) - h_0)^\beta \tag{1}$$

where Q is the river discharge, h the water level, h_0 the null-discharge elevation, and α and β are related to the geometry of the channel cross-section and to the friction coefficient modulating the discharge.

This technique have been successfully applied to estimate river discharge using altimetry-based water levels when it was possible to derive the rating curve a common period of observation. It permits to derive river discharge with accuracy better than 20% (e.g., [18–20,22,46–48]).

As the altimetry crossing over the river does not generally occur at the location of the in-situ station but several tenths of kilometers upstream or downstream, the flow cross-sectional area is likely to vary over these short distances. To avoid errors caused by these changes, previous studies used a linear regression between altimetry and in situ water levels (h_{alti} and h_{insitu} respectively) when the dynamic of the flow is quite similar (e.g., [20,46]):

$$h_{insitu} = ah_{alti} + b \tag{2}$$

where a and b are the coefficients of the linear regression that allows the radar altimetry data to exactly fit the variations from the in situ gauge used to estimate the river discharge.

5. Results

5.1. Altimetry-Based Network of Gauging Stations

An altimetry-based network of 34 virtual stations (VS) was defined across the Ogooué river and its major tributaries (Ivindo and Ngounie rivers in northeast and southwest of the ORB respectively). Water level time series were mostly derived from Envisat, Saral and Sentinel-3A observations. As ENVISAT and SARAL missions were orbiting on the same nominal orbit, 15 of them provide a pluri-annual record from June 2002 to October 2010 and from February 2013 to June 2016, one of them from June 2002 to October 2010 and one of them from February 2013 to June 2016. As this orbit was formerly used by ERS-2, this record was extended from May 1995 to July 2003 for the 6 VS located closer to the mouth. Due to the small width of the rivers in the upstream part of the ORB and the presence of topography, VS could not be defined 125 km upstream Lambaréné because of the narrow width of the Ogooué river and its surrounding rugged topography. This network is completed with 2 VS created using observations from Jason-2 (from July 2008 to August 2016) and Jason-3 (since January 2016). Due to the large cross-track of these missions (315 km at the Equator), very few cross-sections between the rivers and the altimeter ground tracks occur. The network was also completed by 11 VS defined under Sentinel-3A ground tracks, which was launched in February 2016. The locations of the VS from the different altimetry missions are presented in Figure 2 and Table 1.

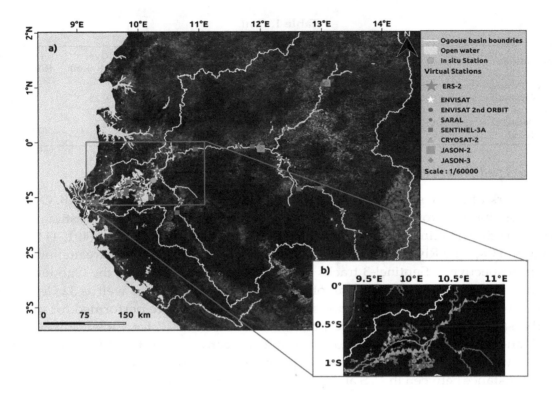

Figure 2. Locations of the altimetry virtual stations in the Ogooué River Basin. VS from ERS-2, ENVISAT, ENVISAT 2nd orbit, SARAL, Sentinel-3A, Cryosat-2, Jason-1, Jason-2, Jason-3 are represented using orange stars, white stars, brown dots, red dots, blue squares, cyan triangles, green squares and orange diamonds respectively. For readability purpose, virtual stations from missions with repeat period shorter than 35 days are presented in (**a**) and virtual stations from Cryosat-2 are presented in (**b**).

Table 1. List of VS where water stages are derived from altimetry measurements.

Virtual Stations	Missions	Longitude (°)	Latitude (°)	River Width (km)	Distance to the River Mouth (km)
SV_229_Ogooué	ENVISAT, SARAL	13.3533	−1.3082	0.22	693.09
SV_272_Ivindo	ENVISAT, SARAL	12.4228	0.2542	0.20	524.91
SV_272_Ogooué	ENVISAT, SARAL	12.3051	−0.2816	0.36	477.59
SV_315_Ogooué	ENVISAT, SARAL	11.6422	−0.0618	0.37	386.57
SV_358_Ngounié	ENVISAT, SARAL	10.6490	−1.2761	0.20	274.948
SV_401_Ngounié	ENVISAT, SARAL	10.3227	−0.5958	0.62	193.531
SV_401_Ogooué	ERS-2, ENVISAT, SARAL	10.3045	−0.5129	1.30	198.222
SV_444_Ogooué	ERS-2, ENVISAT, SARAL	9.2569	−1.0722	0.59	42.582
SV_730_Ogooué	SARAL	12.9154	−0.8316	0.19	607.233
SV_773_Ogooué	ENVISAT, SARAL	12.4700	−0.5595	0.23	494.696
SV_902_lake_Onangué	ERS-2, ENVISAT, SARAL	9.9912	−1.0001		160.97
SV_902_Ogooué	ERS-2, ENVISAT, SARAL	10.0280	−0.8323	1.25	141.66
SV_902_Ogooué_2	ENVISAT, SARAL	10.0551	−0.7091	0.32	152.215
SV_945_lake_Louandé	ENVISAT	9.6497	−0.8047		107.174
SV_945_lake_Ogognié	ENVISAT, SARAL	9.6844	−0.9624		101.671
SV_945_Ogooué	ERS-2, ENVISAT, SARAL	9.6755	−0.9220	1.19	97.413
SV_945_Ogooué_2	ERS-2, ENVISAT, SARAL	9.6571	−0.8382	0.47	103.125
Station 1	ENVISAT 2nd orbit	10.6414	−0.1864	0.49	127.826
Station 2	ENVISAT 2nd orbit	10.0208	−0.8328	0.88	138.166
Station 3	ENVISAT 2nd orbit	9.9445	−0.8082	1.17	252.246
SV_128_Ogooué	SENTINEL-3A	9.2788	−1.0638	1.23	45.271
SV_378_Ogooué	SENTINEL-3A	9.8069	−0.8454	1.24	112.654
SV_185_lake_Onangué	SENTINEL-3A	10.1962	−1.0009		169.463
SV_050_Ogooué	SENTINEL-3A	10.9045	−0.1177	0.30	298.329
SV_107_Ogooué	SENTINEL-3A	11.8457	−0.0803	0.37	412.048
SV_164_Ogooué	SENTINEL-3A	12.6126	−0.8438	0.31	564.01
SV_356_Ogooué	SENTINEL-3A	12.9676	−0.8423	0.30	617.561
SV_164_Ivindo	SENTINEL-3A	13.0280	1.0330	0.19	677.105
SV_050_Ngounié	SENTINEL-3A	10.6477	−1.2728	0.35	301.24

Table 1. *Cont.*

Virtual Stations	Missions	Longitude (°)	Latitude (°)	River Width (km)	Distance to the River Mouth (km)
SV_378_lake_Avanga	SENTINEL-3A	9.7878	−0.9345		112.299
SV_050_Ngounié	SENTINEL-3A	9.8386	−0.7021	0.23	301.24
Lambaréné	CRYOSAT-2	10.2220	−0.7139		
SV_185_Ogooué	JASON-2, JASON-3	12.0035	−0.1148	0.36	430.318
SV_096_Ivindo	JASON-2, JASON-3	13.0790	1.0758	0.18	687.576

5.2. Altimetry-Based Water Levels Validation Using the Lambaréné Gauge Record

The network of gauge stations was not maintained after the 1980s in the ORB. Only the gauge station from Lambaréné provided measurements of water stages and discharge estimates during the period of acquisition of altimetry data. Four ERS-2/ENVISAT/SARAL tracks (401, 444, 902 and 945) are crossing the Ogooué River close to Lambaréné, at 22 and 135.5 km upstream and at 38 and 89 downstream respectively. Sentinel-3 tracks 050, 128 and 378 are crossing the Ogooué River at 121 km upstream and 66 and 133 downstream of Lambaréné respectively, as well as 32 Cryosat-2 tracks. Comparisons between altimetry-based and in situ water stages were performed for measurements acquired the same day. The results are presented in Figures 3–7 for ERS-2/ENVISAT/SARAL, Sentinel-3A and ENVISAT 2nd orbit and Cryosat-2 respectively. For the different missions except Cryosat-2, the water levels are presented from upstream (a) to downstream (c or d), (b) corresponding to the closest distance between the VS and the in situ station of Lambaréné.

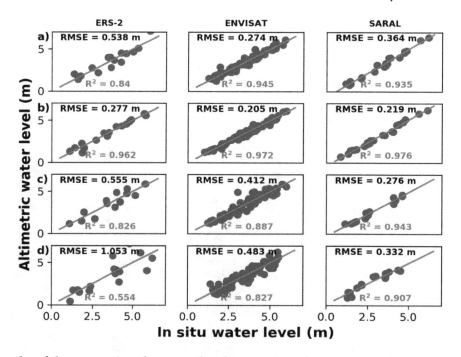

Figure 3. Results of the comparison between the altimetry-based water stages from ERS-2/ENVISAT/SARAL for tracks 401 (**a**), 902 (**b**), 945 (**c**), 444 (**d**) and the in situ ones from Lambaréné gauge station.

Overall very good results are obtained for all these stations. As expected, the quality of the water stage retrieval decreases as the distance to the in situ station increases. Better results were generally obtained using SARAL data, the first mission to operate in Ka-band, than using ERS-2 and ENVISAT ones (Ku-band, Figures 3 and 4). Results obtained using Sentinel-3A data, the first mission to operate in SAR mode on a frequent repetitive orbit, are very encouraging (Figure 5). In spite of the few cycles available (15), altimetry-based water levels obtained using data from this mission already exhibit a similar quality as the ones obtained using data from SARAL when considering the closer distance to the Lambaréné in situ station (Figure 5b). A very good agreement is also found with Cryosat-2 at the

Lambaréné (R^2 = 0.96 and RMSE = 0.25 m) demonstrating the potential of this mission for the retrieval of water stages (Figure 6).

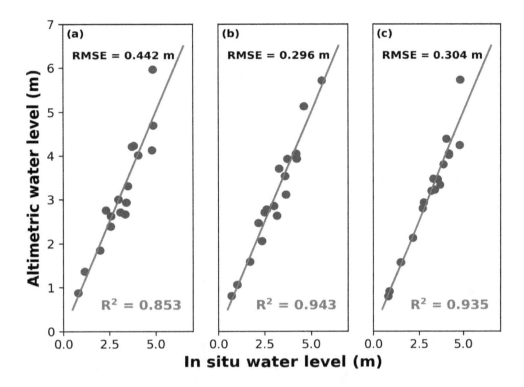

Figure 4. Results of the comparison between the altimetry-based water stages from ENVISAT on its second orbit for (**a**) Station 1, (**b**) Station 2 and (**c**) Station 3 and the in-situ ones from Lambaréné gauge station.

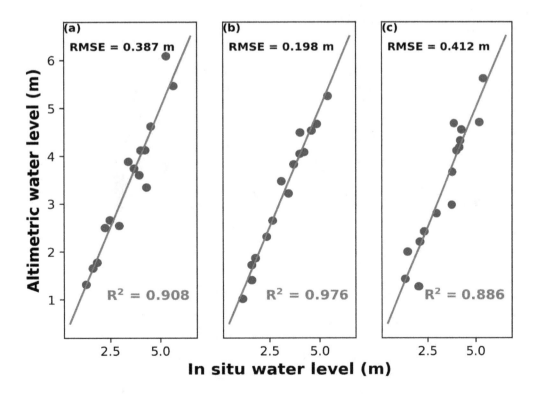

Figure 5. Results of the comparison between the altimetry-based water stages from Sentinel-3A for tracks (**a**) 050, (**b**) 378 and (**c**) 128 and the in-situ ones from Lambaréné gauge station.

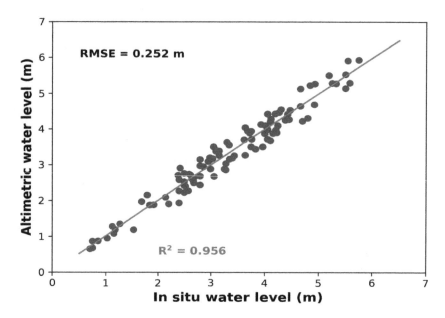

Figure 6. Results of the comparison between the altimetry-based water stages from Cryosat-2 and the in-situ ones from Lambaréné gauge station.

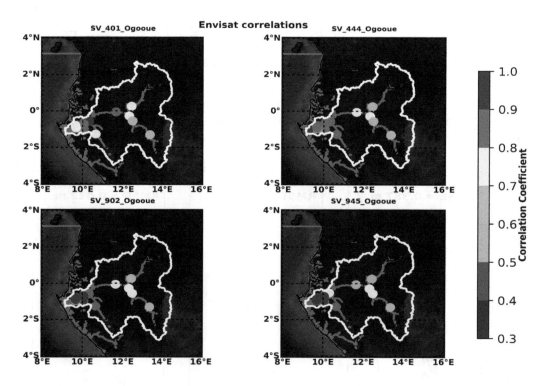

Figure 7. Maps of maximum of cross-correlation between time series from ENVISAT data in the ORB for the four VS around Lambaréné.

5.3. Consistency of the Altimetry-Based Water Levels in the Ogooué River Basin

Only one in-situ gauge station record had water stage measurements during the altimetry period in the ORB. Even if this record can be used to validate 4 SV from ERS-2/ENVISAT/SARAL, 3 from Sentinel-3A, and one combining Cryosat-2 water level estimates at different cross-sections, a consistency check was performed for the other stations in the ORB. It consists in estimating the cross-correlation between the different time-series of water levels from the same mission. Maxima of correlation were reported on Figures 6 and 7 for ENVISAT and SARAL respectively. As the repeat period of these mission is 35 days on their nominal orbit, only time-lags of plus or minus 53 days (one repeat period and a half) were considered due to the relatively small scale of the ORB. Due to the changes of the river features (slope, depth, width, . . .), bias and RMSE were not computed between the time-series of altimetry-based water levels. This consistency-check was applied neither to the ERS-2 VS that are not sufficiently numerous (only 6 located on the downstream part of the river) nor to the ENVISAT VS from the second orbit because of the short period of operation of ENVISAT (17 cycles available).

It can be seen that maximum values of cross-correlation between the time series are greater downstream than upstream of the ORB whatever the VS considered to make the comparisons (Figures 7 and 8). Larger maxima of cross-correlation are obtained with the ENVISAT mission than with the SARAL mission. They are above 0.7 between ENVISAT VS in the downstream part of the basin up to the confluence between its two major tributaries, the Ivindo and the Ngounié Rivers. On these two tributaries, the maximum of cross-correlations generally range from 0.6 to 0.8 (Figure 7). Much lower agreement was found between the SARAL VS. If maxima of cross-correlation are generally above 0.6 on the downstream part of the ORB but they rapidly decrease down to 0.4 upstream. Higher consistency is found using ENVISAT VS on a larger part of the ORB.

No time-lag is present in most of the cases. Nevertheless, a few maxima cross-correlation coefficients between time series occurred with a time-lag of one month, for both ENVISAT and SARAL missions. This situation happened in the case of two stations located upstream of the Ogooué River (ENVISAT SV_229_Ogooué using SV_401_Ogooué, SV_902_Ogooué and SV_945_Ogooué validated against records from Lambaréné in situ gauge as reference and SARAL SV_ 773_Ogooué using SV_401_Ogooué and SV_945_Ogooué as reference) and in the case of the station SV_945_Lake_Ogognié located on the Lake Ogognié downstream of the ORB in cross-correlation with the validated time series from SV_401_Ogooué.

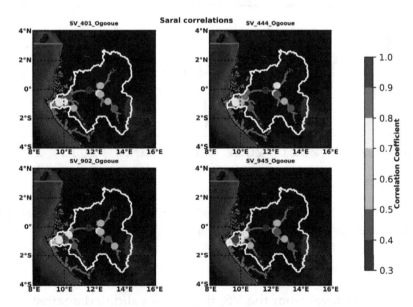

Figure 8. Maps of maximum of cross-correlation between time series from SARAL data in the ORB for the four VS around Lambaréné.

Due to the lack of in situ gauge station in the ORB with the exception of the one from Lambaréné, no validation was performed for the water stages derived from Jason-2 and Jason-3 data. As the two cross-sections between Jason-2 and 3 ground-tracks (185 on the Ogooué and 096 on the Ivindo) are close from cross-sections between Sentinel-3A ground-tracks (107 on the Ogooué and 164 on the Ivindo) (see Table 1 and Figure 2), cross-comparisons between the water levels derived from Jason-2, Jason-3 and Sentinel-3A were performed during the common period of availability of the different datasets. For this purpose, data from the 10-day repeat orbit missions (Jason-2 and Jason-3 that were orbiting 2 minutes appart) were interpolated at the date of acquisition of Sentinel-3A. In both cases, Jason-2 time-series of water levels exhibit a clear seasonal cycle, especially for track 096 on the Ivindo, with high stages observed during the primary peak from October to December and the secondary peak in April-May and low stages observed in January-February (small dry season) and from June to August for the large dry season, which is consistent our knowledge of the hydrological cycle in the ORB (Figure 9). This also the case for Jason-3 time-series that agree well with the ones from Jason-2 (R = 0.88 and 0.87 and RMSE = 0.51 and 0.68 m for tracks 185 on the Ogooué and 096 on the Ivindo respectively). If similar but smoother water levels variations were observed in the Ivindo at the SV built under Sentinel-3A track 164 (R = 0.92 and 0.95, RMSE = 0.71 and 0.56 m, Figure 9a), no realistic water level variations were observed at the SV built under Sentinel-3A track 107 (Figure 9b). This SV was discarded.

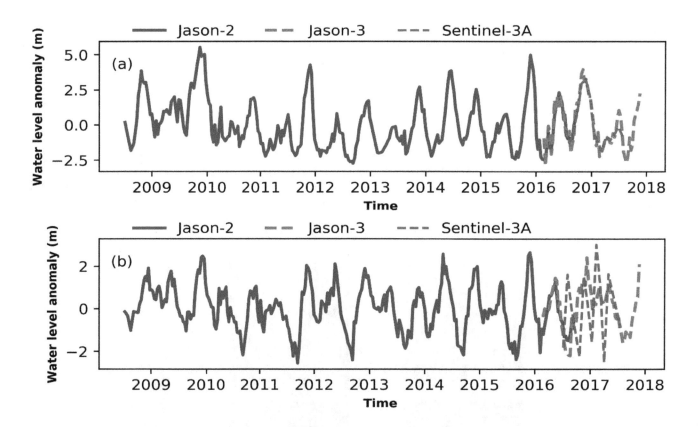

Figure 9. Time series of water level from Jason-2 (blue), Jason-3 (dashed green) and Sentinel-3A (dashed red) on the Ivindo (**a**) and upstream Ogooué (**b**) rivers.

5.4. Multi-Mission Discharge Estimates

The time-series of water level from the VS, that were validated against the stage record from the Lambaréné in situ gauge, were combined using Equation (2) to form a unique time series over May 1995-July 2017 time period. It is presented in Figure 10 along with the time-series of water levels from Lambaréné gauge station. The combined altimetry-based time series of water levels is in good agreement with the in-situ one (R^2 = 0.96 and RMSE = 0.38 m). Contrary to the individual altimetry-based time-series of water levels used in this study that are constrained by the revisit time of the satellite (27 days for ENVISAT on its second orbit, 35 days for ERS-2/ENVISAT and SARAL on their nominal orbit, 369 days for Cryosat-2), the combined time-series present the advantage to have between 3 to 4 (during the ERS-2 and the ENVISAT periods between 1995 and 2010) and 9 (during the SARAL and the Sentinel-3A periods since 2013 when combining with the Cryosat-2 measurements available since 2010) measurements each month.

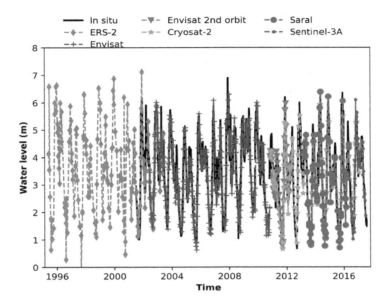

Figure 10. Time series of water level at Lambaréné from the in-situ gauge record (black continuous line), the multi-mission altimetry-based record (ERS-2 data are represented with diamonds, ENVISAT with blue crosses on its nominal orbit and with green triangles on its second orbit, Cryosat-2 with green-blue stars, SARAL with red circles, Sentinel-3 with purple dots).

Applying the rating curve relating water stages and discharge, established for the Lambaréné gauge using in situ measurements of water levels and discharge, an altimetry-based time series of discharge was obtained. Figure 11 shows discharges estimated from altimetry-based and in situ water levels at Lambaréné. A very good agreement between both sources (R = 0.94 and RMSE = 701.6 $m^3 \cdot s^{-1}$ for a mean annual discharge of 4253.4 $m^3 \cdot s^{-1}$ or 16.5% of the mean annual discharge over the observation period). Considering the total annual discharge over the common observation period, altimetry-based and in situ estimates only differs by 1.4 $m^3 \cdot s^{-1}$ (or 0.03% of the mean annual discharge) in average (Table 2).

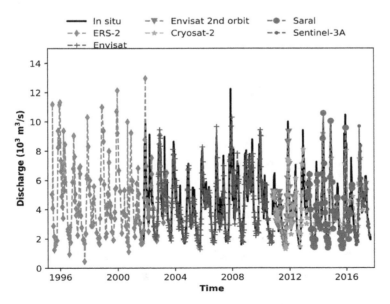

Figure 11. Time series of river discharge at Lambaréné from the in-situ gauge record (black continuous line), the multi-mission altimetry-based record (ERS-2 data are represented with diamonds, ENVISAT with blue crosses on its nominal orbit and with green triangles on its second orbit, Cryosat-2 with green-blue stars, SARAL with red circles, Sentinel-3 with purple dots).

Table 2. Average annual flow derived for the different VS of every altimetry mission and also combining their different records.

Missions	Stations	Estimated Discharge ($m^3 \cdot s^{-1}$)	Relative Error (%)	R(-)	RMSE ($m^3 \cdot s^{-1}$)
In situ	Lambaréné	4253.427	0	1	0
ERS-2	SV_401_Ogooué	4951.475	16.41	0.850	1107.647
	SV_444_0gooué	5465.756	21.45	0.141	2466.262
	SV_902_Ogooué	4498.633	5.76	0.987	332.152
	SV_945_Ogooué	5104.026	19.99	0.886	920.226
	Combined	4848.378	13.99	0.755	1440.860
ENVISAT	SV_401_Ogooué	4311.470	1.36	0.969	431.757
	SV_444_Ogooué	4281.369	0.66	0.884	804.946
	SV_902_Ogooué	4316.212	1.47	0.981	352.028
	SV_945_Ogooué	4342.597	2.09	0.922	722.334
	Combined	4340.594	2.049	0.942	604.321
SARAL	SV_401_Ogooué	3927.343	−7.67	0.975	470.446
	SV_444_Ogooué	3103.856	−27.023	0.947	409.759
	SV_902_Ogooué	3851.170	−9.46	0.986	346.980
	SV_945_Ogooué	2898.672	−31.85	0.973	335.693
	Combined	4060.427	−4.537	0.978	396.820
ENVISAT 2nd Orbit	Station 1	3850.274	−9.47	0.816	868.273
	Station 2	3740.297	−12.06	0.963	450.838
	Station 3	3841.379	−9.69	0.895	654.027
	Combined	3839.296	−9.736	0.898	679496
SENTINEL-3A	SV_050_Ogooué	4454.109	4.72	0.915	757.845
	SV_378_Ogooué	4210.519	−1.01	0.983	316.740
	SV_128_Ogooué	4373.213	2.82	0.931	629.888
	Combined	4262.683	0.22	0.942	597.610
CRYOSAT-2	Lambaréné	4188.220	−1.53	0.971	408.739
SENTINEL-3A + CRYOSAT-2	Combined	4118.9275	−3.162	0.967	462.141
CRYOSAT-2 + SARAL	Combined	4136.3508	−2.752	0.977	405.978
CRYOSAT-2 + ENVISAT 2nd orbit	Combined	4243.2487	−0.239	0.956	515.057
All missions	Combined	4252.052	−0.03	0.936	701.645

6. Discussion

Despite the relatively small size of the ORB and the small length of the Ogooué River (~900 km), the distance between Equatorial cross-tracks of the missions (80 km for ERS-2/ENVISAT/SARAL, and 315 km for Topex-Poseidon/Jason-1/Jason-2/Jason-3 on their nominal orbits), make possible the construction of a dense network of altimetric VS. It is composed of 16 SV using ENVISAT and SARAL, 6 SV using ERS-2 data, 10 SV using Sentinel-3A data, and only 2 SV using Jason-2 and Jason-3 data. In spite of the improvements made in the processing of data from early high precision missions, too few valid data were found in the GDR E data for Jason-1 to provide a continuous monitoring of water stages at two locations where the SV were constructed. These two cross-sections are located in the upstream parts of the basin where the river widths are between 180 and 360 m. Similarly, even if a very good agreement was found between ERS-2 based water levels in the downstream part of the basin, up to several tenths of kilometers upstream Lambaréné in situ station, no SV was defined either on the upstream Ogooué River or on its tributaries. Due to the small width of the river and the presence of higher topography, ERS-2 was affected more by tracking loss than ENVISAT. Furthermore, the accuracy of the data was degraded due to more frequent changes of acquisition modes than ENVISAT (Table 3). ENVISAT was, most of the time, operating in the 320 MHz Ku chirp bandwidth acquisition mode than in the 80 MHz or the 20 MHz modes, that is to say with the better range resolution as the size of the range detection window is 64, 256 and 1024 m respectively. For ERS-2, different cases were observed.

When no valid water level time series were obtained from ERS-2 acquisitions, it can be attributed to either rapid changes in the topography along the track causing changes in Ku chirp bandwidth acquisition modes (with a large number of acquisitions at 82.5 MHz causing a loss of accuracy of the data) or altimeter lock on the hills on top of the river (with acquisitions mostly at 330 MHz but a few tenths of meters above the river, see the example presented in [37] for ENVISAT). When time-series of water levels were derived, the percentage of acquisition at high frequency is high but lower than the one for ENVISAT causing a loss of accuracy compared with ENVISAT (Table 1).

Compared to earlier missions on the same nominal orbit operating at Ku-band (ERS-2 and ENVISAT), SARAL, the first mission to operate at Ka-band, agrees better with gauge records. Higher R and lower RMSE were found between the time-series of water stages from in-situ and from altimetry-based measurements for the four SV close to Lambaréné (Figure 3). In spite of the short period of observation of SARAL on its nominal orbit (35 cycles from February 2013 to June 2016 and even shorter considering that SARAL started to drift since July 2016), the benefits of the Ka-band smaller footprint compared to the Ku-one can be clearly observed as it was reported in other locations [49]. The use of other cycles would have lead to large errors in the water stage retrievals because of (i) slope effects causing large changes of water base levels, (ii) changes in the characteristics of the river (width, depth) responsible changes in amplitude of the water levels, (iii) detection of other water bodies (e.g., lakes) than the Ogooué River.

Only considering ERS-2, ENVISAT and SARAL on their nominal orbit would have resulted in gaps in the monitoring of water levels in the ORB between October 2010 and February 2013 and after July 2016, and especially in Lambaréné where the river discharge is estimated. To complete the time-series, data from the ENVISAT 2nd orbit were considered from October 2010 to April 2012. Three SV were build giving water estimates of equivalent quality as the ones obtained on the nominal orbit (Figure 4). Applying an innovative processing, time-series of water levels were derived from Cryosat-2 measurements, the first altimetry mission to operate in SAR mode. The results show, that the data from this mission, that are of very good quality (R^2 and RMSE of very similar quality as ENVISAT or SARAL, see Figure 6). The monitoring of the water levels in the Ogooué can continue using acquisitions from the Sentinel-3A mission and soon, from the Sentinel-3B. Comparisons performed against the in-situ gauge records from Lambaréné of one year and a half of Sentinel-3A-based water levels show the high quality of the data acquired in SAR mode (R^2 = 0.98 and RMSE < 20 cm for the SV located at the closer distance to Lambaréné station, see Figure 5).

Table 3. Acquisition modes of the ERS-2 and ENVISAT measurements for the VS in the ORB.

Virtual Stations	Rivers or Lakes	Missions	ENVISAT Data Modes			ERS-2 Data Modes	
			320 Hz (%)	80 Hz (%)	20 Hz (%)	330 Hz (%)	82.5 Hz (%)
SV_229_Ogooué	Ogooué	ENVISAT, SARAL	99.786	0.213	0	100	0
SV_272_Ivindo	Ivindo	ENVISAT, SARAL	96.893	1.804	1.302	58.0	42.0
SV_272_Ogooué	Ogooué	ENVISAT, SARAL	98.267	0.533	1.2	6.25	93.75
SV_315_Ogooué	Ogooué	ENVISAT, SARAL	91.896	6.212	1.890	100	0
SV_358_Ngounié	Ngounié	ENVISAT, SARAL	88.76	10.3	0.939	70.270	29.729
SV_401_Ngounié	Ngounié	ENVISAT, SARAL	98.776	1.146	0.077	81.132	18.867
SV_401_Ogooué	Ogooué	ERS-2, ENVISAT, SARAL	98.776	1.146	0.077	81.132	18.867
SV_444_Ogooué	Ogooué	ERS-2, ENVISAT, SARAL	99.887	0.113	0	85.0	15.0
SV_730_Ogooué	Ogooué	SARAL	92.347	6.152	1.5005	34.426	65.573
SV_773_Ogooué	Ogooué	ENVISAT, SARAL	99.178	0.821	0	70.588	29.411
SV_902_lake_Onangué	lake Onangué	ERS-2, ENVISAT, SARAL	99.669	0.259	0.070	87.804	12.195
SV_902_Ogooué	Ogooué	ERS-2, ENVISAT, SARAL	99.669	0.259	0.070	87.804	12.195
SV_902_Ogooué_2	Ogooué	ENVISAT, SARAL	99.669	0.259	0.070	87.804	12.195
SV_945_lake_Louandé	lake Louandé	ENVISAT	99.513	0.487	0	100	0
SV_945_lac_Ogognié	lake Ogognié	ENVISAT, SARAL	99.513	0.487	0	100	0
SV_945_Ogooué	Ogooué	ERS-2, ENVISAT, SARAL	99.513	0.487	0	100	0
SV_945_Ogooué_2	Ogooué	ENVISAT	99.513	0.487	0	100	0

Due to the presence of only one in-situ station with records during the altimetry acquisition period, the consistency of the altimetry-based time-series of water levels was checked for ENVISAT and SARAL through cross-correlation estimates. They were performed choosing as reference the SV close to Lambaréné station whose records were validated. The difference observed for the cross-correlations between ENVISAT and SARAL occur in the upstream parts of the basin (Figures 7 and 8 respectively). The lower scores for SARAL can be attributed to the shorter observation period of SARAL compared with ENVISAT (35 cycles for SARAL against 85 for ENVISAT, that is to say, more or less a ratio of one third). This difference in number of observations has a smaller impact on the downstream part of the ORB as the water stages are mostly driven by the flow from the upstream parts. On the upstream part, changes in rainfall conditions between two VS are likely to strongly modify the water stage at these two locations. Even a few of this kind of changes can reduce the correlation between the two time-series of river levels between these two stations, especially for shorter records. Due to the relatively small length of the basin and the quite long repeat period of ENVISAT and SARAL (35 days), maxima of correlation are generally observed with no time-lag, except for a limited number of stations in the upstream parts of the ORB where differences lower than one month were obtained.

Based on the good agreement found between in situ and altimetry-based water stages for the VS located around Lambaréné gauge station, the stage discharge rating curve was applied to any individual VS but also for the combination of VS for the same mission. The results are presented in Table 2. This combination represents a good compromise between quality of the resulting time series of discharge and number of observations. For all the missions, the results of the combination are very close to the best result obtained with a single one. It also decreases the risk to miss a rapid flood event that will modify the annual discharge. The combination of data from several missions (ENVISAT 2nd orbit + Cryosat-2, SARAL + Cryosat-2, Sentinel-3A + Cryosat-2) also increase the quality of the discharge estimate. When combining the data from the missions (Cryosat-2 period of availability of the data is overlapping the ones from ENVISAT, ENVISAT 2nd orbit, SARAL and Sentinel-3A), very good performances are obtained (see Table 2).

7. Conclusions

This study provides one of the first assessments of the performances of multiple satellite altimetry-based water levels in a river basin, from ERS-2 to Sentinel-3A. Comparisons between altimetry water stages and in situ gauge records from Lambaréné, the unique station still in operation in the ORB showed the improvement in performances of the missions in operation since ERS-2. In spite of good performances in the downstream part of the ORB ($R^2 > 0.82$ and RMSE < 0.6 m for 3 of the four comparisons), no VS using ERS-2 data were created in the upstream part of the basin due to the small width of the rivers and the presence of topography. On the contrary, the whole basin was sampled using ENVISAT ($R^2 > 0.82$ and RMSE < 0.5 m) and SARAL ($R^2 > 0.90$ and RMSE < 0.4 m) on the same orbit. A very good consistency was also found for these two missions when computing cross-correlations between altimetry-based water stages all over the basin. As SARAL orbit control was not as strict as is generally is for high precision altimetry missions, the drifts of several kilometers from the nominal paths lead to lower agreement in the upstream parts of the basin than using ENVISAT. Missions operating in SAR mode also exhibit very good accuracy ($R^2 > 0.88$ and RMSE < 0.4 m for only one year and a half of Sentinel-3A data and $R^2 = 0.96$ and RMSE = 0.25 m for Cryosat-2). For Cryosat-2, the concept of VS was extended: data acquired on close cross-sections were combined to build the time series of water level. The use of Cryosat-2 (and ENVISAT 2nd orbit) data allowed a continuous monitoring of the ORB for all the altimetry period, showing the potential of the data acquired during geodetic (drifting) orbit of the altimetry missions for land hydrology.

The altimetry-based time-series of water levels from the different altimetry missions close to the Lambaréné in situ stations were converted to discharges using the rating curves from the in situ station. Very good agreement was also found, the quality of which is directly linked to the accuracy of the water levels at the VS for the different missions. This study also shows the interest to combine the data from different missions to improve quality of the monitoring of level and discharge in terms of sampling frequency and annual discharge estimates.

The network of altimetry VS built in this study and associated discharge estimates has a strong interest for the scientific community. It will allow (i) the continuous monitoring of water stages in an almost ungauged basin, (ii) the analysis of possible effects of climate variability and anthropogenic effects (i.e., deforestation) on the hydrological cycle of the ORB. It will be very useful for the calibration and validation of the NASA/CNES Surface Water and Ocean Topography (SWOT) mission, that will use the SAR interferometry technique to map surface water elevation at a spatial resolution of 100 m as its one day calibration orbit encompass the downstream part of the Ogooué Basin.

Acknowledgments: This work is dedicated to Gaston Liénou who sadly passed away in 2015. He enormously contributed to the continuity of study operations in the Ogooué river basin. The authors are grateful to the Center for Topographic studies of the Ocean and Hydrosphere (CTOH) at LEGOS (Toulouse, France) for providing the altimetry dataset. We also thank the Société de l'Énergie et de l'Eau du Gabon (SEEG) branch of Lambaréné (Lambaréné, Gabon) for supplying of the Ogooué water level from the Lambaréné hydrological station. Aurélie Flore KOUMBA PAMBO, Scientific Cell coordinator, Agence Nationale des Parcs Nationaux (ANPN) and Centre National de la Recherche Scientifique et Technologique (CENAREST), Libreville, Gabon, is warmly thanked for her assistance with respect to the reasearch authorizations. Other thanks go to the "Service Coopération et de l'Action Culturelle (SCAC)" of the French Embassy in Cameroon and TNC-Gabon, and the project "Centre de Topographie des Océans et de l'Hydrosphère" funded by CNES. This work was supported by the French national programme EC2CO-Biohefect "Régime d'Altération/Érosion en Afrique Centrale (RALTERAC)". Final acknowledgements are for the project "Jeune Équipe Associée Internationale—Réponse du Littoral Camerounais aux Forçages Océaniques Multi-Échelles (JEAI-RELIFOME)" of the University of Douala and the Laboratoire Mixte International (LMI) Dynamique des écosystèmes continentaux d'Afrique Centrale en contexte de changements globaux (DYCOFAC) for its support in the accomplishment of this work. The authors want to thank two anonymous Reviewers for their helpful comments.

Author Contributions: Sakaros Bogning and Frédéric Frappart conceived and designed the study; Sakaros Bogning performed the altimetry data processing with the help of Fabien Blarel, Frédéric Frappart, Fernando Niño and Frédérique Seyler; Jean-Pierre Briquet and Gil Mahé processed the in situ data; all the authors analyzed the results and contributed to the writing of the paper.

References

1. Younger, P.L. *Water*; Hodder & Stoughton: London, UK, 2012.
2. Vörösmarty, C.J.; Green, P.; Salisbury, J.; Lammers, R.B. Global water resources: Vulnerability from climate change and population growth. *Science* **2000**, *289*, 284–288. [CrossRef] [PubMed]
3. Oki, T.; Kanae, S. Global hydrological cycles and world water resources. *Science* **2006**, *313*, 1068–1072. [CrossRef] [PubMed]
4. Gleick, P.H. Global freshwater resources: Soft-path solutions for the 21st century. *Science* **2003**, *302*, 1524–1528. [CrossRef] [PubMed]
5. Alsdorf, D.E.; Rodríguez, E.; Lettenmaier, D.P. Measuring surface water from space. *Rev. Geophys.* **2007**, *45*, RG2002. [CrossRef]
6. Alsdorf, D.; Beighley, E.; Laraque, A.; Lee, H.; Tshimanga, R.; O'Loughlin, F.; Mahé, G.; Dinga, B.; Moukandi, G.; Spencer, R.G.M. Opportunities for hydrologic research in the Congo Basin. *Rev. Geophys.* **2016**, *54*, 378–409. [CrossRef]
7. Stammer, D.; Cazenave, A. *Satellite Altimetry over Oceans and Land Surfaces*; Taylor & Francis: Boca Raton, FL, USA, 2017.
8. Crétaux, J.-F.; Nielsen, K.; Frappart, F.; Papa, F.; Calmant, S.; Benveniste, J. Hydrological applications of satellite altimetry: Rivers, lakes, man-made reservoirs, inundated areas. In *Satellite Altimetry over Oceans and Land Surfaces*; Stammer, D., Cazenave, A., Eds.; Earth Observation of Global Changes; CRC Press: Boca Raon, FL, USA, 2017; pp. 459–504.
9. Morris, C.S.; Gill, S.K. Variation of Great Lakes water levels derived from Geosat altimetry. *Water Resour. Res.* **1994**, *30*, 1009–1017. [CrossRef]
10. Birkett, C.M. The contribution of TOPEX/POSEIDON to the global monitoring of climatically sensitive lakes. *J. Geophys. Res.* **1995**, *100204*, 179–225. [CrossRef]

11. Koblinsky, C.J.; Clarke, R.T.; Brenner, A.C.; Frey, H. Measurement of river level variations with satellite altimetry. *Water Resour. Res.* **1993**, *29*, 1839–1848. [CrossRef]

12. Birkett, C.M. Contribution of the TOPEX NASA Radar Altimeter to the global monitoring of large rivers and wetlands. *Water Resour. Res.* **1998**, *34*, 1223. [CrossRef]

13. Frappart, F.; Calmant, S.; Cauhopé, M.; Seyler, F.; Cazenave, A. Preliminary results of ENVISAT RA-2-derived water levels validation over the Amazon basin. *Remote Sens. Environ.* **2006**, *100*, 252–264. [CrossRef]

14. Baup, F.; Frappart, F.; Maubant, J. Use of satellite altimetry and imagery for monitoring the volume of small lakes. In Proceedings of the International Geoscience and Remote Sensing Symposium (IGARSS), Quebec City, QC, Canada, 13–18 July 2014.

15. Sulistioadi, Y.B.; Tseng, K.-H.; Shum, C.K.; Hidayat, H.; Sumaryono, M.; Suhardiman, A.; Setiawan, F.; Sunarso, S. Satellite radar altimetry for monitoring small rivers and lakes in Indonesia. *Hydrol. Earth Syst. Sci.* **2015**, *19*, 341–359. [CrossRef]

16. Frappart, F.; Papa, F.; Malbeteau, Y.; León, J.G.; Ramillien, G.; Prigent, C.; Seoane, L.; Seyler, F.; Calmant, S. Surface freshwater storage variations in the orinoco floodplains using multi-satellite observations. *Remote Sens.* **2015**, *7*, 89–110. [CrossRef]

17. Bjerklie, D.M.; Lawrence Dingman, S.; Vorosmarty, C.J.; Bolster, C.H.; Congalton, R.G. Evaluating the potential for measuring river discharge from space. *J. Hydrol.* **2003**, *278*, 17–38. [CrossRef]

18. Kouraev, A.V.; Zakharova, E.A.; Samain, O.; Mognard, N.M.; Cazenave, A. Ob' river discharge from TOPEX/Poseidon satellite altimetry (1992–2002). *Remote Sens. Environ.* **2004**, *93*, 238–245. [CrossRef]

19. Zakharova, E.A.; Kouraev, A.V.; Cazenave, A.; Seyler, F. Amazon River discharge estimated from TOPEX/Poseidon altimetry. *Comptes Rendus Geosci.* **2006**, *338*, 188–196. [CrossRef]

20. Papa, F.; Durand, F.; Rossow, W.B.; Rahman, A.; Bala, S.K. Satellite altimeter-derived monthly discharge of the Ganga-Brahmaputra River and its seasonal to interannual variations from 1993 to 2008. *J. Geophys. Res. Ocean.* **2010**, *115*, C12013. [CrossRef]

21. Birkinshaw, S.J.; Moore, P.; Kilsby, C.G.; O'Donnell, G.M.; Hardy, A.J.; Berry, P.A.M. Daily discharge estimation at ungauged river sites using remote sensing. *Hydrol. Process.* **2014**, *28*, 1043–1054. [CrossRef]

22. Leon, J.G.; Calmant, S.; Seyler, F.; Bonnet, M.-P.; Cauhopé, M.; Frappart, F.; Filizola, N.; Fraizy, P. Rating curves and estimation of average water depth at the upper Negro River based on satellite altimeter data and modeled discharges. *J. Hydrol.* **2006**, *328*, 481–496. [CrossRef]

23. Tarpanelli, A.; Barbetta, S.; Brocca, L.; Moramarco, T. River Discharge Estimation by Using Altimetry Data and Simplified Flood Routing Modeling. *Remote Sens.* **2013**, *5*, 4145–4162. [CrossRef]

24. Paris, A.; Dias de Paiva, R.; Santos da Silva, J.; Medeiros Moreira, D.; Calmant, S.; Garambois, P.-A.; Collischonn, W.; Bonnet, M.-P.; Seyler, F. Stage-discharge rating curves based on satellite altimetry and modeled discharge in the Amazon basin. *Water Resour. Res.* **2016**, *52*, 3787–3814. [CrossRef]

25. Wilson, M.D.; Bates, P.; Alsdorf, D.; Forsberg, B.; Horritt, M.; Melack, J.; Frappart, F.; Famiglietti, J. Modeling large-scale inundation of Amazonian seasonally flooded wetlands. *Geophys. Res. Lett.* **2007**, *34*. [CrossRef]

26. Getirana, A.; Bonnet, M.; Calmant, S.; Roux, E.; Rotunno Filho, O.C.; Mansur, W.J. Hydrological monitoring of poorly gauged basins based on rainfall-runoff modeling and spatial altimetry. *J. Hydrol.* **2009**, *379*, 205–219. [CrossRef]

27. Milzow, C.; Krogh, P.E.; Bauer-Gottwein, P. Combining satellite radar altimetry, SAR surface soil moisture and GRACE total storage changes for model calibration and validation in a large ungauged catchment. *Hydrol. Earth Syst. Sci. Discuss.* **2010**, *7*, 9123–9154. [CrossRef]

28. Pereira-Cardenal, S.; Riegels, N.D.; Berry, P.A.M.; Smith, R.G.; Yakovlev, A.; Siegfried, T.U.; Bauer-Gottwein, P. Real-time remote sensing driven river basin modeling using radar altimetry. *Hydrol. Earth Syst. Sci.* **2011**, *15*, 241–254. [CrossRef]

29. De Paiva, R.C.D.; Buarque, D.C.; Collischonn, W.; Bonnet, M.P.; Frappart, F.; Calmant, S.; Bulhões Mendes, C.A. Large-scale hydrologic and hydrodynamic modeling of the Amazon River basin. *Water Resour. Res.* **2013**, *49*, 1226–1243. [CrossRef]

30. Mignard, S.L.A.; Mulder, T.; Martinez, P.; Charlier, K.; Rossignol, L.; Garlan, T. Deep-sea terrigenous organic carbon transfer and accumulation: Impact of sea-level variations and sedimentation processes off the Ogooue River (Gabon). *Mar. Pet. Geol.* **2017**, *85*, 35–53. [CrossRef]

31. Mahe, G.; Lerique, J.; Olivry, J.-C. Le fleuve Ogooué au Gabon: Reconstitution des débits manquants et mise en évidence de variations climatiques à l'équateur. *Hydrol. Cont.* **1990**, *5*, 105–124.

32. Lambert, T.; Darchambeau, F.; Bouillon, S.; Alhou, B.; Mbega, J.D.; Teodoru, C.R.; Nyoni, F.C.; Massicotte, P.; Borges, A.V. Landscape Control on the Spatial and Temporal Variability of Chromophoric Dissolved Organic Matter and Dissolved Organic Carbon in Large African Rivers. *Ecosystems* **2015**, *18*, 1224–1239. [CrossRef]

33. Lienou, G.; Mahe, G.; Paturel, J.E.; Servat, E.; Sighomnou, D.; Ekodeck, G.E.; Dezetter, A.; Dieulin, C. Evolution des régimes hydrologiques en région équatoriale camerounaise: Un impact de la variabilité climatique en Afrique équatoriale? *Hydrol. Sci. J.* **2008**, *53*, 789–801. [CrossRef]

34. Home—CTOH. Available online: http://ctoh.legos.obs-mip.fr/ (accessed on 24 October 2017).

35. Frappart, F.; Legrésy, B.; Niño, F.; Blarel, F.; Fuller, N.; Fleury, S.; Birol, F.; Calmant, S. An ERS-2 altimetry reprocessing compatible with ENVISAT for long-term land and ice sheets studies. *Remote Sens. Environ.* **2016**, *184*, 558–581. [CrossRef]

36. Blarel, F.; Frappart, F.; Legrésy, B.; Blumstein, D.; Fatras, C.; Mougin, E.; Papa, F.; Prigent, C.; Rémy, F.; Niño, F.; et al. Radar altimetry backscattering signatures at Ka, Ku, C and S bands over land. In Proceedings of the Living Planet Symposium, Prague, Chech Republic, 9–13 May 2016.

37. Biancamaria, S.; Frappart, F.; Leleu, A.-S.; Marieu, V.; Blumstein, D.; Desjonquères, J.-D.; Boy, F.; Sottolichio, A.; Valle-Levinson, A. Satellite radar altimetry water elevations performance over a 200 m wide river: Evaluation over the Garonne River. *Adv. Space Res.* **2017**, *59*, 128–146. [CrossRef]

38. Frappart, F.; Roussel, N.; Biancale, R.; Martinez Benjamin, J.J.; Mercier, F.; Perosanz, F.; Garate Pasquin, J.; Martin Davila, J.; Perez Gomez, B.; Gracia Gomez, C.; et al. The 2013 Ibiza Calibration Campaign of Jason-2 and SARAL Altimeters. *Mar. Geodesy* **2015**, *38*, 219–232. [CrossRef]

39. Vu, P.; Frappart, F.; Darrozes, J.; Marieu, V.; Blarel, F.; Ramillien, G.; Bonnefond, P.; Birol, F. Multi-Satellite Altimeter Validation along the French Atlantic Coast in the Southern Bay of Biscay from ERS-2 to SARAL. *Remote Sens.* **2018**, *10*, 93. [CrossRef]

40. Salameh, E.; Frappart, F.; Marieu, V.; Spodar, A.; Parisot, J.-P.; Hanquiez, V.; Turki, I.; Laignel, B. Monitoring Sea Level and Topography of Coastal Lagoons Using Satellite Radar Altimetry: The Example of the Arcachon Bay in the Bay of Biscay. *Remote Sens.* **2018**, *10*, 297. [CrossRef]

41. Frappart, F.; Papa, F.; Marieu, V.; Malbeteau, Y.; Jordy, F.; Calmant, S.; Durand, F.; Bala, S. Preliminary Assessment of SARAL/AltiKa Observations over the Ganges-Brahmaputra and Irrawaddy Rivers. *Mar. Geodesy* **2015**, *38*, 568–580. [CrossRef]

42. Wingham, D.J.; Rapley, C.G.; Griffiths, H. New Techniques in Satellite Altimeter Tracking Systems. In Proceedings of the International Geoscience and Remote Sensing Symposium (IGARSS), Zurich, Switzerland, 1986; pp. 1339–1344.

43. Nielsen, K.; Stenseng, L.; Andersen, O.; Knudsen, P. The Performance and Potentials of the CryoSat-2 SAR and SARIn Modes for Lake Level Estimation. *Water* **2017**, *9*, 374. [CrossRef]

44. Schneider, R.; Godiksen, P.N.; Villadsen, H.; Madsen, H.; Bauer-Gottwein, P. Application of CryoSat-2 altimetry data for river analysis and modelling. *Hydrol. Earth Syst. Sci.* **2017**, *21*, 751–764. [CrossRef]

45. Jiang, L.; Schneider, R.; Andersen, O.; Bauer-Gottwein, P. CryoSat-2 Altimetry Applications over Rivers and Lakes. *Water* **2017**, *9*, 211. [CrossRef]

46. Birkinshaw, S.J.; O'Donnell, G.M.; Moore, P.; Kilsby, C.G.; Fowler, H.J.; Berry, P.A.M.M. Using satellite altimetry data to augment flow estimation techniques on the Mekong River. *Hydrol. Process.* **2010**, *24*, 3811–3825. [CrossRef]

47. Papa, F.; Bala, S.K.; Pandey, R.K.; Durand, F.; Gopalakrishna, V.V.; Rahman, A.; Rossow, W.B. Ganga-Brahmaputra river discharge from Jason-2 radar altimetry: An update to the long-term satellite-derived estimates of continental freshwater forcing flux into the Bay of Bengal. *J. Geophys. Res. Ocean.* **2012**, *117*, C11021. [CrossRef]

48. Getirana, A.C.V.; Peters-Lidard, C. Estimating water discharge from large radar altimetry datasets. *Hydrol. Earth Syst. Sci.* **2013**, *17*, 923–933. [CrossRef]

49. Bonnefond, P.; Verron, J.; Aublanc, J.; Babu, K.; Bergé-Nguyen, M.; Cancet, M.; Chaudhary, A.; Crétaux, J.-F.; Frappart, F.; Haines, B.; et al. The Benefits of the Ka-Band as Evidenced from the SARAL/AltiKa Altimetric Mission: Quality Assessment and Unique Characteristics of AltiKa Data. *Remote Sens.* **2018**, *10*, 83. [CrossRef]

Evolution of the Performances of Radar Altimetry Missions from ERS-2 to Sentinel-3A over the Inner Niger Delta

Cassandra Normandin [1,*], Frédéric Frappart [2,3], Adama Telly Diepkilé [4], Vincent Marieu [1], Eric Mougin [2], Fabien Blarel [3], Bertrand Lubac [1], Nadine Braquet [5] and Abdramane Ba [6]

[1] Oceanic and Continental Environments and Paleoenvironments (EPOC), Mixed Research Unit (UMR) 5805, University of Bordeaux, Allée Geoffroy Saint-Hilaire, 33615 Pessac, France; vincent.marieu@u-bordeaux.fr (V.M); bertrand.lubac@u-bordeaux.fr (B.L)

[2] Geosciences Environment Toulouse (GET), University of Toulouse, National Center for Scientific Reaseach (CNRS), Institute for Research and Development (IRD), UPS. Observatory Midi-Pyrénées (OMP), 14 Av. E. Belin, 31400 Toulouse, France; frederic.frappart@legos.obs-mip.fr (F.F.); eric.mougin@get.omp.eu (E.M.)

[3] Laboratory of Studies on Spatial Geophysics and Space Oceanography (LEGOS), University of Toulouse, National Center for Space Studies (CNES), CNRS, IRD, UPS. OMP, 14 Av. E. Belin, 31400 Toulouse, France; fabien.blarel@legos.obs-mip.fr

[4] Department of Education and Research (DER) Math-Informatics, Faculty of Sciences and Technology (FST)/University of Sciences, Techniques and Technologies of Bomako (USTTB), Bamako 3206, Mali; Adama.Diepkile@USherbrooke.ca

[5] National Research Institute of Science and Technology for Environment and Agriculture (IRSTEA), IRD, 361 rue Jean-François Breton, 34196 Montpellier, France; nadine.braquet@ird.fr

[6] Laboratory of Optics, Spectroscopy and Atmospheric Sciences (LOSSA), Department of Education and Research (DER) Physics, Faculty of Sciences and Technology (FST)/University of Sciences, Techniques and Technologies of Bomako (USTTB), Bamako 3206, Mali; abdramaneba55@yahoo.fr

* Correspondence: cassandra.normandin@u-bordeaux.fr

Abstract: Radar altimetry provides unique information on water stages of inland hydro-systems. In this study, the performance of seven altimetry missions, among the most commonly used in land hydrology (i.e., European Remote-Sensing Satellite-2 (ERS-2), ENVIronment SATellite (ENVISAT), Satellite with Argos and ALtika (SARAL), Jason-1, Jason-2, Jason-3 and Sentinel-3A), are assessed using records from a dense in situ network composed of 19 gauge stations in the Inner Niger Delta (IND) from 1995 to 2017. Results show an overall very good agreement between altimetry-based and in situ water levels with correlation coefficient (R) greater than 0.8 in 80% of the cases and Root Mean Square Error (RMSE) lower than 0.4 m in 48% of cases. Better agreement is found for the recently launched missions such as SARAL, Jason-3 and Sentinel-3A than for former missions, indicating the advance of the use of the Ka-band for SARAL and of the Synthetic-aperture Radar (SAR) mode for Sentinel-3A. Cross-correlation analysis performed between water levels from the same altimetry mission leads to time-lags between the upstream and the downstream part of the Inner Niger Delta of around two months that can be related to the time residence of water in the drainage area.

Keywords: altimetry; water levels; validation; Inner Niger Delta

1. Introduction

Surface waters, which are part of the continental branch of the terrestrial water cycle, play an essential role in supplying fresh water for basic human and economic needs. They are strongly impacted by climate changes and anthropogenic pressures caused by population growth and changes in agricultural practices [1–3]. Despite the importance of their monitoring for addressing integrated water resource management, use in operational flood forecasting or disaster mitigation, reliable in situ measurements of water stage and discharge has become increasingly scarce information due to either the disappearance of the gauge networks or the difficulty to get access to data [4,5].

Satellite radar altimetry, initially developed for the measurement of the ocean surface topography through the measurement of the distance between the Earth's surface and the spaceborne radar altimeter [6], has demonstrated its efficiency for deriving water levels of inland water bodies (see Crétaux et al. [7] for a recent review). Radar altimetry, was initially used over land to retrieve water levels over homogeneous surfaces such as large lakes and enclosed seas [8,9], but also at cross-sections between rivers and altimetry ground-tracks of several kilometers of width in large river basins [10,11]. These early results were obtained using Geosat and Topex/Poseidon (T/P) ranges (i.e., the distance between the satellite and the surface) derived from the Ocean retracking algorithm. Root Mean Square Errors (RMSE) lower than 0.05 m and 1.1 m through comparisons with in situ water stages were obtained over lakes and rivers respectively. The comparisons are performed at the so-called Virtual Stations. Virtual stations (VS) are defined as the cross-sections of an altimetry ground-track and a water body (i.e., lake, reservoir river channel, floodplain, or wetland) where the temporal variations of the height from one cycle to the next can be associated with changes in water level [7]. With the launch of ENVISAT in 2002, ranges processed using other retracking algorithms were included in the Geophysical Data Records (GDR) made available by space agencies. Among them, the Offset Center Of Gravity (OCOG, also known as Ice-1) was found to provide, most of the time, the most accurate estimate of river water levels (with RMSE generally lower than 0.3 m and correlation coefficient R greater than 0.9) [12]. Combined with the availability of land-dedicated corrections of the ionosphere, wet troposphere delays and improvements in the data processing, this allowed the generalization of the use of radar altimetry for the monitoring of inland waters [7,13].

Thanks to these different improvements and the use of high-frequency data (10, 18, 20 or 40 Hz depending on the altimetry mission) instead of 1 Hz data (~7 km of sampling along the track) as over the open ocean water bodies of a few or below one hundred meters of width can now be monitored with very good accuracy (e.g., [14–16]). As radar altimetry data have global coverage and are freely available, they are now commonly used in a wide range of hydrological applications (see Crétaux et al. [7] for a recent review), and, even in support for the management of in situ networks [17]. Until now, no study provided a systematic assessment of the performance over rivers of different altimetry missions that were operating since the beginning of the high-precision altimetry era, which started with the launch of Topex/Poseidon (T/P) in 1992, contrary to that done over lakes [18,19].

The goal of this study is to evaluate the quality of altimetry-based water levels for all missions in repetitive orbits whose data contained in the GDR were processed using the OCOG retracking algorithm. For this purpose, comparisons between altimetry-based water stages and in situ measurements from a dense gauges network were performed in the Inner Niger Delta (IND).

Several studies already used altimetry-based water levels to better understand spatio-temporal dynamics of the flood in this region [20–22] and to estimate river discharges [23,24]. The IND was chosen as study area as it is densely covered with (i) in situ gauge stations whose records are available over the whole high-precision altimetry era and (ii) cross-sections between altimetry ground-tracks from different missions and rivers of various widths.

A dense network of VS was built in the IND, composed of 52, 63, 623 VS for European Remote-Sensing Satellite-2 (ERS-2), ENVIronment SATellite (ENVISAT), Satellite with Argos and ALtika (SARAL) respectively, 31 for Sentinel-3A and 8, 8, 9 for Jason-1, Jason-2 and Jason-3 respectively. At each virtual station, time variations of river levels from radar altimetry are constructed.

In this study, comparisons between altimetry-based water stages derived from acquisitions of Jason-1, Jason-2 and Jason-3, ERS-2, ENVISAT, SARAL and Sentinel-3A and in situ water levels from 19 gauge stations located in the IND are presented in terms of RMSE and R. Intra-mission results consistency were also assessed through cross-correlations between virtual stations along the river.

2. Method

2.1. Principle of Radar Altimetry and Data Processing

2.1.1. Principle of Altimetry Measurement

The principle of radar altimetry is the following: a radar altimeter emits an electromagnetic wave in the nadir direction and measures its round-trip time. The distance between the satellite and the Earth surface—the altimeter range (R_0)—is derived with a precision of a few centimeters. The satellite altitude (H) referred to an ellipsoid is determined from precise orbitography technique with accuracy better than 2 cm. Taking into account propagation corrections caused by delays resulting from interactions of electromagnetic wave with the atmosphere, and geophysical corrections, the height of the reflecting surface (h) with reference to an ellipsoid can be estimated as [25,26]:

$$h = H - \left(R_0 + \sum \left(\Delta R_{propagation} + \Delta R_{geophysical} \right) \right) \tag{1}$$

where H is the height of the center of mass of the satellite above the ellipsoid estimated using precise orbit determination (POD) technique, R_0 is the nadir altimeter range from the center of mass of the satellite to the sea surface taking into account instrumental corrections.

$$\sum \Delta R_{propagation} = \Delta R_{ion} + \Delta R_{dry} + \Delta R_{wet} \tag{2}$$

where ΔR_{ion} is the atmospheric refraction range correction due to the free electron content associated with the dielectric properties of the ionosphere, ΔR_{dry} is the atmospheric refraction range correction due to the dry gas component of the troposphere, ΔR_{wet} is the atmospheric refraction range correction due to the water vapor and the cloud liquid water content of the troposphere.

$$\sum \Delta R_{geophysical} = \Delta R_{solid\ Earth} + \Delta R_{pole} \tag{3}$$

where $\Delta R_{solid\ Earth}$ and ΔR_{pole} are the corrections respectively accounting for crustal vertical motions due to the solid Earth and pole tides.

2.1.2. Time Variations of River Levels from Radar Altimetry Measurements

In this study, the Multi-mission Altimetry Processing Software (MAPS), developed by Frappart et al. [27] was used to visualize and process the altimetry data over land [22,28–30] and ocean [31,32] to build the VS in the IND. Data processing is composed of three main steps: (i) a coarse delineation of the VS using Google Earth; (ii) a refined selection of the valid altimetry data based on visual inspection; and (iii) the computation of the time series of water level. The altimetry-based water level is computed for each cycle using the median of the selected altimetry heights, along with their respective deviation (i.e., mean absolute deviation). This process is repeated each cycle to construct the water level time series at the virtual stations and illustrated in Figure 1.

Altimetry datasets are referenced either to WGS84 ellipsoid or to Topex/Poseidon ellipsoid. A datum conversion from T/P ellipsoid to WGS84 is automatically performed using Equation (4) adapted from Jekeli et al. [33] and implemented in the version of MAPS used in Salameh et al. [34]:

$$\Delta h = \frac{a' \left(1 - e'^2\right)}{\sqrt{1 - e'^2 \sin^2 \varphi}} - \frac{a \left(1 - e^2\right)}{\sqrt{1 - e^2 \sin^2 \varphi}} \tag{4}$$

where Δh is the variation of height at latitude φ due to the change of ellipsoid from T/P to WGS84 datum, $a = 6,378,137$ m and $e = 0.081819190842621$ are the semi-major axis and the eccentricity of the WGS84 datum, $a' = 6,378,136.3$ m and $e' = 0.081819221456$ are the semi-major axis and the eccentricity of the T/P datum.

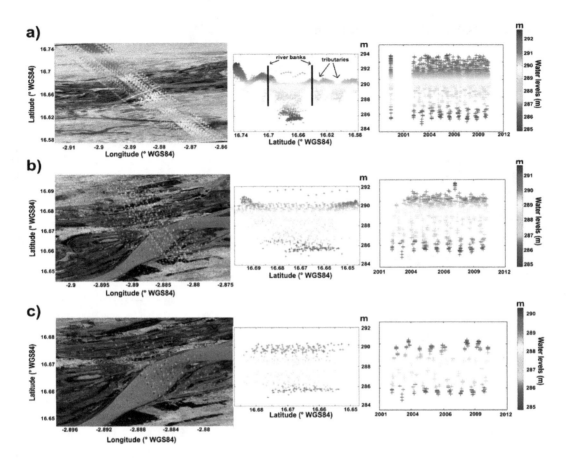

Figure 1. The different steps of the altimetry data using Multi-mission Altimetry Processing Software (MAPS). First, a rough selection of the altimetry data (represented with crosses of color) is performed: (**a**) all altimetry data located a few kilometers away from the center of the river are selected (left panel), the major topographic features, such hills, river banks, tributaries, etc. can be identified (central panel), temporal variations of the altimetry signal cannot be related to any hydrological signal (right panel). Then, a more accurate selection is made; (**b**) the number of data is decreasing and outliers are removed (left panel), the shape of the river and its temporal variations in width (central panel) and height (left panel) becomes clearer. This process is repeated until final selection is achieved; (**c**) all subfigures are derived from the MAPS Graphical User Interface (GUI).

Some along-track altimetry profiles exhibit a parabolic shape caused by non-nadir reflections known as hooking effect (see Figure 2). Hooking effect is corrected as follows:

$$h(s_0) = h(s_i) + \frac{1}{2R_{corr}(s_0)}\left(1 + \left(\frac{\partial H}{\partial s}(s_i)\right)^2\right)ds^2 \tag{5}$$

where s is the along-track coordinate, $h(s_0)$ is the altimeter height at nadir, $R_{corr}(s_0)$ the altimeter range at nadir corrected from the geophysical and environmental effects, s_0 the location of the nadir along the altimeter track, s_i the coordinates of the slant measurements, $\partial H/\partial s$ the rate of altitude variation of the satellite along the orbital segment, and ds the along track difference between s_0 and s_i. Then, the altimeter height at nadir is computed using the summit of the parabola representing the actual water level:

$$h(s_0) = as_0^2 + bs_0 + c \tag{6}$$

where a, b and c are parabola coefficients calculating using a least-square fitting of the altimeter data affected by hooking.

Finally, s_0 and $h(s_0)$ are defined as follow:

$$s_0 = -\frac{b}{2a} \text{ and } h(s_0) = c - \frac{b}{4a} \tag{7}$$

$$h(s_0) = as_0^2 + bs_0 + c \tag{8}$$

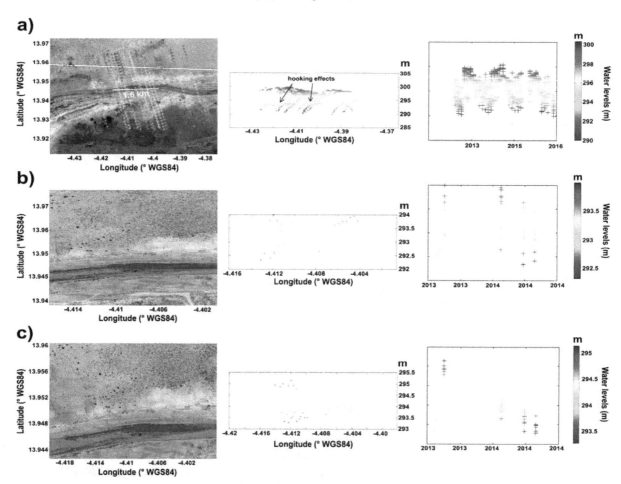

Figure 2. Example of the hooking correction. (**a**) On a rough selection, parabolic profiles in different cycles were identified. (**b**) Zooming on them, it appears that they are responsible for deviation of several tenths of centimeters of the river levels. (**c**)Once the correction of the hooking effect is applied, the deviation is reduced to a maximum of a couple of tenths cm).

2.2. Validation of the Altimetry-Based Water Levels

Validation of the altimetry-based water levels is performed against records from the closest in situ gauge stations. The along-stream distance between a VS and the closest in situ gauge stations is generally lower than 100 km (see Tables S1–S5). Root Mean Square Error (RMSE), R and R^2 values were estimated between altimetry-based water levels and in situ ones measured the same day using the classical formulas:

$$\text{RMSE} = \left(\frac{1}{n} \sum_{i=1}^{n} (h_{alti}(t_i) - h_{in \, situ}(t_i))^2 \right)^{1/2} \tag{9}$$

$$R = \frac{\sum_{i=1}^{n} (h_{alti}(t_i) - \langle h_{alti}(t_i) \rangle)(h_{in \, situ}(t_i) - \langle h_{in \, situ}(t_i) \rangle)}{\left(\sum_{i=1}^{n} (h_{alti}(t_i) - \langle h_{alti}(t_i) \rangle)^2 \right)^{1/2} \left(\sum_{i=1}^{n} (h_{in \, situ}(t_i) - \langle h_{in \, situ}(t_i) \rangle)^2 \right)^{1/2}} \tag{10}$$

$$R^2 = \frac{\sum_{i=1}^{n}(h_{alti}(t_i) - \langle h_{in\ situ}(t_i)\rangle)^2}{\sum_{i=1}^{n}(h_{in\ situ}(t_i) - \langle h_{in\ situ}(t_i)\rangle)^2} \tag{11}$$

where h_{alti} and $h_{in\ situ}$ are the altimetry-based and the in situ water stages respectively, t_i is the measurement time and n the number of common observations. The average of a variable x is written $<x>$.

As the in situ gauge stations are leveled against a reference unavailable to us, no bias estimates were computed between the in situ and the altimetry-based water levels, but they were between the different missions in the same orbit as follows:

$$\text{Bias} = \frac{1}{n}\sum_{i=1}^{n}(h_{alti1}(t_i) - h_{alti2}(t_i)) \tag{12}$$

while h_{alti1} is the more recent mission in the orbit and h_{alti2} is the older one.

The consistency of the intra-mission altimetry-based water levels as well as likely time-lag between water stages in the IND were estimated using the maximum of the cross-correlation function R_{hh} and the argument of the maximum:

$$R_{hh}(\tau) = \frac{\sum_{i=1}^{n}(h_{alti}(t_i) - \langle h_{alti}(t_i)\rangle)(h_{in\ situ}(t_i - \tau) - \langle h_{in\ situ}(t_i)\rangle)}{\left(\sum_{i=1}^{n}(h_{alti}(t_i) - \langle h_{alti}(t_i)\rangle)^2\right)^{1/2}\left(\sum_{i=1}^{n}(h_{in\ situ}(t_i - \tau) - \langle h_{in\ situ}(t_i)\rangle)^2\right)^{1/2}} \tag{13}$$

where τ is the time displacement.

3. Study Area and Datasets

3.1. Study Area

The IND is an extensive Sahelian floodplain located between longitudes 3–5° W and latitudes 13–17° N in Central Mali (Figure 3a). It is encompassed between the in situ gauge stations of Macina (−5.37° W, 13.95° N), on the Niger River, and Douna (−5.9° W, 13.22 °N), on the Bani River, upstream, and Diré (−3.38° W, 16.27° N), downstream. Its drainage area represents a surface of 73,000 km^2 [35]. The flooded area extent depends on the intensity of the West African Monsoon and can reach 35,000 km^2 during the wettest rainy seasons [36–40]. The flooding period ranges from August to December and during the dry season, from March to May, the area dries out with the exception of the rivers mainstem and the permanent lakes3.2. Radar Altimetry Data

The data used in this study come from the acquisitions of the following radar altimetry missions in their nominal orbit: Jason-1 (2002–2008), Jason-2 (2008–2016), Jason-3 (since 01/2016), ERS-2 (05/1995–06/2003), ENVISAT (03/2002–10/2010), SARAL (02/2013–2016), Sentinel-3A (since 02/2016). The main characteristics of these missions are presented below.

3.1.1. Missions with a 35-Day Repeat Period (European Remote-Sensing Satellite-2 (ERS-2), ENVIronment SATellite (ENVISAT), Satellite with Argos and ALtika (SARAL))

ERS-2, ENVISAT and SARAL orbited at an average altitude of 790 km, with an inclination of 98.54°, in a sun-synchronous orbit with a 35-day repeat cycle. They provided observations of the Earth surface (ocean, land, and ice caps) from 82.4° latitude north to 82.4° latitude south. This orbit was formerly used by ERS-1 mission, with an equatorial ground-track spacing of about 85 km.

ERS-2 was launched in 1995 by the European Space Agency (ESA) as ERS-1 follow-on mission. The satellite carries, among other instruments, a radar altimeter (RA) operating at Ku-band (13.8 GHz) developed for measuring height over ocean, land and ice caps. ERS-2 data are available from 17 May 1995 to 9 August 2010. After 22 June 2003, the dataset coverage is limited to ground station visibility.

ENVISAT mission was launched on 1 March 2002 by ESA. It carried 10 instruments including the advanced radar altimeter (RA-2). It was based on the heritage of the sensor on-board the ERS-1 and

2 satellites. RA-2 was a nadir-looking pulse-limited radar altimeter operating at two frequencies at Ku-(13.575 GHz), as ERS-1 and 2, and S-(3.2 GHz) bands [41]. ENVISAT remained in its nominal orbit until October 2010 and its mission ended on 8 April 2012. RA-2 stopped operating correctly at S-band in January 2008.

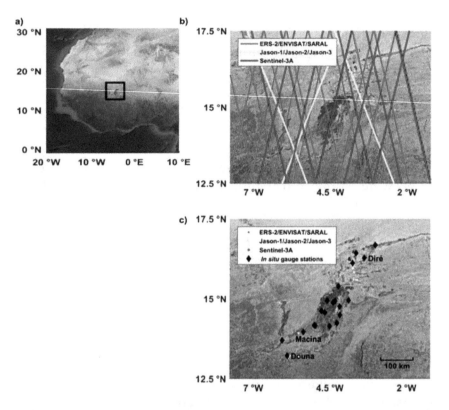

Figure 3. (**a**) Location of the IND in Africa; (**b**) Altimetry tracks over the IND from European Remote-Sensing Satellite-2 (ERS-2), ENVIronment SATellite (ENVISAT), Satellite with Argos and ALtika (SARAL) (blue dots), Jasons-1/Jason-2/Jason-3 (yellow dots) and Sentinel-3A (red dots); and (**c**) Location of virtual stations to calculate water levels (colored circles) using altimetry data and in situ gauge stations (black diamonds) in IND.

SARAL mission was launched on 25 February 2013. SARAL is a new collaboration between Centre National d'Etudes Spatiales (CNES) and Indian Space Research Organization (ISRO). Its payload comprises the AltiKa radar altimeter and bi-frequency radiometer, and a triple system for precise orbit determination: the real-time tracking system Détermination Immédiate d'Orbite par Doris embarqué (DIODE) of the Doppler Orbitography and Radio-positioning Integrated by Satellite (DORIS) instrument, a Laser Retroflector Array (LRA), and the Advance Research and Global Observation Satellite (ARGOS-3). AltiKa radar altimeter is a solid-state mono-frequency altimeter that provides accurate range measurements. It is the first altimeter to operate in the Ka-band (35.75 GHz) [42]. It has been put in a drifting orbit since July 2016.

3.1.2. Missions with a 10-Day Repeat Period (Jason-1, Jason-2 and Jason-3)

Jason-1, Jason-2 and Jason-3 orbit at an altitude of 1336 km, with an inclination of 66°, on a 10-day repeat cycle, providing observations of the Earth surface (ocean and land) from 66° latitude North to 66° latitude South, with an equatorial ground-track spacing of about 315 km. This orbit was formerly used by Topex/Poseidon mission.

Jason-1 mission was launched on 7 December 2001 by a cooperation between CNES and National Aeronautics and Space Administration (NASA). Jason-1 sensors are based on the former Topex/Poseidon missions, composed of the Poseidon-2 altimeter which is a two-frequency altimeter with C (5.3 GHz) and Ku (13.575 GHz) -bands. Its payload is also composed of the Jason Microwave Radiometer from NASA and a triple system for precise orbit determination: DORIS instrument from the CNES, Black Jack Global Positioning System receiver from NASA and a LRA from NASA/Jet Propulsion Laboratory (JPL) [43]. Jason-1 remained in its nominal orbit until 26 January 2009 and was decommissioned on 21 June 2013.

Jason-2 mission was launched on 20 June 2008 as a cooperation between CNES, the European Organization for the Exploitation of Meteorological Satellites (EUMETSAT), NASA and the National Oceanic and Atmospheric Administration (NOAA). Its payload is mostly composed of the Poseidon-3 radar altimeter from CNES, the Advanced Microwave Radiometer (AMR) from JPL/NASA, and a triple system for precise orbit determination: the real-time tracking system DIODE of DORIS instrument from CNES, a Global Navigation Satellite System (GNSS) receiver and a LRA from NASA/JPL. Poseidon-3 radar altimeter is a two-frequency solid-state altimeter that measures accurately the distance between the satellite and the surface (range) and provides ionospheric corrections over the ocean [44]. It operates at Ku and C bands. Raw data are processed by SSALTO (Segment Sol multimissions d'ALTimétrie, d'Orbitographie). Jason-2 remained in its nominal orbit until 3 July 2016.

Jason-3 mission was launched on 17 January 2016 as cooperation between CNES, EUMETSAT, NASA and NOAA. This satellite is composed of Poseidon-3B radar altimeter with a Precise Orbit Determination (POD) package with a Global Positioning System (GPS) receiver, DORIS and a LRA from NASA/JPL.

3.1.3. Mission with a 27-Day Repeat Period (Sentinel-3A)

Sentinel-3A mission was launched on 16 February 2016 by ESA to an orbit of altitude 814 km. The satellite caries one altimeter radar called SRAL (SAR Radar ALtimeter), a dual-frequency SAR altimeter (Ku-band at 13.575 GHz and C-band at 5.41 GHz). Its payload comprises also a Microwave Radiometer (MWR) instrument for wet path delay measurements and a triple system for precise orbit determination: a POD including a GPS receiver, a LRA and a DORIS instrument [45].

All this information is summarized in Table 1.

Table 1. Major characteristics of the high-precision radar altimetry missions used in this study.

Mission	Jason-1/2/3	ERS-2 ENVISAT	SARAL	Sentinel-3A
Instrument	Poseidon-2 Poseidon-3 Poseidon-3B	Radar Altimeter (RA) Radar Altimeter (RA-2)	AltiKa	Sar Radar Altimeter (SRAL)
Space agency	Centre National d'Etudes Spatiales (CNES), National Aeronautics and Space Administraion (NASA)	European Space Agency (ESA)	CNES, Indian Space Research Organization (ISRO)	European Space Agency (ESA)
Operation	2001–2013 Since 2008 Since 2016	1995–2003 2002–2012	Since 2013	Since 2016
Acquisition mode	Low Resolution Mode (LRM)	LRM	LRM	Pseudo Low Resolution Mode (PLRM), SAR
Acquisition	Along-track	Along-track	Along-track	Along-track
Frequency (GHz)	13.575 (Ku) 5.3 (C)	13.8 (Ku) 13.575 (Ku) 3.2 (S)	35.75 (Ka)	13.575 (Ku) 5.41 (C)
Altitude (km)	1315	800	800	814.5
Orbit inclination (°)	66	98.55	98.55	98.65
Repetitively (days)	9.9156	35	35	27
Equatorial cross-track separation (km)	315	75	75	104

The data used in this study are summarized in Table 2. Ranges used to derive altimeter heights and backscattering coefficients are those processed with OCOG/Ice-1/Ice retracking algorithm [46]. Previous studies showed that Ice-1-derived altimetry heights are the more suitable for hydrological studies in terms of accuracy of water levels and availability of the data (e.g., [12,47,48]) among the commonly available retracked data present in the GDRs.

Table 2. Major characteristics of the high-precision radar altimetry missions used in this study.

Altimetry Mission	Jason-1	Jason-2	Jason-3	ERS-2	ENVISAT	SARAL	Sentinel-3A
GDR	E	D	D	Centre de Topographie des Océans et de l'Hydrosphère (CTOH) [13]	V2.1	T	ESA IPF 06.07 land
Along-track sampling	20 Hz	20 Hz	20 Hz	20 Hz	18 Hz	40 Hz	20 Hz
Retracker	ICE	ICE	ICE	ICE-1	ICE-1	ICE-1	Offset Centre of gravity (OCOG)
ΔR_{iono}	GIM-based						
ΔR_{dry}	European Centre for Medium-Range Weather Forecasts (ECMWF)-based using Digital Elevation Model (DEM)			ECMWF-based using h from altimeter	ECMWF-based using DEM		
ΔR_{wet}	ECMWF-based using DEM						
$\Delta R_{solid\ Earth}$	Based on Catwright et al. [49]						
ΔR_{pole}	Based on Wahr et al. [50]						

3.2. In Situ Water Levels

Daily stage records from 19 in situ gauge stations located in the IND were used in this study to validate altimetry-based water levels (see Table 3 for their names, locations and periods of data availability and Figure 3c for their locations). Measurements were acquired at 12:00 a.m. local time. They are made available by the Malian water agency (Direction Nationale de l'Hydraulique—DNH).

Table 3. List of in situ gauge stations in the IND used this study.

In Situ Gauge Station	Longitude (°)	Latitude (°)	Validation Period
Akka	−4.23	15.39	1992–2017
Diondiori	−4.78	14.61	2008–2010
Diré	−3.38	16.27	1991–2017
Douna	−5.90	13.22	1991–2004
Goundam	−3.65	16.42	2009–2017
Kakagnan	−4.33	14.93	2008–2010
Kara	−5.01	14.16	1992–2011
Kirango	−6.07	13.7	2015–2017
Konna	−3.9	14.95	1992–1999
Koryoumé	−3.03	16.67	1992–2017
Macina	−5.29	14.14	1991–2017
Mopti	−4.18	14.48	1991–2017
Sévéri	−4.19	14.75	2008–2010
Sormé	−4.4	14.87	2008–2010
Sossobé	−4.67	14.56	2008–2010
Tilembeya	−4.98	14.15	1991–2006
Toguéré Kou	−4.59	14.93	2008–2010
Tonka	−3.76	16.11	1991–2017
Tou	−4.52	14.13	2008–2010

4. Results

4.1. Direct Validation of the Altimetry-Based Water Stages

The nominal altimetry ground-tracks from ERS-2, ENVISAT, SARAL (35-day repeat orbit), Sentinel-3A (27-day repeat orbit), Jason-1, Jason-2 and Jason-3 (10-day repeat orbit) missions present a large number of cross-sections with river streams and floodplains in the IND (see Figure 3b). A dense network of virtual stations from different missions was defined in the IND (see Table 4). Virtual station (VS) locations in the IND are presented in Figure 3c.

Table 4. Number of virtual stations defined in the IND for each mission.

Mission	ERS-2	ENVISAT	SARAL	Sentinel-3A	Jason-1	Jason-2	Jason-3
Number of virtual stations (VS)	52	63	62	31	8	8	9

Altimetry-based water levels were compared to water stage records from close in situ gauge. These comparisons were performed for VS located on the rivers and not on the floodplains for distances between the in situ gauge and the VS lower than 100 km. In situ gauge records from 19 stations were used to perform the 89 following comparisons:

- 19 against ERS-2-based water stages;
- 32 against ENVISAT-based water stages;
- 14 against SARAL-based water stages;
- 3 against Jason-1 and Jason-2-based water stages;
- 2 against Jason-3-based water stages;
- 16 against Sentinel-3A-based water stages.

The complete results of these comparisons (distance between the in situ gauge and the VS, number of data used for comparisons (N), RMSE, R and R^2) are presented in Tables S1 to S5 for ERS-2, ENVISAT, SARAL, Jason-1/2/3 and Sentinel-3A missions (in supplementary information). The results of these comparisons are also presented as maps in Figure 4 for the altimetry missions that were launched before 2010 (ERS-2, ENVISAT, Jason-1 and 2) and in Figure 5 for the most recent missions. The number of data used for the comparison is, most of the time, statistically significant, except for 15 comparisons against ENVISAT (less than 20 common observations), 4 against SARAL (less than 15 common observations) as well as the whole comparisons against Sentinel-3A as only 16 cycles were used:

- between 28 and 70 for the 19 ERS-2-based time series of water level (out of 85 available cycles);
- between 7 and 81 for 32 ENVISAT-based time series of water level (out of 89 available cycles);
- between 6 and 28 for the 14 SARAL-based time series of water level (out of 35 available cycles);
- between 46 and 147 for the 3 Jason-1-based time series of water level (out of 262 available cycles);
- between 37 and 72 for the 3 Jason-2-based time series of water level (out of 303 available cycles);
- between 45 and 50 for the 2 Jason-3-based time series of water level (out of 55 available cycles);
- between 3 and 15 for the 16 Sentinel-3A-based time series of water level (out of 16 available cycles).

Very good agreements were generally found between altimetry-based and in situ water stages for all the missions over a total of 89 comparisons performed. Values of R greater than 0.95 were obtained 41 times (45%), between 0.95 and 0.9, 18 times (20%), and between 0.8 and 0.9, 18 times (20%). Correlation coefficients R lower than 0.7 were obtained only 4 times (4%) (Figure 6a). The minimum R value is 0.57. RMSE lower than 0.3 m were obtained 12 times (13%), between 0.3 and 0.5 m, 17 times (19%), between 0.5 and 0.75 m, 29 times (32%) and above 1 m, 19 times (21%) (Figure 6b).

Better agreement was found for the recent missions such as SARAL, Jason-3 and Sentinel-3A than for the older ones (ERS-2, ENVISAT, Jason-1 and Jason-2). Focusing on the correlations, for instance R was greater than 0.9 in 10 out of 19 comparisons for ERS-2 (52%), in 19 out of 32 (59%) for ENVISAT, in 10 out of 14 (71%) for SARAL, in 14 out of 16 (88%) for Sentinel-3A, 1 (R = 0.89) out of 3 (33%) for Jason-1, 3 out of 3 for Jason-2 (100%) and 3 out of 3 (100%) for Jason-3 (Tables S1–S5). The first results of the use of Sentinel-3A are very encouraging as only very few cycles were available, keeping in mind that among the three correlation coefficients lower than 0.9, three were already higher than 0.85.

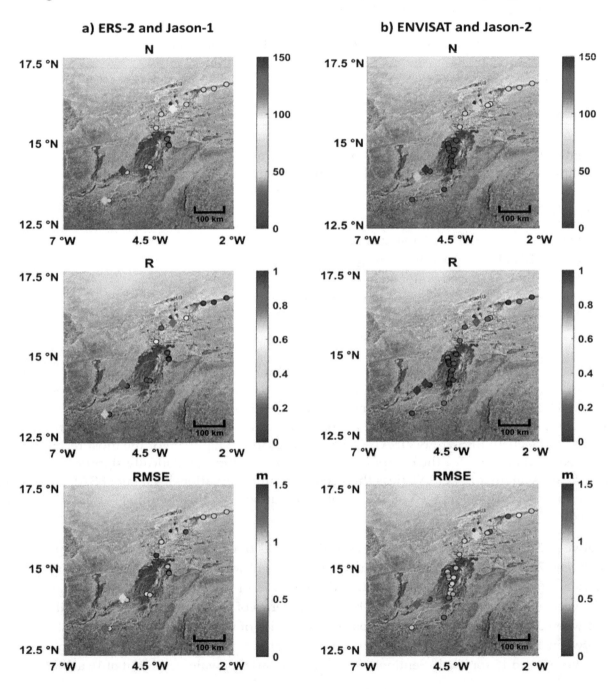

Figure 4. Comparisons between in situ and altimetry water levels for several missions (**a**) ERS-2 and Jason-1 data and (**b**) ENVISAT and Jason-2. For each comparison, the number of samples (N), correlation (R), and RMSE is presented. Diamonds points correspond to Jason data and circles with black contours to ERS-2/ENVISAT.

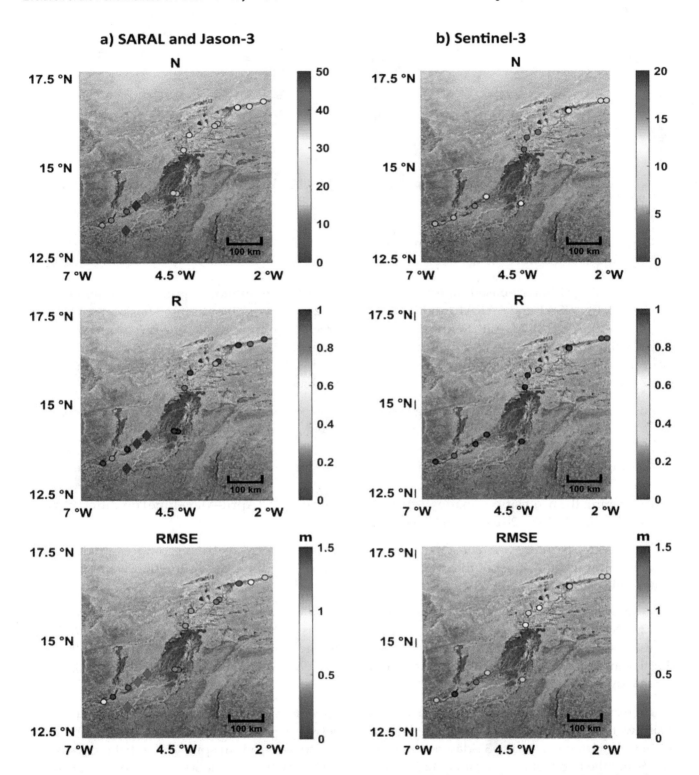

Figure 5. Comparisons between in situ and altimetry water levels for several missions (**a**) SARAL and Jason-3 data and (**b**) Sentinel-3A. For each comparison, the number of samples (N), correlation (R), and RMSE is presented. Diamonds points correspond to Jason data and circles with black contours to SARAL/Sentinel-3A.

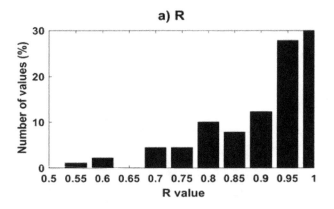
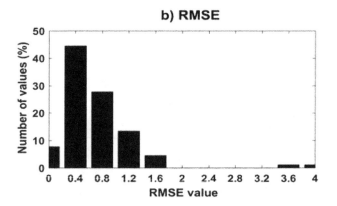

Figure 6. Histograms of (**a**) correlations and (**b**) RMSE between altimetry-based and in situ water stages for all missions.

Combining altimetry-based time series of water levels from missions in the same nominal orbits (i.e., ERS-2, ENVISAT and SARAL, and Jason-1, Jason-2 and Jason-3) or at inter-mission cross-overs (ERS-2/ENVISAT/SARAL with Sentinel-3A ground-tracks), multi-mission time series of water levels were obtained. In this latter case, the maximum difference in time between SARAL and Sentinel-3A acquisitions is half the length of the shortest repeat period of the two satellites (i.e., 13 days considering the 27 days of Sentinel-3A repeat period). Inter-mission biases were removed:

- using the acquisitions made during tandem phases when two missions were in the same orbit a few seconds or minutes apart from each other (e.g., Jason-1 and Jason-2, Jason-2 and Jason-3, ERS-2 and ENVISAT);
- averaging the acquisitions made during the common period of observations at low water stages (April–May–June) for Sentinel-3A and SARAL;
- averaging the acquisitions made during low water periods (April–May–June) on different years for ENVISAT (2003–2010) and SARAL (2013–2016).

Low water periods were chosen, rather than high water periods or the complete hydrological cycle, to minimize the effect of the difference in temporal sampling, assuming that water levels are more stable during low water stages.

Examples of multi-mission time series of water levels are presented in Figures 7 and 8 for ERS-2/ENVISAT/SARAL and Sentinel-3A (when there is a cross-over) and for Jason-1, Jason-2 and Jason-3 respectively along with in situ gauge station gauge records of Diré upstream part of the IND, Mopti, central part and Macina, downstream part, over 1995–2017 (Figure 7a–c respectively), and of Macina, over 2002–2017 (Figure 8).

In examples presented in Figure 7, a very good agreement is found between altimetry-based and in situ water stages. There is a gap in the time series between November 2010 and January 2013 as no altimetry mission was in the 35-day repeat orbit during this period. In spite of the distance between the VS and the two first in situ gauge stations (77 and 40 km, with Diré and Mopti stations respectively, but only 1 km from the Macina station under the ERS-2/ENVISAT and SARAL ground-tracks, and 8 km from Macina station under Sentinel-3A ground-tracks), better results were found in the first examples than in the latter one, with higher R^2 and lower RMSE for ERS-2 and ENVISAT and similar ones for SARAL and Sentinel-3A. An underestimation of the annual amplitude of the water levels is observed during the ERS-2 observation period. In the example presented in Figure 8, the VS and the Macina in situ gauge station are separated by only 1 km. The quality of the water stage retrieval is increasingly better from Jason-1 to Jason-3. It is important to mention that Jason-1 data contained in the GDR E, released in May 2016, allow the accurate and continuous estimation of water stages over the IND contrary to the previous GDR versions that contained few useful data over land (except over large lakes, see [18,51,52] for instance).

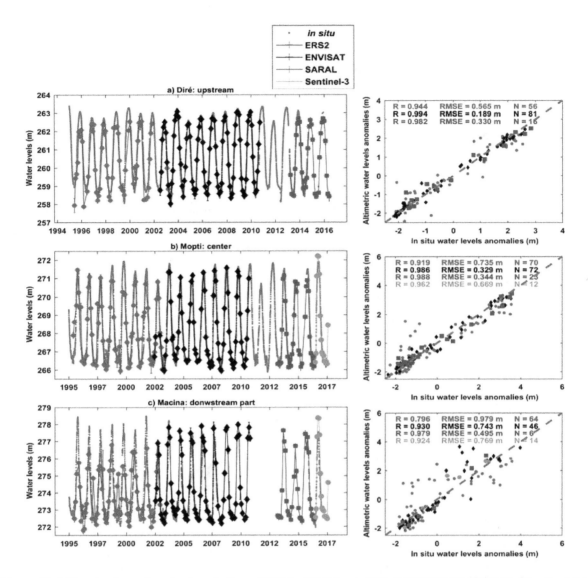

Figure 7. Altimetry-based water levels from 1995 to 2017 from ERS-2/ENVISAT/SARAL/Sentinel-3A (red/black/green respectively) and in situ (grey) data (left). Scatter plots of water levels anomalies from radar altimetry and in situ gauge stations at (**a**) Diré (upstream IND), (**b**) Mopti (center IND) and (**c**) Macina (downstream IND).

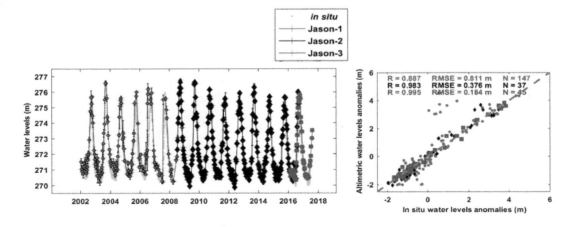

Figure 8. Altimetry-based water levels from 2002 to 2017 from Jason-1/Jason-2/Jason-3 (red/black/blue respectively) and in situ water levels (light green) data (left). Scatter plots of water levels anomalies from radar altimetry and in situ gauge stations at Macina (right figure, downstream IND).

4.2. Intermission Water Stage Comparison

Comparisons between water levels estimated by altimetry missions during their tandem phase (i.e., few cycles during which two missions were orbiting a few minutes apart from one another) were performed between ERS-2 and ENVISAT from June 2002 to July 2003 (11 cycles in common), Jason-1 and Jason-2 from July 2008 to January 2009 (21 cycles in common), Jason-2 and Jason-3 from January to September 2016 (23 cycles in common). They allow the increase of the number of comparisons in the IND, not only on the rivers but also on the wetlands that are monitored using in situ gauges.

A total of 48 comparisons between ERS-2 and ENVISAT-based water levels were performed in the IND with several samples (N) varying from 3 to 11 (Figure 9). On the total number of comparisons between ERS-2 and ENVISAT missions, 22 were performed on more than 8 samples (45%) (Figure 9a). Very good agreement was generally obtained between altimetry-based water stages from the two missions (Figure 9a,b). The value R greater than 0.95 was obtained 24 times (50%), between 0.95 and 0.9, 7 times (15%), and between 0.8 and 0.9, 7 times (15%) (Figure 10a). Correlation coefficients (R) lower than 0.5 were obtained 5 times (10%) (Figure 9b). Values of RMSE lower than 0.3 m were obtained 13 times (27%), between 0.3 and 0.5 m, 10 times (21%), between 0.5 and 0.75 m, 13 times (27%) and above 1 m, 8 times (17%) (Figure 10b). Large biases are observed between ERS-2 and ENVISAT (-1.16 ± 0.38 m on average) (Figure 9d).

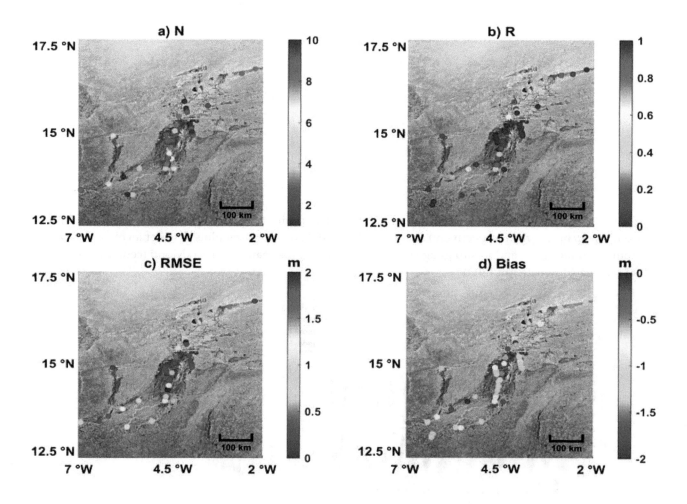

Figure 9. Comparisons between ERS-2 and ENVISAT water levels in terms of (**a**) number of samples (N), (**b**) correlation (R), (**c**) RMSE, and (**d**) bias.

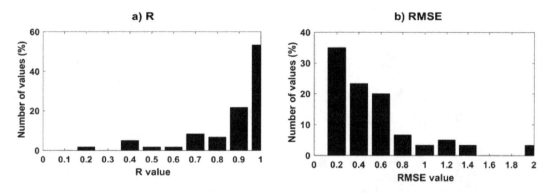

Figure 10. Histograms of (**a**) correlation coefficients and (**b**) RMSE for ERS-2/ENVISAT intermission water stages comparisons.

A total of 5 and 7 comparisons between Jason-1 and Jason-2, and Jason-2 and Jason-3-based water levels were performed in the IND with several samples (N) varying from 5 to 15 for Jason-1/Jason-2, and from 13 to 22 for Jason-2/Jason-3. Three of these comparisons were performed on more than 10 samples (60%) for Jason-1 (Figure 11a). Very good agreement was generally found between altimetry-based water stages from Jason-1 and 2. Values of R greater than 0.95 were obtained 3 times (60%). In the two other cases, R equal to 0.80 and 0.69 were found (Figure 11b). Values of RMSE lower than 0.3 m were obtained 3 times (60%). In the two other cases, RMSE lower than 0.35 m were found (Figure 11c). Biases between Jason-1 and 2 ranged between 0.34 and 1.06 m (0.75 ± 0.28 m) (Figure 11d).

Better agreement is found between altimetry-based water stages from Jason-2 and 3. Correlation coefficients R greater than 0.95 were obtained in all the cases except one (86%) for which R equals to 0.84 (Figure 12b). Values of RMSE lower than 0.30 m were obtained in all the cases except one (86%) for which RMSE equals to 0.40 m (Figure 12c). Biases between Jason-1 and 2 are very similar, ranging from −0.36 to −0.14 m (−0.27 ± 0.08 m) (Figure 12d).

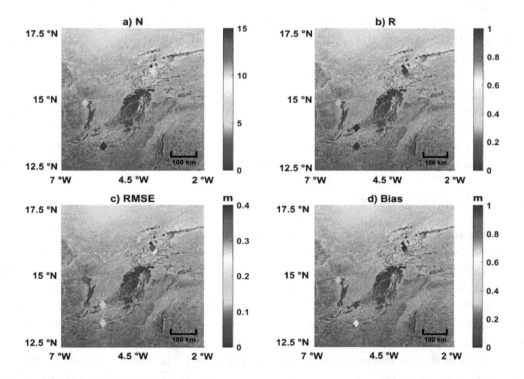

Figure 11. Comparisons between Jason-1 and Jason-2 water levels in terms of (**a**) number of samples (N); (**b**) correlation (R); (**c**) RMSE; and (**d**) bias.

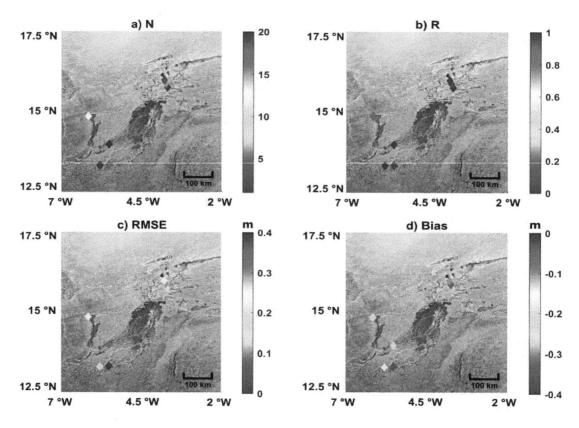

Figure 12. Comparisons between Jason-2 and Jason-3 water levels in terms of (**a**) number of samples (N); (**b**) correlation (R); (**c**) RMSE and (**d**) bias.

Examples of time series combining ERS-2 and ENVISAT, Jason-1 and 2, and Jason-2 and 3 are presented on Figure 13a–c. Virtual stations from close locations were chosen: ERS-2/ENVISAT VS 0545-a and Jason-1/Jason-2/Jason-3 VS 046-d. They are located in the upstream part of the IND at an approximate along-stream distance of 45 km. Seasonal amplitudes of between 4 and 5 m are observed during the common period of availability of the different altimetry-based water levels. The agreement is better for the recent missions than for the older ones: R increases from 0.92 to 0.99 whereas RMSE decreases from 0.62 to 0.22 m as well as the bias from -1.35 to 0.36 m. Lower deviations are generally observed on the time series from the more recent missions than other the older ones: 0.13 ± 0.11, 0.13 ± 0.15, 0.19 ± 0.22, 0.21 ± 0.16, 0.15 ± 0.11 m were obtained averaging the mean absolute deviation from individual cycles for ERS-2, ENVISAT, Jason-1, 2 and 3 respectively. Please note that, for readability purpose, biases between the time series were removed in Figure 13.

4.3. Multi-Mission Time Series on Floodplains

Floodplains and wetlands are generally not monitored using in situ gauges. Radar altimetry is a unique tool for the long-term observations of the changes in water levels over inundated areas [53–57]. Among the VS defined in the IND, 16 were built in floodplains under ERS-2/ENVISAT/SARAL ground-tracks and 9 under Sentinel-3A ground-tracks, but none under Jason-1, Jason-2 and Jason-3 ground-tracks. Their consistency was checked during the tandem phase between ERS-2 and ENVISAT (Figure 14). Values of R greater than 0.95 were obtained 10 times (63%), between 0.95 and 0.9, twice (12%), and between 0.8 and 0.9, 3 times (19%). However, R lower than 0.5 was obtained once (6%) (Figure 14a). Values of RMSE lower than 0.3 m were obtained 7 times (44%), between 0.3 and 0.5 m, 5 times (31%), between 0.5 and 0.75 m, 3 times (19%) and above 1 m, once (6%) (Figure 14b).

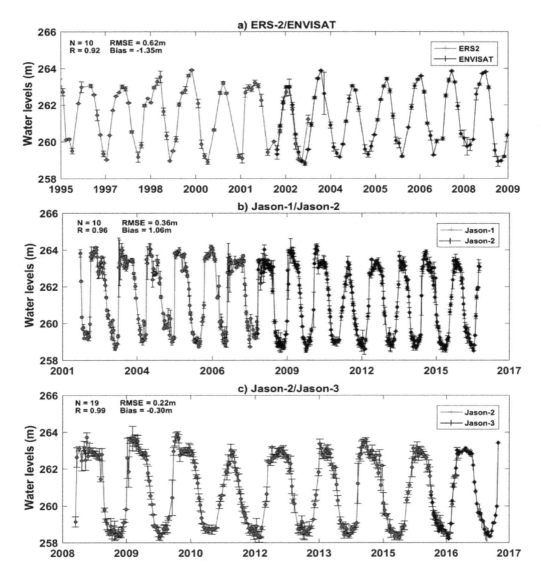

Figure 13. Intermission quality time series: (**a**) for the VS 0545-a located in the upstream part in the IND between ERS-2/ENVISAT; (**b**) for the VS 046-d (upstream part) between Jason-1/Jason-2; and (**c**) for the VS 046-d between Jason-2/Jason-3. The two VS are separated by 45 km.

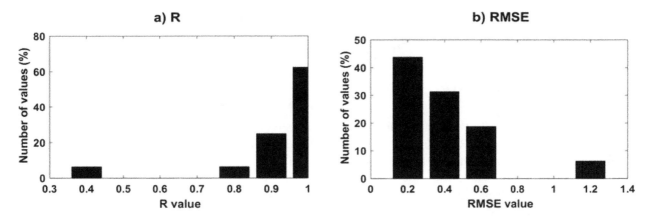

Figure 14. Histograms of (**a**) R and (**b**) RMSE for ERS-2/ENVISAT intermission water stages comparisons on floodplains virtual stations.

Three examples of multi-mission time series of water levels (corrected for inter-mission bias) over the IND floodplains are presented for VS located in the center, in the north east and in the south (Figure 15a–c respectively). They provide time-variations of water stages between 1995–2016 with the exception of a gap between November 2010 and January 2013.

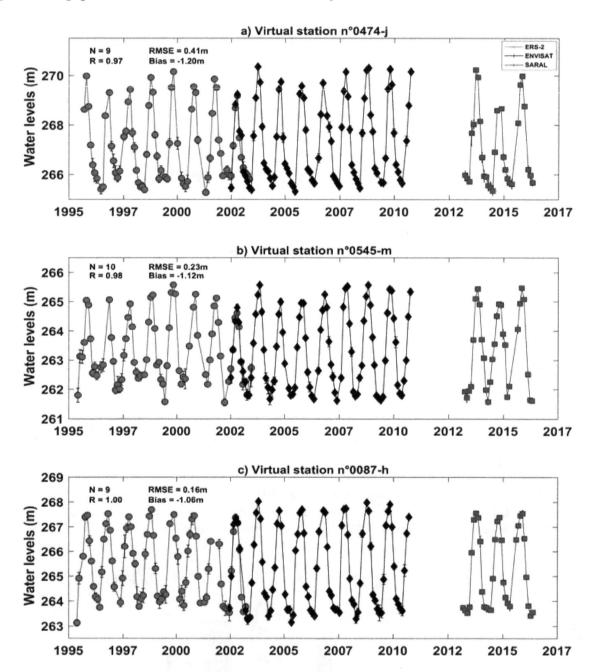

Figure 15. Water levels time series on floodplains derived from radar altimetry multi-mission (ERS-2/ENVISAT and SARAL) from 1995 to 2016 at 3 different locations in the Inner Niger Delta: the center of the delta (**a**); the north eastern (**b**) and the southern (**c**) parts.

4.4. Consistency of the Altimetry-Based Water Levels in the Inner Niger Delta (IND)

A consistency check was performed between a reference VS located either in Niger River mainstem or in the Bani major tributary and the other VS from the same altimetry mission located upstream and downstream on the same river course estimating the maximum of correlation using the cross-correlation function and the associated time-lag, similarly as in Bogning et al. [29].

Cross-correlation function maxima and associated time-lags are presented for ERS-2 and ENVISAT in Figure 16 and Figure S1 respectively and for SARAL and Sentinel-3A in Figure 17 and Figure S2 respectively for three different VS locations in the IND: near Douna, over the Bani tributary, near Mopti, in the downstream part of the IND, and near Diré in the upstream part of the IND. As the repeat period of these missions is either 35 or 27 days in their nominal orbit, only time-lags of plus or minus one repeat period and a half (i.e., 53 and 41 days) were considered due to the relatively small scale of the IND. Due to the changes of the river features (slope, depth, width, etc.), biases and RMSEs were not computed between the time series of altimetry-based water levels.

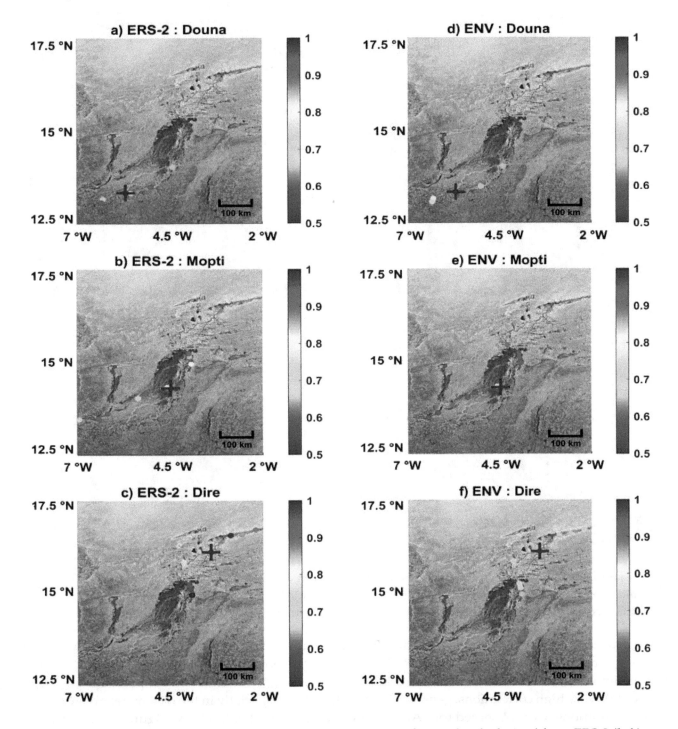

Figure 16. Maxima of cross-correlation between time series of water levels derived from ERS-2 (**left**) and ENVISAT (**right**) data in the IND for three different VS near Douna (**a,d**); Mopti (**b,e**) and Diré (**c,f**). Red crosses symbolize the VS chosen as reference (auto-correlation of 1).

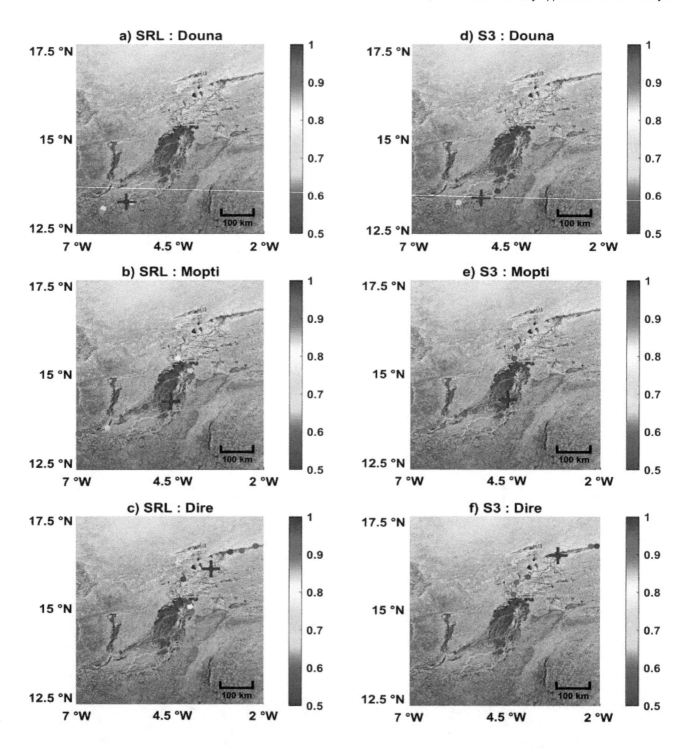

Figure 17. Maxima of cross-correlation between time series of water levels derived from SARAL (left) and Sentinel-3A (right) data in the IND for three different VS near Douna (**a,d**); Mopti (**b,e**) and Diré (**c,f**). Red crosses symbolize the VS chosen as reference (auto-correlation of 1).

Correlation coefficients (R) higher than 0.7 are generally observed for ERS-2 and ENVISAT, except in the upstream part of the delta for ERS-2, with correlation ranging from 0.5 to 0.7 (Figure 16c). Correlations higher than 0.8 are observed in the IND downstream and central parts for ENVISAT (Figure 16). Very high correlations, generally higher than 0.85, especially in the downstream part, except on a few locations, were observed for SARAL and, especially, for Sentinel-3A (Figure 17). No time-lag was observed over the Bani River for none of the altimetry missions (Figures S1a, d and S2a,d). Time lags of plus or minus one cycle were respectively found for the upper and lower stations along the Niger mainstem (Figures S1 and S2).

5. Discussion

The coverage of altimetry tracks from missions of the 35-day (ERS-2/ENVISAT/SARAL) and the 27-day (Sentinel-3A) repeat orbits allows the construction of a dense network of VS on both rivers and floodplains in the IND, completed by a few VS from missions in the 10-day repeat orbit (Figure 3c). Yet, a larger number (more than 15%) of ENVISAT (6) and SARAL-based (62) VS than ERS-2-based (52) ones were built on the ground-tracks of the 35-day repeat orbit. These VS with no valid ERS-2-based water levels are located on the upstream part of the IND for 5 of them, central part for 3 others and downstream part for the last 3 ones.

Due to the relative flatness of the IND, the no-construction of VS cannot be attributed to data losses caused by tracking issues, but to incorrect range estimates. They were likely to be caused by the narrow width of the river streams, especially in the upstream part, where they are generally lower than 300 m. They can also be accounted for the possible complexity of the waveforms (e.g., multi-peaked over areas with several bright targets encompassed in the altimeter footprint such as several river streams that can reach 1 km of width as in the downstream part of the IND, or several floodplain lakes as in the central part).

Comparisons with a high number of in situ gauge records show an overall very good agreement between in situ and altimetry-based water levels, increasingly better for the most recent missions than the older ones. Contrary to what was found in other study areas with variations of topography (e.g., [28,29]), the data used to build the VS were mostly acquired at 330 and 320 MHz Ku chirp bandwidth operation mode by both ERS-2 and ENVISAT (i.e., with the better range resolution in more than 95% of the cases for any VS—Figure 18). This is the reason quite similar results when comparing ERS-2 and ENVISAT-based water stages with in situ water levels were found in terms of R, with generally higher RMSE for ERS-2 (see Figure 4, Tables S1 and S2), and also a good agreement when comparing ERS-2 and ENVISAT-based water levels during their tandem phase, except in terms of bias (Figure 10). In the same orbit, better results were found using SARAL data (Figure 5 and Table S3), as already observed in other river basins, resulting from the use of the Ka-band with its smaller footprint and higher chirp frequency of 500 MHz [29,58].

Very similar numbers of VS were built using Jason-1, Jason-2 and Jason-3 data. The release of the Jason-1 GDR E, which contained valid data over land and not mostly only over large lakes as in the previous versions, allowed the extension of the duration of the time series of water levels over rivers for the missions in the 10-day repeat orbit down to 2001, at the expense of lower accuracy than using Jason-2 and Jason-3 data (Figures 4 and 5 and Table S4). This lower accuracy can be accounted for the small river widths, most of the time lower than 500 m, except in some locations in the downstream part of the IND.

On the contrary, Jason-2 and Jason-3 confirm their great capability for detecting and accurately estimating water levels over narrow rivers, especially over flat areas [28,30]. Comparisons performed during the tandem phase logically confirm the lower agreement between Jason-1 and Jason-2 than between Jason-2 and Jason-3 (Figures 11 and 12). In terms of bias, if high and variable biases were found between Jason-1 and Jason-2-based time series of water levels, much lower ones with low variability were found between Jason-2 and Jason-3 ones during their tandem period. Poseidon-3, on-board Jason-3, is nominally operating in open loop or DIODE/DEM tracking mode over land surfaces, meaning that the reception window is controlled by an a-priori elevation from an along-track DEM loaded in the altimeter [30].

Very good results were also found using one year and a half of measurements from Sentinel-3A, the first altimeter to operate in SAR over all types of surfaces. During this short time period, the results obtained are almost as good as the ones obtained using SARAL, confirming the strong potential of this technique for monitoring inland water stages. This is likely due to the sharper waveform obtained in SAR mode compared with Low Resolution Mode (LRM) [59] as OCOG is particularly well adapted for very specular echoes from a single reflector (e.g., a river stream). On the contrary, less accurate estimates can be expected from multi-peak waveforms.

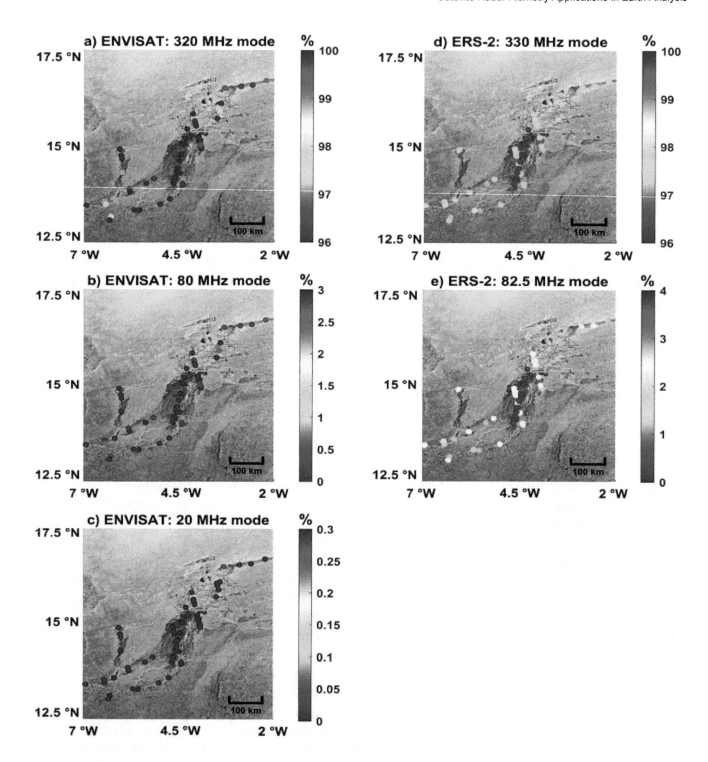

Figure 18. Percentage of acquisition in the different Ku chirp bandwidth operation mode for each VS in the IND for ENVISAT at 320 MHz (**a**); at 80 MHz; (**b**) and at 20 MHz; (**c**) ERS-2 at 330 MHz (**d**) and at 82.5 MHz (**e**).

The consistency of intra-mission water levels derived from radar altimetry measurements was analyzed using cross-correlations for ERS-2, ENVISAT, SARAL and Sentinel-3A. In this case, too, higher correlations were found for the most recent missions operating in (LRM) at Ka band (SARAL) and in SAR at Ku band (Sentinel-3A). Time-lags corresponding to the maximum of cross-correlation (Figures S1 and S2) also showed that there is a time-lag of around two months between the upstream and the downstream part of the IND, in accordance with the water residence time estimated using satellite images [39].

6. Conclusions

An extensive assessment of the performance of almost all the missions put in orbit from ERS-2 to Sentinel-3A was performed over the IND. Around 90 comparisons between in situ and altimetry-based were performed as well as more than 75 intermission comparisons between ERS-2 and ENVISAT, Jason-1 and Jason-2, and Jason-2 and Jason-3. Results of these comparisons show (i) a better agreement between altimetry-derived and in situ water stages over rivers of width varying from a few hundreds of meters to ~1.5 km and floodplains, (ii) an increase in accuracy in water level estimates for the most recent missions such as SARAL, Jason-3 and Sentinel-3A benefiting from the use of the Ka-band for SARAL and of the SAR acquisition mode for Sentinel-3A.

Due to the small number of VS defined on Jason-3 ground-tracks, the DIODE/DEM was not evaluated. Intra-mission consistency check performed on ERS-2, ENVISAT, SARAL and Sentinel-3A missions exhibits a time-lag for the maximum of around 2 months that can be related to the residence time of the water in the IND.

Very good performance of the recently launched altimeter SARAL onboard Sentinel-3A ensures the continuity of the monitoring of the IND whose data density should increase with the launch, in 2018, of Sentinel-3B, in the same orbit as Sentinel-3A but Sentinel-3B flies at $\pm140°$ out of phase with Sentinel-3A.

This network of VS will present a strong interest for (i) assessing the impacts of climate variability and human effects (e.g., dam operation and rice production), and (ii) validating the measurement of the future Surface Water and Ocean Topography (SWOT) mission, the first mission to operate in close nadir interferometry SAR for providing elevations in two swaths, to be launched in 2021, on both rivers and floodplains.

Supplementary Materials
Figure S1: Time-lag corresponding to maxima of cross-correlation between time series of water levels derived from SARAL (left) and SENTINEL-3A (right) data in the IND for three different VS near Douna (a and d), Mopti (b and e) and Diré (c and f). Light blue crosses symbolize the VS chosen as reference (auto-correlation of 1 and time-lag null), Figure S2: Time-lag corresponding to maxima of cross-correlation between time series of water levels derived from SARAL (left) and SENTINEL-3A (right) data in the IND for three different VS near Douna (a and d), Mopti (b and e) and Diré (c and f). Light blue crosses symbolize the VS chosen as reference (auto-correlation of 1 and time-lag null), Table S1: Results of the comparisons between in situ and altimetry water levels derived from ERS-2 mission: number of sample (N), correlation coefficient (R), determination coefficient (R^2) and Root Mean Square Error (RMSE), Table S2: Results of the comparisons between in situ and altimetry water levels derived from ENVISAT mission: number of sample (N), correlation coefficient (R), determination coefficient (R^2) and Root Mean Square Error (RMSE), Table S3: Results of the comparisons between in situ and altimetry water levels derived from SARAL mission: number of sample (N), correlation coefficient (R), determination coefficient (R^2) and Root Mean Square Error (RMSE), Table S4: Results of the comparisons between in situ and altimetry water levels derived from Jason-1, Jason-2 and Jason-3 missions: number of sample (N), correlation coefficient (R), determination coefficient (R^2) and Root Mean Square Error (RMSE), Table S5: Results of the comparisons between in situ and altimetry water levels derived from Sentinel-3A mission: number of sample (N correlation coefficient (R), determination coefficient (R^2) and Root Mean Square Error (RMSE).

Author Contributions: C.N., F.F., A.T.D. and E.M. designed the study. C.N. processed the altimetry data to produce the time series of water levels. A.T.D. and N.B. provided the in situ data and gave important information on their quality. All of the authors of the present work contributed to the discussion of the results, as well as the writing of the manuscript.

Acknowledgments: We thank four anonymous reviewers who helped us improving our manuscript.

References

1. Vörösmarty, C.J.; Green, P.; Salisbury, J.; Lammers, R.B. Global water resources: vulnerability from climate change and population growth. *Science* **2000**, *289*, 284–288. [CrossRef] [PubMed]
2. Oki, T.; Kanae, S. Global hydrological cycles and world water resources. *Science* **2006**, *313*, 1068–1072. [CrossRef] [PubMed]

3. Haddeland, I.; Heinke, J.; Biemans, H.; Eisner, S.; Flörke, M.; Hanasaki, N.; Konzmann, M.; Ludwig, F.; Masaki, Y.; Schewe, J.; et al. Global water resources affected by human interventions and climate change. *Proc. Natl. Acad. Sci. USA* **2014**, *111*, 3251–3256. [CrossRef] [PubMed]

4. Gleick, P.H. Global freshwater resources: soft-path solutions for the 21st century. *Science* **2003**, *302*, 1524–1528. [CrossRef] [PubMed]

5. Alsdorf, D.E.; Rodríguez, E.; Lettenmaier, D.P. Measuring surface water from space. *Rev. Geophys.* **2007**, *45*, RG2002. [CrossRef]

6. Stammer, D.; Cazenave, A. *Satellite Altimetry over Oceans and Land Surfaces*; Taylor & Francis: Boca Raton, FL, USA, 2017; ISBN 978-1-4987-4345-7.

7. Crétaux, J.-F.; Nielsen, K.; Frappart, F.; Papa, F.; Calmant, S.; Benveniste, J. Hydrological applications of satellite altimetry: rivers, lakes, man-made reservoirs, inundated areas. In *Satellite Altimetry Over Oceans and Land Surfaces*; Earth Observation of Global Changes; Stammer, D., Cazenave, A., Eds.; CRC Press: Boca Raton, FL, USA, 2017; pp. 459–504.

8. Morris, C.S.; Gill, S.K. Variation of Great Lakes water levels derived from Geosat altimetry. *Water Resour. Res.* **1994**, *30*, 1009–1017. [CrossRef]

9. Birkett, C.M. The contribution of TOPEX/POSEIDON to the global monitoring of climatically sensitive lakes. *J. Geophys. Res.* **1995**, *100204*, 25179–25204. [CrossRef]

10. Koblinsky, C.J.; Clarke, R.T.; Brenner, A.C.; Frey, H. Measurement of river level variations with satellite altimetry. *Water Resour. Res.* **1993**, *29*, 1839–1848. [CrossRef]

11. Birkett, C.M. Contribution of the TOPEX NASA Radar Altimeter to the global monitoring of large rivers and wetlands. *Water Resour. Res.* **1998**, *34*, 1223. [CrossRef]

12. Frappart, F.; Calmant, S.; Cauhopé, M.; Seyler, F.; Cazenave, A. Preliminary results of ENVISAT RA-2-derived water levels validation over the Amazon basin. *Remote Sens. Environ.* **2006**, *100*. [CrossRef]

13. Frappart, F.; Legrésy, B.; Niño, F.; Blarel, F.; Fuller, N.; Fleury, S.; Birol, F.; Calmant, S. An ERS-2 altimetry reprocessing compatible with ENVISAT for long-term land and ice sheets studies. *Remote Sens. Environ.* **2016**, *184*. [CrossRef]

14. Baup, F.; Frappart, F.; Maubant, J. Use of satellite altimetry and imagery for monitoring the volume of small lakes. In *International Geoscience and Remote Sensing Symposium (IGARSS)*; 2014.

15. Sulistioadi, Y.B.; Tseng, K.-H.; Shum, C.K.; Hidayat, H.; Sumaryono, M.; Suhardiman, A.; Setiawan, F.; Sunarso, S. Satellite radar altimetry for monitoring small rivers and lakes in Indonesia. *Hydrol. Earth Syst. Sci.* **2015**, *19*, 341–359. [CrossRef]

16. Frappart, F.; Papa, F.; Malbeteau, Y.; León, J.G.J.G.; Ramillien, G.; Prigent, C.; Seoane, L.; Seyler, F.; Calmant, S. Surface freshwater storage variations in the orinoco floodplains using multi-satellite observations. *Remote Sens.* **2015**, *7*, 89–110. [CrossRef]

17. da Silva, J.S.; Calmant, S.; Seyler, F.; Moreira, D.M.; Oliveira, D.; Monteiro, A. Radar Altimetry Aids Managing Gauge Networks. *Water Resour. Manag.* **2014**, *28*, 587–603. [CrossRef]

18. Birkett, C.; Reynolds, C.; Beckley, B.; Doorn, B. From Research to Operations: The USDA Global Reservoir and Lake Monitor. In *Coastal Altimetry*; Springer Berlin Heidelberg: Berlin/Heidelberg, Germany, 2011; pp. 19–50.

19. Ričko, M.; Birkett, C.M.; Carton, J.A.; Crétaux, J.-F. Intercomparison and validation of continental water level products derived from satellite radar altimetry. *J. Appl. Remote Sens.* **2012**, *6*, 61710. [CrossRef]

20. Cretaux, J.-F.; Berge-Nguyen, M.; Leblanc, M.; Rio, R.A.D.; Delclaux, F.; Mognard, N.; Lion, C.; Pandey, R.-K.; Tweed, S.; Calmant, S.; et al. Flood mapping inferred from remote sensing data. *Int. Water Technol. J.* **2011**, *1*, 48–62.

21. Goita, K.; Diepkile, A.T. Radar altimetry of water level variability in the Inner Delta of Niger River. In Proceedings of the IEEE International Geoscience and Remote Sensing Symposium, Munich, Germany, 22–27 July 2012; pp. 5262–5265.

22. Frappart, F.; Fatras, C.; Mougin, E.; Marieu, V.; Diepkilé, A.T.; Blarel, F.; Borderies, P. Radar altimetry backscattering signatures at Ka, Ku, C, and S bands over West Africa. *Phys. Chem. Earth* **2015**, *83–84*, 96–110. [CrossRef]

23. Tarpanelli, A.; Amarnath, G.; Brocca, L.; Massari, C.; Moramarco, T. Discharge estimation and forecasting by MODIS and altimetry data in Niger-Benue River. *Remote Sens. Environ.* **2017**, *195*, 96–106. [CrossRef]

24. Tourian, M.J.; Schwatke, C.; Sneeuw, N. River discharge estimation at daily resolution from satellite altimetry over an entire river basin. *J. Hydrol.* **2017**, *546*, 230–247. [CrossRef]

25. Chelton, D.B.; Ries, J.C.; Haines, B.J.; Fu, L.-L.; Callahan, P.S. Chapter 1 Satellite Altimetry. In *Satellite Altimetry and Earth Sciences A Handbook of Techniques and Applications*; Elsevier, 2001; Volume 69, pp. 1–131. ISBN 0074-6142.

26. Frappart, F.; Blumstein, D.; Cazenave, A.; Ramillien, G.; Birol, F.; Morrow, R.; Rémy, F. Satellite Altimetry: Principles and Applications in Earth Sciences. In *Wiley Encyclopedia of Electrical and Electronics Engineering*; John Wiley & Sons, Inc.: Hoboken, NJ, USA, 2017; pp. 1–25. ISBN 047134608X.

27. Frappart, F.; Papa, F.; Marieu, V.; Malbeteau, Y.; Jordy, F.; Calmant, S.; Durand, F.; Bala, S. Preliminary Assessment of SARAL/AltiKa Observations over the Ganges-Brahmaputra and Irrawaddy Rivers. *Mar. Geod.* **2015**, *38*. [CrossRef]

28. Biancamaria, S.; Frappart, F.; Leleu, A.S.; Marieu, V.; Blumstein, D.; Desjonquères, J.D.; Boy, F.; Sottolichio, A.; Valle-Levinson, A. Satellite radar altimetry water elevations performance over a 200 m wide river: Evaluation over the Garonne River. *Adv. Sp. Res.* **2017**, *59*, 128–146. [CrossRef]

29. Bogning, S.; Frappart, F.; Blarel, F.; Niño, F.; Mahé, G.; Bricquet, J.P.; Seyler, F.; Onguéné, R.; Etamé, J.; Paiz, M.C.; Braun, J.J. Monitoring water levels and discharges using radar altimetry in an ungauged river basin: The case of the Ogooué. *Remote Sens.* **2018**, *10*, 350. [CrossRef]

30. Biancamaria, S.; Schaedele, T.; Blumstein, D.; Frappart, F.; Boy, F.; Desjonquères, J.D.; Pottier, C.; Blarel, F.; Niño, F. Validation of Jason-3 tracking modes over French rivers. *Remote Sens. Environ.* **2018**, *209*, 77–89. [CrossRef]

31. Frappart, F.; Roussel, N.; Biancale, R.; Martinez Benjamin, J.J.; Mercier, F.; Perosanz, F.; Garate Pasquin, J.; Martin Davila, J.; Perez Gomez, B.; Gracia Gomez, C.; et al. The 2013 Ibiza Calibration Campaign of Jason-2 and SARAL Altimeters. *Mar. Geod.* **2015**, *38*. [CrossRef]

32. Vu, P.; Frappart, F.; Darrozes, J.; Marieu, V.; Blarel, F.; Ramillien, G.; Bonnefond, P.; Birol, F. Multi-Satellite Altimeter Validation along the French Atlantic Coast in the Southern Bay of Biscay from ERS-2 to SARAL. *Remote Sens.* **2018**, *10*, 93. [CrossRef]

33. Jekeli, C. *Geometric Reference System in Geodesy*; Division of Geodesy and Geospatial Science School of Earth Sciences, Ohio State University: Columbus, OH, USA, 2006.

34. Salameh, E.; Frappart, F.; Marieu, V.; Spodar, A.; Parisot, J.P.; Hanquiez, V.; Turki, I.; Laignel, B. Monitoring sea level and topography of coastal lagoons using satellite radar altimetry: The example of the Arcachon Bay in the Bay of Biscay. *Remote Sens.* **2018**, *10*, 297. [CrossRef]

35. Mahé, G.; Bamba, F.; Soumaguel, A.; Orange, D.; Olivry, J.C. Water losses in the inner delta of the River Niger: water balance and flooded area. *Hydrol. Process.* **2009**, *23*, 3157–3160. [CrossRef]

36. De Noray, M.-L. Delta intérieur du fleuve Niger au Mali—quand la crue fait la loi: l'organisation humaine et le partage des ressources dans une zone inondable à fort contraste. *VertigO-la Rev. électronique en Sci. l' Environ.* **2003**, *4*, 1–9. [CrossRef]

37. Zwarts, L. *The Niger, A Lifeline: Effective Water Management in the Upper Niger Basin*; RIZA: Lelystad, 2005; ISBN 978-90-807150-6-6.

38. Jones, K.; Lanthier, Y.; van der Voet, P.; van Valkengoed, E.; Taylor, D.; Fernández-Prieto, D. Monitoring and assessment of wetlands using Earth Observation: The GlobWetland project. *J. Environ. Manag.* **2009**, *90*, 2154–2169. [CrossRef] [PubMed]

39. Bergé-Nguyen, M.; Crétaux, J.-F. Inundations in the Inner Niger Delta: Monitoring and Analysis Using MODIS and Global Precipitation Datasets. *Remote Sens.* **2015**, *7*, 2127–2151. [CrossRef]

40. Ogilvie, A.; Belaud, G.; Delenne, C.; Bailly, J.-S.; Bader, J.-C.; Oleksiak, A.; Ferry, L.; Martin, D. Decadal monitoring of the Niger Inner Delta flood dynamics using MODIS optical data. *J. Hydrol.* **2015**, *523*, 368–383. [CrossRef]

41. Benveniste, J.; Roca, M.; Levrini, G.; Vincent, P.; Baker, S.; Zanife, O.; Zelli, C.; Bombaci, O. The radar altimetry mission: RA-2, MWR, DORIS and LRR. *ESA Bull.* **2001**, *106*, 25101–25108.

42. Steunou, N.; Desjonquères, J.D.; Picot, N.; Sengenes, P.; Noubel, J.; Poisson, J.C. AltiKa Altimeter: Instrument Description and In Flight Performance. *Mar. Geod.* **2015**, *38*, 22–42. [CrossRef]

43. Taylor, P.; Perbos, J.; Escudier, P.; Parisot, F.; Zaouche, G.; Vincent, P.; Menard, Y.; Manon, F.; Kunstmann, G.; Royer, D.; et al. Jason-1: Assessment of the System Performances Special Issue: Jason-1 Calibration/Validation. *Mar. Geod.* **2003**, *26*, 37–41. [CrossRef]

44. Desjonquères, J.D.; Carayon, G.; Steunou, N.; Lambin, J. Poseidon-3 Radar Altimeter: New Modes and In-Flight Performances. *Mar. Geod.* **2010**, *33*, 53–79. [CrossRef]

45. Donlon, C.; Berruti, B.; Buongiorno, A.; Ferreira, M.H.; Féménias, P.; Frerick, J.; Goryl, P.; Klein, U.; Laur, H.; Mavrocordatos, C.; et al. The Global Monitoring for Environment and Security (GMES) Sentinel-3 mission. *Remote Sens. Environ.* **2012**, *120*, 37–57. [CrossRef]

46. Wingham, D.J.; Rapley, C.G.; Griffiths, H. New Techniques in Satellite Altimeter Tracking Systems. *Proc. IGARSS Symp. Zurich* **1986**, 1339–1344.

47. Frappart, F.; Do Minh, K.; L'Hermitte, J.; Cazenave, A.; Ramillien, G.; Le Toan, T.; Mognard-Campbell, N. Water volume change in the lower Mekong from satellite altimetry and imagery data. *Geophys. J. Int.* **2006**, *167*. [CrossRef]

48. Santos da Silva, J.; Calmant, S.; Seyler, F.; Rotunno Filho, O.C.; Cochonneau, G.; Mansur, W.J. Water levels in the Amazon basin derived from the ERS 2 and ENVISAT radar altimetry missions. *Remote Sens. Environ.* **2010**, *114*, 2160–2181. [CrossRef]

49. Cartwright, D.E.; Edden, A.C. Corrected Tables of Tidal Harmonics. *Geophys. J. R. Astron. Soc.* **1973**, *33*, 253–264. [CrossRef]

50. Wahr, J.M. Deformation induced by polar motion. *J. Geophys. Res.* **1985**, *90*, 9363. [CrossRef]

51. Crétaux, J.F.; Jelinski, W.; Calmant, S.; Kouraev, A.; Vuglinski, V.; Bergé-Nguyen, M.; Gennero, M.C.; Nino, F.; Abarca Del Rio, R.; Cazenave, A.; Maisongrande, P. SOLS: A lake database to monitor in the Near Real Time water level and storage variations from remote sensing data. *Adv. Sp. Res.* **2011**, *47*, 1497–1507. [CrossRef]

52. Frappart, F.; Biancamaria, S.; Normandin, C.; Blarel, F.; Bourrel, L.; Aumont, M.; Azemar, P.; Vu, P.-L.; Le Toan, T.; Lubac, B.; et al. Influence of recent climatic events on the surface water storage of the Tonle Sap Lake. *Sci. Total Environ.* **2018**, *636*, 1520–1533. [CrossRef]

53. Frappart, F.; Papa, F.; Famiglietti, J.S.; Prigent, C.; Rossow, W.B.; Seyler, F. Interannual variations of river water storage from a multiple satellite approach: A case study for the Rio Negro River basin. *J. Geophys. Res. Atmos.* **2008**, *113*. [CrossRef]

54. Lee, H.; Shum, C.K.; Yi, Y.; Ibaraki, M.; Kim, J.W.; Braun, A.; Kuo, C.Y.; Lu, Z. Louisiana wetland water level monitoring using retracked TOPEX/POSEIDON altimetry. *Mar. Geod.* **2009**, *32*, 284–302. [CrossRef]

55. Frappart, F.; Papa, F.; Güntner, A.; Werth, S.; Santos da Silva, J.; Tomasella, J.; Seyler, F.; Prigent, C.; Rossow, W.B.; Calmant, S.; Bonnet, M.-P. Satellite-based estimates of groundwater storage variations in large drainage basins with extensive floodplains. *Remote Sens. Environ.* **2011**, *115*. [CrossRef]

56. da Silva, J.S.; Seyler, F.; Calmant, S.; Filho, O.C.R.; Roux, E.; Araújo, A.A.M.; Guyot, J.L. Water level dynamics of Amazon wetlands at the watershed scale by satellite altimetry. *Int. J. Remote Sens.* **2012**, *33*, 3323–3353. [CrossRef]

57. Zakharova, E.A.; Kouraev, A.V.; Rémy, F.; Zemtsov, V.A.; Kirpotin, S.N. Seasonal variability of the Western Siberia wetlands from satellite radar altimetry. *J. Hydrol.* **2014**, *512*, 366–378. [CrossRef]

58. Bonnefond, P.; Verron, J.; Aublanc, J.; Babu, K.; Bergé-Nguyen, M.; Cancet, M.; Chaudhary, A.; Crétaux, J.-F.; Frappart, F.; Haines, B.; et al. The Benefits of the Ka-Band as Evidenced from the SARAL/AltiKa Altimetric Mission: Quality Assessment and Unique Characteristics of AltiKa Data. *Remote Sens.* **2018**, *10*, 83. [CrossRef]

59. Keith Raney, R. The delay/doppler radar altimeter. *IEEE Trans. Geosci. Remote Sens.* **1998**, *36*, 1578–1588. [CrossRef]

9

Independent Assessment of Sentinel-3A Wet Tropospheric Correction over the Open and Coastal Ocean

Maria Joana Fernandes [1,2,*] and Clara Lázaro [1,2]

[1]   Faculdade de Ciências, Universidade do Porto, 4169-007 Porto, Portugal; clazaro@fc.up.pt
[2]   Centro Interdisciplinar de Investigação Marinha e Ambiental (CIIMAR/CIMAR), Universidade do Porto,
      4050-123 Porto, Portugal
*   Correspondence: mjfernan@fc.up.pt

Abstract: Launched on 16 February 2016, Sentinel-3A (S3A) carries a two-band microwave radiometer (MWR) similar to that of Envisat, and is aimed at the precise retrieval of the wet tropospheric correction (WTC) through collocated measurements using the Synthetic Aperture Radar Altimeter (SRAL) instrument. This study aims at presenting an independent assessment of the WTC derived from the S3A MWR over the open and coastal ocean. Comparisons with other four MWRs show Root Mean Square (RMS) differences (cm) of S3A with respect to these sensors of 1.0 (Global Precipitation Measurement (GPM) Microwave Imager, GMI), 1.2 (Jason-2), 1.3 (Jason-3), and 1.5 (Satellite with ARgos and ALtika (SARAL)). The linear fit with respect to these MWR shows scale factors close to 1 and small offsets, indicating a good agreement between all these sensors. In spite of the short analysis period of 10 months, a stable temporal evolution of the S3A WTC has been observed. In line with the similar two-band instruments aboard previous European Space Agency (ESA) altimetric missions, strong ice and land contamination can be observed, the latter mainly found up to 20–25 km from the coast. Comparisons with the European Centre for Medium-Range Weather Forecasts (ECMWF) and an independent WTC derived only from third party data are also shown, indicating good overall performance. However, improvements in both the retrieval algorithm and screening of invalid MWR observations are desirable to achieve the quality of the equivalent WTC from Jason-3. The outcome of this study is a deeper knowledge of the measurement capabilities and limitations of the type of MWR aboard S3A and of the present WTC retrieval algorithms.

Keywords: Sentinel-3; satellite altimetry; microwave radiometer; wet tropospheric correction; wet path delay; sensor calibration

1. Introduction

Starting with European Remote Sensing 1 (ERS-1) in 1991 and TOPography Experiment (TOPEX)/Poseidon in 1992, satellite missions carrying radar altimeters have resulted in invaluable continuous measurements being obtained over the ocean, ice, rivers and lakes for more than 25 years. This has been made possible through the joint effort of various space agencies: the European Space Agency (ESA), the Centre National d'Études Spatiales (CNES), European Organisation for the Exploitation of Meteorological Satellites (EUMETSAT), the National Aeronautics and Space Administration (NASA), the Jet Propulsion Laboratory (JPL), the National Oceanic and Atmospheric Administration (NOAA), and the Indian Space Research Organization (ISRO). ESA supported the 35-day repeat missions ERS-1 (1991), ERS-2 (1995), and Envisat (2002), the geodetic mission CryoSat-2 (2010), and more recently, the 27-day repeat mission Sentinel-3A (2016), the latter in collaboration with EUMETSAT. Satellite with ARgos and ALtika (SARAL), a 35-day mission follow-up of Envisat, is a joint effort between the CNES and ISRO. Partly or in collaboration, NASA, NOAA, JPL, CNES, and EUMETSAT have been responsible for the so-called reference 10-day repeat missions: TOPEX/Poseidon (1991), Jason-1 (2002), Jason-2 (2008), and Jason-3 (2016) [1]. For a detailed and update review of satellite altimetry, see e.g., [2].

These missions have been designed to provide complementary spatial and temporal resolutions, allowing e.g., a better characterisation of the mesoscale oceanic circulation [3,4], the determination of accurate global and regional sea level trends [5–9], and unprecedented observations over river, lakes and reservoirs [10,11] as well as ice surfaces [12].

Satellite altimetry allows the determination of the height of e.g., the sea or lake surface above a reference ellipsoid by means of the following equation [13]:

$$h = H - R = H - R_{obs} - \Delta R \tag{1}$$

where H is the spacecraft height above a reference ellipsoid provided by precise orbit determination, referred to as an International Terrestrial Reference Frame, R_{obs} is the measured altimeter range corrected for all instrument effects, and R is the corresponding range corrected for all instrument, range, and geophysical effects. The term ΔR includes all corrections that need to be applied to the observed range due to the signal propagation delay through the atmosphere, its interaction with the sea surface and terms related with specific geophysical phenomena being given by Equation (2):

$$\Delta R = \Delta R_{dry} + \Delta R_{wet} + \Delta R_{iono} + \Delta R_{SSB} + \Delta R_{tides} + \Delta R_{DAC} \tag{2}$$

The first four terms in Equation (2) are the range corrections, accounting for the interaction of the altimeter radar signal with the atmosphere (dry, wet and ionospheric corrections) and with the sea surface (sea state bias). The last two terms refer to geophysical phenomena, dynamic atmospheric correction, and tides (ocean, load, solid earth and pole tides), which should be removed if they are not part of the signals of interest. A detailed description of the range and geophysical corrections can be found e.g., in [13,14].

Equation (1) evidences that the determination of an accurate surface height requires the knowledge of the spacecraft orbit, the range measurement, and of all correction terms in Equation (2) with the same accuracy. Amidst these terms is the wet tropospheric correction (WTC), due to the water vapour and liquid water content in the atmosphere. Since these variables have large space–time variability, if not properly modelled, the WTC is one of the major sources of uncertainty in many satellite altimetry applications [13,15,16].

Due to its variability, the most precise way to account for the WTC in satellite altimetry is by means of collocated measurements from a microwave radiometer (MWR), obtained on the same spacecraft. Two main types of radiometers have been deployed on the altimeter satellites: three-band radiometers on board the reference missions and two-band radiometers on all ESA missions and SARAL. Over the open ocean they provide accurate WTC retrievals within 1 cm [17].

The algorithms adopted in the WTC retrieval from the measured brightness temperatures (TBs) in the various spectral bands have been designed for the open ocean, assuming a constant ocean emissivity. In the presence of other surfaces such as land or ice, the measurements lay outside the predicted validity interval and the observations become unusable. Moreover, these instruments possess large footprints from 10 to 40 km depending on the frequency [18,19]. Therefore, in the coastal regions they sense land well before the altimeter, originating bands of invalid measurements around the coastline of a width of 10–40 km. In recent years, several methods have been developed to improve the MWR-derived WTC in the coastal zones, extending the validity of the correction up to the coast. An overview of these methods is given in [16].

Building directly on the proven heritage of ERS-1, ERS-2, and Envisat, launched on 16 February 2016, Sentinel-3A (S3A) carries a suite of innovative instruments that include [20]: (1) the Sea and Land Surface Temperature Radiometer (SLSTR) based on Envisat's Advanced Along Track

Scanning Radiometer (AATSR), to determine global sea surface temperatures to an accuracy of better than 0.3 K; the Ocean and Land Colour Instrument (OLCI), based on Envisat's Medium Resolution Imaging Spectrometer (MERIS), with 21 bands, (compared to 15 bands on MERIS), a design optimised to minimise Sun-glint, and a resolution of 300 m over all surfaces; a dual-frequency (Ku and C-band) advanced Synthetic Aperture Radar Altimeter (SRAL) developed from Envisat RA-2, CryoSat Synthetic Aperture Interferometric Radar Altimeter (SIRAL) and Jason-2/Poseidon-3, providing accurate surface topography measurements; and a microwave radiometer for accurate measurement of the wet pat delay of the SRAL observations [20].

SRAL and the MWR constitute the topography package, providing accurate measurements of sea surface height, significant wave height, and wind speed, essential for ocean forecasting systems and climate monitoring. Accurate SRAL measurements also extend to sea-ice, ice sheets, rivers, and lakes. Being the first altimeter mission to operate, globally, a radar altimeter in the SAR closed burst mode, with an improved along track resolution of about 300 m [21], Sentinel-3A is a pioneer and challenging mission in many aspects, pushing the experts of the various fields to the limit to tune the retrieval algorithms to its instruments and exploit the derived geophysical parameters.

As mentioned above, amongst its suite of instruments, Sentinel-3A carries a dual channel (23.8-GHz and 36.5-GHz) microwave radiometer aiming at the retrieval of the wet path delay, which provides important support for the quality of the SRAL measurements. Due to their instrumental characteristics and retrieval algorithms, the two-band MWRs deployed on ESA altimeter missions are known for their good performance in the open ocean [22]. However, when they approach the coast, the retrieval algorithm, which was designed for surfaces with ocean emissivity, generates very noisy values, as the footprint encounters surfaces with different levels of emissivity. The same happens at high latitudes in regions covered with ice [23,24].

This work aims at performing an independent assessment of S3A MWR-based WTC, in the open and coastal ocean, in support of S3A data improvement and exploitation.

The validation is performed by means of comparisons with independent data sets namely: wet path delays derived from the Global Precipitation Measurement (GPM) Microwave Imager (GMI); Global Navigation Satellite System (GNSS)-derived path delays determined at coastal stations; and wet path delays from the MWR on board Jason-2 (J2), Jason-3 (J3), and SARAL/AltiKa. In addition, the overall along-track performance is compared against independent estimates obtained from the GNSS-derived Path Delay Plus (GPD+) algorithm [24] and from atmospheric models. From this thorough analysis, a deeper knowledge of the measurement capabilities and limitations of the dual-frequency radiometer aboard S3A and of the present WTC retrieval algorithms is expected.

This paper is organised in five sections. Section 2 describes the data sets and the methodology used in the S3A WTC assessment. Section 3 describes the results obtained with the various datasets. Finally, Sections 4 and 5 present the discussions of the results and conclusions, respectively.

2. Dataset Description

2.1. Sentinel-3 Data and the Radar Altimeter Database System (RADS)

Sentinel-3A flies in a near-polar Sun-synchronous orbit with a Local Time of the Descending Node (LTDN) at 10:00 h. It has a high inclination of 98.65°, altitude of about 814.5 km, a 27-day repeat cycle (385 orbits per cycle), and 104-km inter-track spacing at the equator [20]. Sentinel-3B will be on a similar orbit, with an 180° phase difference, the two satellites ensuring twice the spatial and temporal coverage.

The Sentinel-3 data used in this study are from the Sentinel-3 Surface Topography Mission (STM) Level-2 Non-Time Critical (NTC) products from the so-called "Spring 2017 Reprocessing Campaign" [25,26]. The product reference is SRAL/MWR L2 Instrument Processing Facility (IPF) (SM-2): version 06.07 from Processing Baseline 2.15. These data have been available in the Radar Altimeter Database System (RADS, http://rads.tudelft.nl/rads/rads.shtml) [27] as reprocessed version

1 (rep1), since mid-August 2017. In addition to the native S3A products, the website includes the orbit and state-of-the-art range and geophysical corrections present in RADS for all other altimeter missions. The data span approximately 10 months, from 15 June 2016 (cycle 05) to 15 April 2017 (cycle 16).

As mentioned above, Sentinel-3A MWR operates in two frequencies. The first, located at a water vapour absorption line (23.8 GHz), is the primary water vapour sensing frequency; the second (36.5 GHz) is an atmospheric window sensitive to surface emissivity and to cloud liquid water. In the dual-frequency microwave radiometers, which do not include any low frequency channel, the effect of surface roughness is often taken into account through the altimeter-derived wind speed or the backscatter coefficient.

The relation between the measured brightness temperatures of the various MWR channels and the wet path delay (WPD) or WTC has been empirically established by means of statistical regression methods. Two main types of algorithms have been used in the retrieval of the WTC from the TB of the various MWR channels: (1) a parametric log-linear algorithm adopted in the reference missions [17]; and (2) a neural network algorithm first developed for Envisat, later used in SARAL/AltiKa and Sentinel-3 [28].

The function adopted in the retrieval of the WPD from the MWR brightness temperatures is usually derived from TB simulations, obtained from a database of atmospheric and sea-surface scenes using a radiative transfer model. Several databases can be used. Traditionally, in the ESA missions the database is built from European Centre for Medium-Range Weather Forecasts (ECMWF) analyses, while in the reference missions, it is usually built from radiosonde observations (for the atmospheric profiles) and from microwave imaging radiometers such as the Special Sensor Microwave Imager (SSM/I), the SSM/I Sounder (SSM/IS), or the Advanced Very High Resolution Radiometer (to provide sea-surface parameters) [29]. In summary, a WPD retrieval algorithm consists in the empirical establishment of the inverse function to get the WPD from the simulated TB.

In [29], the parametric log-linear algorithm is compared with the neural network formalism. These authors also analyse the advantages of the three-band MWR flying in the reference missions versus the two-band MWR on board the ESA missions, as well as how well additional parameters such as the backscatter coefficient can replace the missing third channel in the ESA missions.

The MWR-derived WTC present in this product has been computed by a neural network algorithm based on five inputs [30]: brightness temperatures at 23.8 GHz and 36.5 GHz, Ku-band ocean backscatter coefficient (not corrected for the atmospheric attenuation), sea surface temperature (four seasonal tables, 2° resolution), and the lapse rate (decreasing rate of atmospheric temperature with altitude, from a climatological table, 1° resolution) [30]. This algorithm is an evolution of the one developed for Envisat that only used the first three inputs described above [28].

The S3A products also include composite wet tropospheric correction [30], based on both radiometer and model-based corrections over areas where the radiometer WTC is missing or invalid due to the proximity of land (coastal areas and/or radiometer gaps in open oceans) [30,31]. A first assessment of this correction is also performed.

In the sea level anomaly (SLA) variance analyses performed in Section 3, S3A SRAL data and all required range and geophysical corrections to compute SLA were extracted from RADS.

2.2. The Global Precipitation Measurement (GPM) Microwave Imager (GMI)

The GMI is a dual-polarization, multi-channel, conical-scanning, passive microwave imaging radiometer on board the GPM satellite, launched on 27 February 2014. GPM has a non-Sun-synchronous orbit, at a mean altitude of 407 km and inclination of 65°, covering the latitude band between ±65°, allowing a full sampling of the Earth approximately every 2 weeks. The orbit plane completes half (180°) a rotation relative to the Sun every 41.1 days. The spacecraft undergoes yaw manoeuvres every ~40 days to compensate for the Sun's changing position and prevent the side of the spacecraft facing the Sun from overheating [32].

The GMI has been designed with a strict calibration accuracy requirement, enabling the instrument to serve as a microwave radiometric standard. The GMI features, ensuring its high calibration accuracy, include: protection of the hot load from Sun intrusion, noise diodes on the low-frequency channels for a dual calibration system, and a reflective antenna coating [33].

The GMI data used in this study are gridded products of total column water vapour (TCWV), in binary format provided by Remote Sensing Systems (RSS, http://www.remss.com/). Two $0.25° \times 0.25°$ global grids per day are provided, one containing the ascending and the other the descending GPM passes. The RSS products based on the Version-8.2 algorithm and the Radiative Transfer Model have been used [32].

As described in [34], the WTC has been computed from these TCWV products using the expression by [35], deduced from temperature and humidity profiles from ECMWF model fields:

$$WTC = -\left(a_0 + a_1 WV + a_2 WV^2 + a_3 WV^3\right) WV \times 10^{-2} \tag{3}$$

with $a_0 = 6.8544$, $a_1 = -0.4377$, $a_2 = 0.0714$, and $a_3 = -0.0038$. In Equation (3), WV is in centimetres, as provided in the TCWV products, and WTC results are in metres.

In [24] the WTC derived from a previous version of these products (Version 7, V7) was compared against the WTC derived from the SSM/IS on board the Defense Meteorological Satellite Program satellite series F16 and F17. The SSM/IS sensors can be regarded as a reference, due to their stability and independent calibration [36,37]. This comparison shows that the scale factor and offset between SSM/IS and GMI are 0.99 and -0.26 cm, respectively, while the Root Mean Square (RMS) of the differences between the two data sets is 0.79 cm and 0.77 cm, before and after the adjustment to SSM/IS, respectively. According to RSS, the Version 8 (V8) brightness temperatures from GMI are slightly different from the V7 brightness temperatures; however there are no significant differences between the V7 and V8 ocean products. In the sequel, in this work the GMI V8.2 products have been adopted without applying any calibration parameters. It should be noted that, in spite of the fact that SSM/IS are considered stable sensors, the S3A WTC was not directly compared against the corresponding WTC derived from the SSM/IS TCWV products, also available from RSS, since both S3A, F16, and F17 are in Sun-synchronous orbits with a phase difference of 5–6 h (S3A/F16) and 3–5 h (S3A/F17).

2.3. Microwave Radiometers on Board Jason-2, Jason-3, and SARAL

Launched on 8 June 2008, the Ocean Surface Topography Mission (OSTM) Jason-2 carries the Advanced Microwave Radiometer (AMR) operating at 18.7, 23.8 and 34 GHz. In comparison with the dual-frequency MWR aboard the ESA missions, the additional low frequency channel improves the WTC retrieval, particularly by adding the ability to reduce land effects near the coast [38].

During the first years of the mission, like its predecessors TOPEX/Poseidon and Jason-1, Jason-2 flew in the so-called reference orbit: non-Sun-synchronous, inclination of 66°, altitude of 1336 km, 10-day repeat cycle (127 orbits per cycle), and 315 km inter-track spacing at the equator [39]. Jason-3 was launched to the reference orbit on 17 January 2016. During the calibration phase of approximately 6 months (20 repeat cycles) it shared the same orbit with Jason-2, with a time difference of about 1 min. In October 2016, Jason-2 moved to the interleaved orbit (same characteristics but with the ground tracks at half distance between those of the reference orbit) and later on, on 20 June 2017, to a long-repeat orbit, at an altitude of roughly 1309.5 km.

Jason-3 carries the Advanced Microwave Radiometer 2 (AMR-2), which is similar to the AMR but with improvements in instrument thermal control and stability [40]. Jason-3 is the first altimeter mission to implement special spacecraft cold sky calibration manoeuvres (CSCMs), combined with vicarious on-Earth ocean and land target references, for improving the long-term climate calibration of the radiometer [41]. Currently, Jason-2 also performs routine CSCMs.

Launched on 25 February 2013, SARAL/AltiKa flies on the same orbit as ERS-2 and Envisat: Sun-synchronous with an LTDN at 18:00 h and a Local Time of Ascending Node (LTAN) at 06:00 h.

It has an inclination of 98.5°, altitude of about 800 km, a 35-day repeat cycle (501 orbits per cycle), and 80 km inter-track spacing at the equator [42]. SARAL caries the AltiKa instrument, provided by CNES, consisting of a Ka-band altimeter and an embedded dual-frequency Microwave Radiometer (23.8 GHz/37 GHz) similar to that of Envisat but with a smaller footprint (8 km in Ka-band and 12 km in Ku-band) [19].

2.4. WTC from GNSS at Coastal Stations

Zenith total delays (ZTDs) from a network of 60 stations with a good global coverage have been used. The details of this computation are given in [43], with only the main features being outlined here. These stations have been chosen to cover the various regions of the world with different WTC variability conditions (Figure 1) and to ensure that all zenith wet delays (ZWDs) have been obtained using the same computation parameters, thus ensuring their long-term stability. All stations are located near the coast, up to 100 km from the coast and with an orthometric height <1000 m. The second condition aims at reducing the errors due to the height dependence of the WTC.

As illustrated in the background map of Figure 1, derived from ECMWF operational model fields, the RMS of the WPD has a clear zonal dependence, ranging from only a few cm near the poles up to 35 cm at low latitudes. The patterns of the RMS of WPD are dominated by the zonal dependence, associated with the dependence of the WPD on temperature and the increase in water vapour in the equatorial regions. The variance of the WPD (not shown) also has a strong dependency on latitude, reaching a maximum near the tropics, over regions where the conditions for strong monsoons prevail, and a minimum near the poles, where the content of water vapour in the atmosphere is minimal. Maximum values occur in the Northern Hemisphere, where the Indian and Pacific oceans meet; in the Southern Hemisphere, the highest variability is found northwards of Australia.

The GNSS provides accurate (4–6 mm) values of the ZTD, the sum of the dry and wet components of the tropospheric delay, at station height, while the quantity of interest in this study is the zenith wet delay (ZWD), symmetric with respect to the WTC, at sea (zero height) level. The latter is obtained from the ZTD at station level by subtracting the dry correction or zenith hydrostatic delay (ZHD) derived from the ECMWF sea-level pressure (SLP) field using the modified Saastamoinen model [44] and reducing ZHD and ZWD fields to sea level using the expression by [45], with modifications introduced by [46].

It has been shown [46] that the hydrostatic component of the tropospheric delay can be estimated from global grids of sea level pressure available from ECMWF models with an accuracy of 1 to 3 mm at a global scale, provided an adequate model for the height dependence of atmospheric pressure is adopted. Therefore, using ZHD computed from ECMWF model fields according to [46], the ZWD can be determined from the GNSS with an accuracy greater than 1 cm.

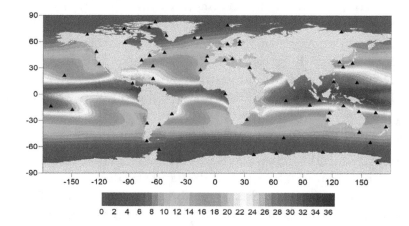

Figure 1. Location of a set of 60 Global Navigation Satellite System (GNSS) stations used in this study (adapted from [43]). The background map represents the Root Mean Square (RMS) of wet path delay (WPD) in cm.

2.5. GPD+ Wet Tropospheric Corrections

GNSS-derived Path Delay Plus (GPD+) provides wet tropospheric corrections derived by data combination, using space–time objective analysis, of all available wet path delays in the neighbourhood of a given point on the Earth. It has been designed: (1) to correct the WTC observations derived from the on-board MWR flagged as invalid due to various error sources: contamination by land, ice or rain, or instrument malfunction; and (2) to estimate improved WTC for satellites such as CryoSat-2, with no on-board MWR, for which only the model-derived WTC was available. In the first case, for all points with good MWR data, GPD+ keeps these values unchanged while for all invalid MWR measurements, a new estimate is obtained from the available observations. In the second case, a new estimate is obtained for all along-track points. Whenever the number of observations is null, GPD+ assumes the value of the first guess, which is the WTC from an atmospheric model: ECMWF operational for the most recent missions [47] and EMWF ReAnalysis (ERA) Interim for all missions with data prior to 2004 [23,24,48].

An important step in the GPD+ retrievals is the efficiency of the criteria to depict all invalid observations. This is done by using various flags provided with satellite data, when available, for example regarding the presence of land, ice, or rain in the radiometer footprint, or the distance from coast. Moreover, statistical criteria are established based on the comparison of MWR values and the corresponding model values, not only on the same point but also in the neighbouring along-track points. Figure 2 illustrates the invalid MWR points for Sentinel-3A cycle 06 (28.6% of the total ocean points). When only points with valid SLA are considered, the percentage of invalid MWR points varies, from cycle to cycle, from 11% to 17%.

For this study, two types of GPD+ WTC have been computed for S3A: (1) using only third-party data, i.e., wet path delays derived from a set of more than 800 GNSS stations and WTC from scanning imaging MWRs on board various remote sensing missions, hereafter designated GPD1; and (2) using all data including the S3A MWR measurements, provided they are valid, and only computing new estimates for the invalid MWR points, hereafter named GPD2.

The available WTC observations used in the GPD+ estimations, spanning the S3A mission period, include: valid observations from S3A´s on-board MWR; scanning imaging MWRs (SI-MWRs) from 11 different satellites; and GNSS-derived WTCs from more than 800 coastal stations.

Details on these data sets and on the GPD+ algorithm can be found in [24].

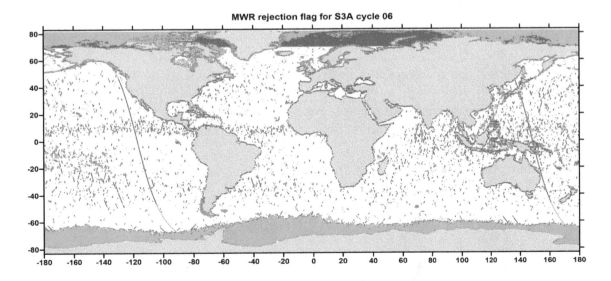

Figure 2. Sentinel-3A (S3A) points for cycle 06 with invalid microwave radiometer (MWR) observations: green—land contamination; blue—ice contamination; pink—rain, outliers, or additional condition such as all points above latitude 70°N or below 70°S.

2.6. The ECMWF Atmospheric Model

The atmospheric model used in this study is the ECMWF operational model. Global $0.125° \times 0.125$ grids of three single-level atmospheric parameters (sea level pressure (SLP), surface temperature (2-m temperature, 2T) and total column water vapour (TCWV)), available every 6 h have been used [49].

These model parameters are used both in the ZWD computations referred in Section 2.4 and to determine a model derived WTC for each S3A along-track position by space–time interpolation from the two closest grids, 6-h apart. In spite of its poor temporal sampling, the ECMWF model is used from here on as an additional point of comparison.

3. Results

Section 3 presents the various analyses performed aimed at a detailed assessment of the MWR-based wet tropospheric correction present on the Sentinel-3A products described in Section 2.1. In this section only the results are presented, their full discussion being described in Section 4.

3.1. Comparison with Other Microwave Radiometers

This section presents the comparison of Sentinel-3A MWR with other microwave radiometers on board the GMI, J2, J3 and SARAL. In these comparisons with other sensors, the wet path delay (WPD, symmetric with respect to the WTC) has been used. Unless otherwise mentioned, all statistical parameters detailed in this section (scale factor, offset, mean, and RMS of differences) refer to WPD. This is done to facilitate the representation and interpretation of the results.

3.1.1. Comparison with GMI

For the comparison with GMI, match points between GPM and S3A with a time difference ΔT < 45 min and within a distance ΔD < 50 km were computed. Only S3A points considered valid by the GPD+ algorithm [24] have been used (see Section 2.5). These criteria aim at removing S3A MWR observations contaminated by e.g., land, ice, rain, and outliers (Figure 2).

The use of match or collocated points is common practice in the comparison of a pair of sensors, thus comparing the measurements from both sensors in points with a given space–time difference. When analysing long periods of data, other approaches are possible such as e.g., in [50] where the authors organised the data sets in monthly bin boxes of three degrees, claiming that this approach is more favourable in e.g., the analyses of temporal variations of geophysical variables such as the WPD. Considering the short time span of the S3A dataset under inspection, match points are considered in this study.

Figure 3 illustrates the match points between GPM and S3A for the 10-month period of this study. The colour scale represents WPD differences between GMI and S3A in cm. Red colours mean that GMI measurements indicate wetter conditions than S3A, while blue colours indicate the opposite. The overall mean and RMS of the differences WPD(GMI)-WPD(S3A) are 0.17 cm and 0.95 cm, respectively (see Table 1).

Figure 4 illustrates the scattergram of the WPD from S3A against the WPD from the GMI (left) and against WPD difference between the GMI and S3A (right), using the whole set of match points (~219,000 points). The scale factor is 1.004 and the offset is 0.12 cm, indicating a very good overall agreement between the two radiometers, with the GMI measurements indicating slightly wetter conditions than S3A.

Figures 5 and 6 illustrate the time evolution of the WPD from both sensors and the respective differences. The top panel of Figure 5 represents the time evolution of the WPD from the GMI (blue) and S3A (pink) while the bottom panel shows the corresponding WPD differences between the GMI and S3A. A strong periodic signal of approximately 41 days (41.1 days) can be observed, due to GPM orbit plane rotation with respect to the Sun. It can be seen that this periodic pattern is due to the changes in the spatial sampling of the match points with time (cf. Figures 5 and 7 and discussion in Section 5).

When daily and 27-day RMS differences are computed (Figure 6), the same periodic pattern is shown, now at 82 days, and seem stable.

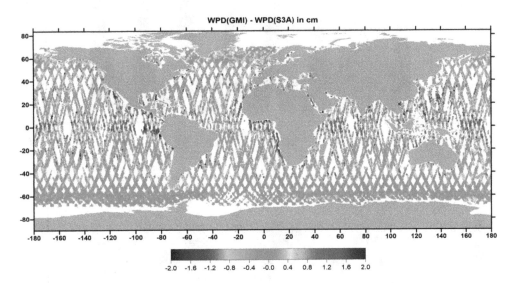

Figure 3. Spatial coverage of match points between S3A and the Global Precipitation Measurement (GPM) Microwave Imager (GMI) with time difference $\Delta T < 45$ min and distance $\Delta D < 50$ km, for S3A cycles 05–16, used in this study (~219,000 points). Colour scale indicates WPD differences between the GMI and S3A in cm.

Table 1. Statistical parameters of the comparison between the WPD from Sentinel-3A and other WPD sources. ECMWF: European Centre for Medium-Range Weather Forecasts; GPD: GNSS-derived Path Delay Plus; WTC: wet tropospheric correction; J2: Jason-2; J3: Jason-3; SARAL: Satellite with ARgos and ALtika.

WTC/S3A	Scale Factor	Offset (cm)	Mean Diff. (cm)	RMS Diff. (cm)
GMI/S3A	1.004	0.115	0.171	0.947
J2/S3A	1.010	−0.660	−0.556	1.213
J3/S3A	1.012	−0.831	−0.698	1.292
SARAL/S3A	0.955	0.557	0.029	1.536
ECMWF/S3A	1.013	−0.086	0.098	1.256
GPD1/S3A	1.005	0.173	0.251	0.967

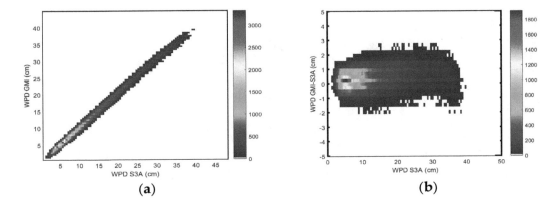

Figure 4. (a) Wet path delay from S3A versus WPD from the GMI; (b) WPD from S3A versus WPD differences between the GMI and S3A (~219,000 points).

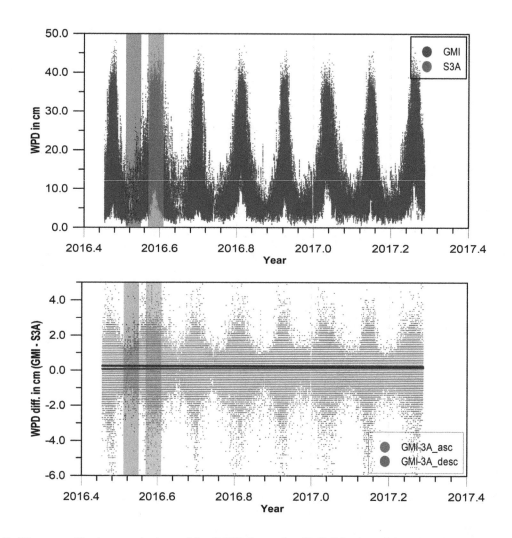

Figure 5. (Top panel): time evolution of the WPD from the GMI (blue) and S3A (pink); **(Bottom Panel):** time evolution of WPD differences between the GMI and S3A for ascending passes (blue) and descending passes (red). Colour bars refer to periods when the GMI/S3A match points are all located at high latitudes (green points in Figure 7) or low latitudes (orange points in Figure 7) to which correspond smaller or larger WPD variability.

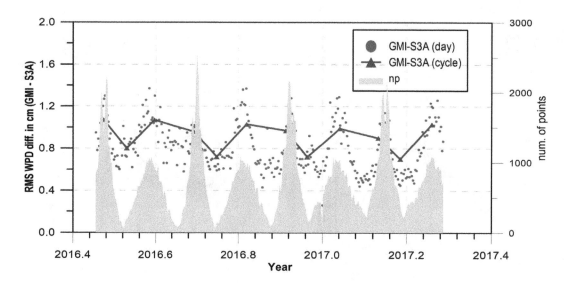

Figure 6. Time evolution of the daily RMS of the WPD differences between the GMI and S3A. The number of points is represented by "np".

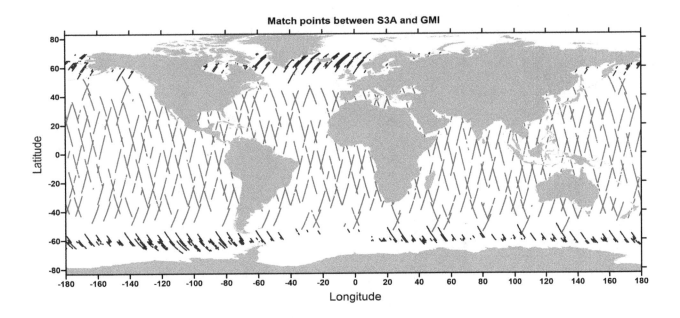

Figure 7. Match points between the GMI and S3A. Colours correspond to different time periods, indicated in Figure 5 by the corresponding colour bars.

3.1.2. Comparison with the MWR on Board Jason-2, Jason-3, and SARAL

To compare the WPD from S3A with that from Jason-2 (J2), Jason-3 (J3), and SARAL, the differences between the corresponding WPD at the crossovers between these missions have been computed. For this purpose, crossovers between S3A and Jason-2 and between S3A and Jason-3 with time difference $\Delta T < 180$ min have been computed. Note that both S3A and SARAL have Sun-synchronous orbits, with S3A LTAN at 22:00 h (LTDN at 10:00 h), and SARAL LTAN at 06:00 h (LTDN at 18:00). Thus, 240 min (4 h) is the minimum time interval to get crossovers between the two missions at low latitudes. With a 4-h difference, every ascending/descending S3A pass crosses a descending/ascending SARAL track. If a smaller period is considered, crossovers are all located at high latitudes, the corresponding values not being representative of the whole set of WTC variability conditions. For example, within a time interval of 180 min, all crossovers are above latitude $40°$ and below latitude $-40°$. In the sequel, for SARAL crossovers with time difference $\Delta T < 240$ min have been considered.

Figure 8 illustrates the spatial pattern of the WPD differences between J2, J3, SARAL, and S3A. The mean and RMS differences between J2, J3, SARAL, and S3A are presented in Table 1. Mean values are -0.56 cm, -0.70 cm and 0.03 cm, respectively, while the corresponding RMS are 1.21 cm, 1.29 cm, and 1.54 cm. The overall agreement with J2 and J3 is better than with SARAL, somehow expected since the crossovers with SARAL have a larger time difference.

Figure 9 shows the scattergrams of the WPD from S3A against the WPD from the various sensors (left panels) as well as against the difference between the WPD from each sensor and that from S3A (right panels). The scale factors and offsets for J2 and J3 are very similar: same scale factor (1.01) and offsets of -0.66 cm and -0.83 cm, respectively. The scale factor for SARAL is slightly different (0.96) with an offset of 0.56 cm. These scatterplots indicate that the best agreement is with J3 and the worst is with SARAL, due to the large time difference between the S3A/SARAL crossover points. Both Figures 8 and 9 indicate that J2 and J3 measure drier than S3A by about 0.6–0.8 cm, while SARAL measurements indicate wetter conditions than S3A by about 0.6 cm.

Figure 10 depicts the time evolution of the WPD differences between J2, J3, SARAL and S3A (daily and 27-day RMS values in centimetres). These results indicate that in spite of the small span of the data, there is an overall agreement between all analysed sensors.

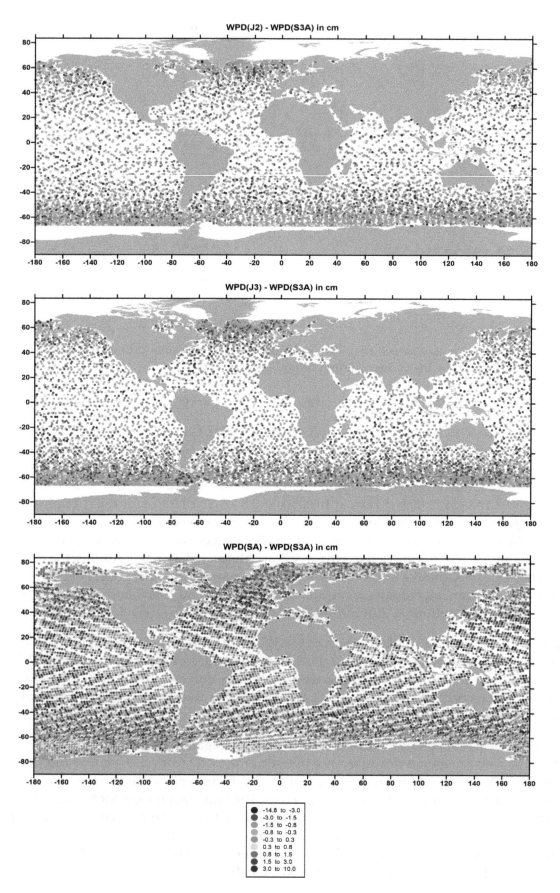

Figure 8. (**Top panel**): Crossover points between J2 and S3A with $\Delta T < 180$ min (~11,600 points); (**Middle panel**): crossovers between J3 and the S3A with $\Delta T < 180$ min (~12,600 points); (**Bottom panel**): crossovers between SARAL and S3A with $\Delta T < 240$ min (~13,700 points).

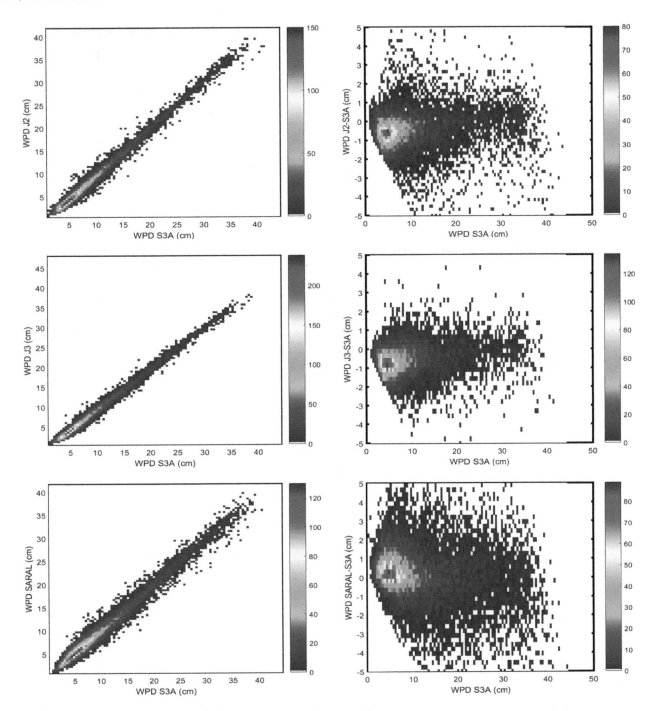

Figure 9. (Left panels): WPD from S3A versus WPD from J2 (**top**), J3 (**middle**), and SARAL (**bottom**), in cm; (**Right panels**): WPD from S3A versus WPD difference between J2 and S3A (**top**), between J3 and S3A (**middle**) and between SARAL and S3A (**bottom**).

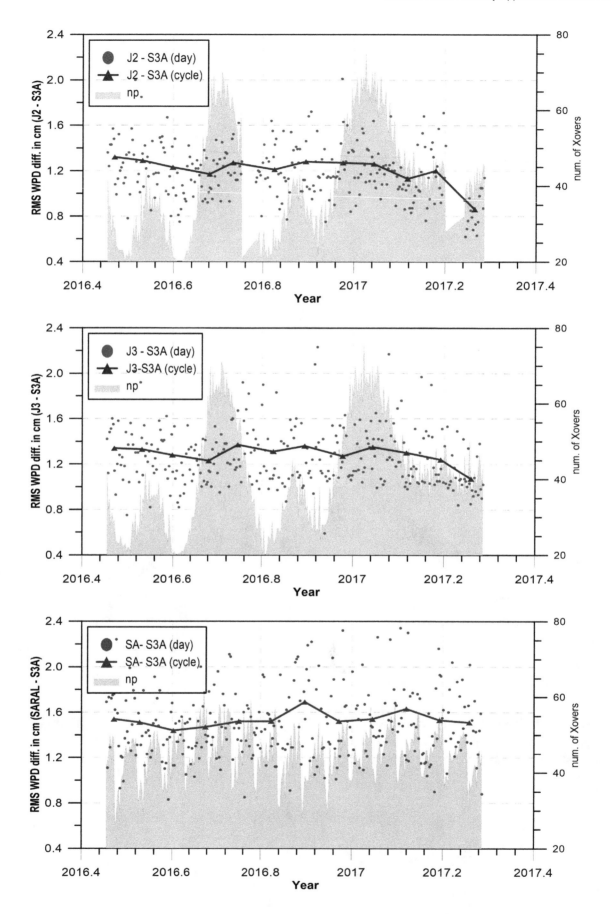

Figure 10. Time evolution of daily and 27-day RMS WPD differences between J2 and S3A (**Top**), between J3 and S3A (**Middle**), and between SARAL and S3A (**Bottom**), in cm.

3.2. Comparison with GNSS

Figure 11 shows the non-collocated comparison with the WTC derived at 60 GNSS stations, as described in Section 2.4. In the comparisons with S3A MWR, only valid MWR observations, except those related with the criterion for land contamination, have been selected. This means that the observations flagged as invalid due to all other error sources (ice and rain contamination and outliers) have been removed. In the comparisons with GPD+ all points have been used. For each epoch of the S3A measurement, a WTC from each GNSS station is linearly interpolated in time. Then, the WTC differences from all stations are binned in classes of distance from coast and the corresponding RMSs are computed.

As the GNSS stations are over land and the MWR measurement points are over the ocean, the distance from coast is also directly related with the distance between the points under comparison. Therefore, the RMS differences between the WTC from either the S3A MWR or GPD+ generally increase with the distance from coast, reaching a minimum at the maximum distance at which land contamination occurs in the MWR observations. Differences between the WTC computed at GNSS stations and derived from S3A MWR (or GPD+) are expected to increase as the distance to the GNSS station increases, as both measurements start to become decorrelated.

The RMSs of the differences between the WTC from GNSS and the WTC from S3A MWR (or the corresponding WPD) show that contamination is observed up to 20–25 km, in line with instrument specifications, the RMS of differences at these distances being about 1.8 cm. On the other hand, no land contamination is observed in the RMS of differences with respect to GPD+ (GPD1 mentioned above), being always smaller than those with respect to the MWR by about 0.3 cm, reaching 1.2 cm near the coast.

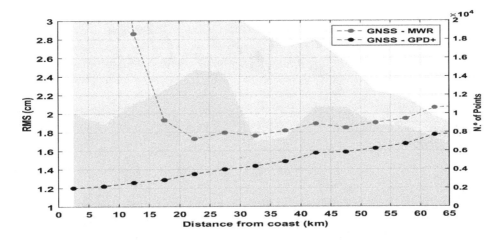

Figure 11. RMS differences between the WTC from the GNSS at coastal stations and the WTC from the S3A MWR, in cm. The grey and red coloured bars represent the number of points in each class of distance for GPD1 and the MWR, respectively.

3.3. Comparison with an Independent GPD+ WTC, ECMWF Operational Model and the Composite WTC

As mentioned in Section 2.5, two types of GPD+ WTC have been computed for S3A cycles 06 to 16: (1) GPD1—using only third-party data; and (2) GPD2—using all data including the S3A MWR. GPD1 WTC can be used as an external dataset, allowing an independent assessment of the S3A WTC. In the second case, GPD2 preserves the S3A MWR-derived WTC, whenever flagged as valid. Only new estimates are computed for the invalid points, using all available observations. The comparison between these two WTCs and between them and the S3A MWR-derived WTC (the latter comparison being performed only at valid MWR observations) gives further insight into the quality of the MWR observations, since it is performed using all S3A points and not only a subset (match points), as is the case of the comparison with the other MWR sensors.

Figure 12 shows an example where the different WTCs from S3A MWR, ECMWF operational model, GPD1 (top) and GPD2 (bottom) are compared for S3A pass 340, cycle 06. Figure 13 shows the corresponding plots for pass 462, same cycle. In these plots, only points with valid SLA are shown. Ice, land, and rain contamination can be observed in the MWR-derived WTC, more pronounced in pass 462 than in pass 340. Figure 14 illustrates the spatial distribution of the RMS of the WPD differences (cm) between GPD1 and S3A MWR for the whole period of study (cycles 05 to 16).

Figure 15 shows the time evolution of the RMS of the differences between the S3A MWR WTC and those from GPD1 and ECMWF, only for the valid MWR points. Note that for these points, GPD2 is equal to the MWR, so these differences are not shown.

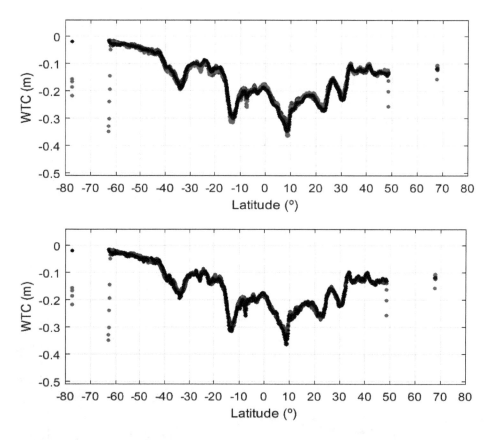

Figure 12. WTC (in metres) for S3A pass 340, cycle 06: ECMWF Operational (blue), MWR (red) and GPD (black, GPD1 in the top plot, GPD2 in the bottom plot) functions of latitude. The plot order is as mentioned in this caption. Thus, whenever the blue points cannot be seen, they are overlaid by the red and/or black points.

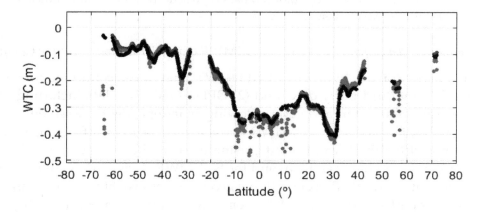

Figure 13. *Cont.*

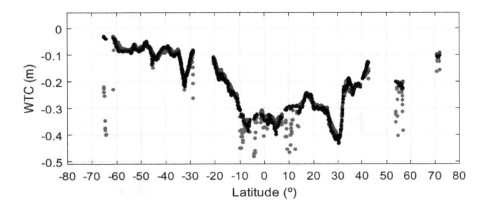

Figure 13. Same as in Figure 12 for S3A pass 462, cycle 06.

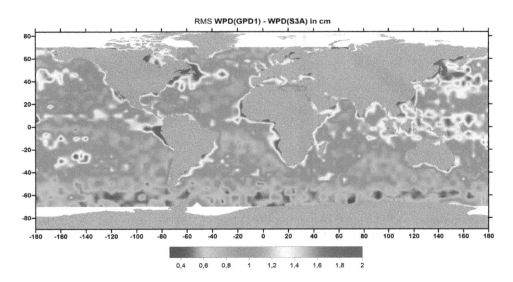

Figure 14. Spatial distribution of the RMS values of the WPD differences (cm) between GPD1 and the S3A MWR, for cycles 05 to 16.

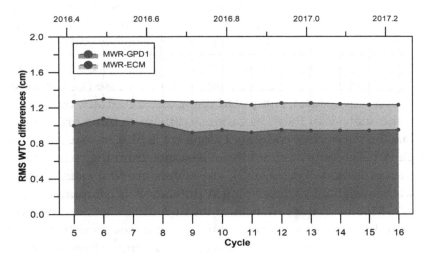

Figure 15. Time evolution of the RMS differences between the S3A MWR-based WTC and those from GPD1 and ECMWF.

The left panel of Figure 16 presents the scatterplots of the WPD from S3A versus the WPD from ECMWF (top) and GPD1 (bottom) in cm, while the right panel shows the corresponding scatterplots against the WPD differences between ECMWF and S3A (top) and between GPD1 and S3A (bottom), evidencing the good agreement between all WPD, better with respect to GPD1 than to the model, in line with the statistical parameters shown in Table 1.

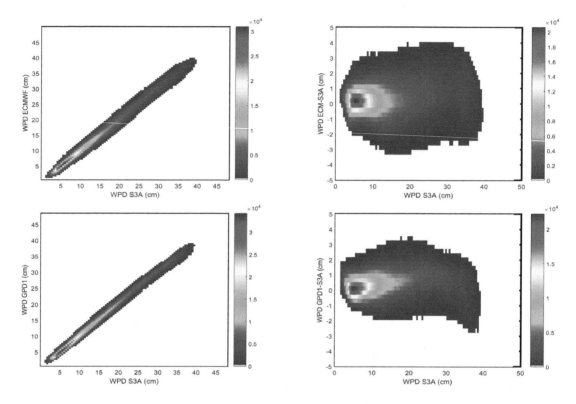

Figure 16. (**Left panels**): WPD from S3A versus WPD from ECMWF (**top**) and GPD1 (**bottom**), in cm; (**Right panels**): WPD from S3A versus WPD difference between ECMWF and S3A (**top**) and between GPD1 and S3A (**bottom**).

Aiming at inspecting the impact of the S3A WTC on the computation of SRAL-derived sea level anomalies in comparison with other WTC such as ECMWF, GPD1, and GPD2, various SLA variance analyses have been performed that can be divided into: (1) SLA along-track variance differences (weighted mean values per cycle and at collocated points, function of distance from coast and function of latitude); and (2) SLA analysis at crossovers (weighted mean cycle values and spatial pattern).

For this purpose, SLA datasets are first derived, for each S3A cycle, using the different WTC under comparison, a pair each time being compared. Then, for case (1), the difference between the weighted variance of each SLA dataset, computed using all along-track points, is estimated for each cycle. In addition, for the analysis of the SLA variance difference function of the distance from coast and function of latitude, the variance of co-located along-track SLA measurements for the whole study period and using each WTC is computed in bins of distance from the coast or latitude, respectively, and the differences are computed. For case (2), crossovers are first estimated using a pair of SLA datasets each time and the weighted variance of SLA differences is computed at cross-over points and estimated for each cycle. Moreover, the variances of the SLA differences at crossovers are computed in regular latitude × longitude grids ($4° × 4°$) and subtracted. Crossovers with a time difference less than 14 days, the minimum period to obtain S3A crossovers at all latitude bands, have been considered.

The results are shown in Figures 17–21, their discussion being described in Section 4.

The S3A products used in this study also include the composite WTC. Although [26] indicates that this correction is still not calibrated, a first assessment is also included here. Figure 22 depicts some representative examples of S3A passes where this WTC is compared against those from ECMWF, the MWR, and GPD2, showing that the present implementation of this correction evidences significant problems.

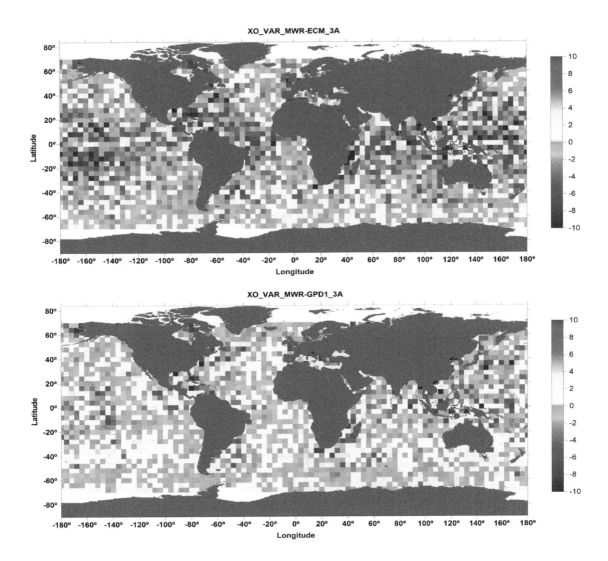

Figure 17. Spatial distribution of the weighted sea level anomaly (SLA) variance differences at crossovers, for SLA datasets computed using the MWR- and ECMWF-derived WTC (**Top**) and those using the WTC from the MWR and GPD1 (**Bottom**) for the period corresponding to S3A cycles 05 to 16. Only points with valid observations have been used.

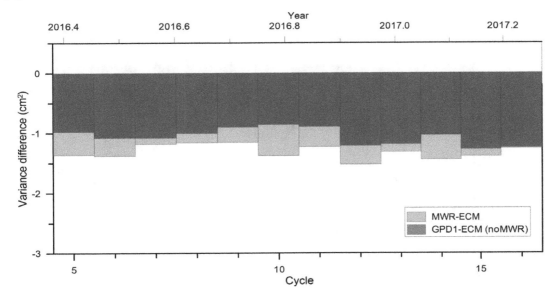

Figure 18. *Cont.*

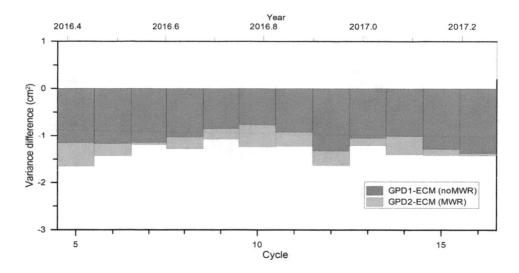

Figure 18. (**Top panel**): temporal evolution of weighted SLA variance differences at crossovers for SLA datasets computed using the WTC from the S3A MWR and ECMWF (orange) and the WTC from GPD1 and ECMWF (blue) using only points with a valid MWR; (**Bottom panel**): temporal evolution of weighted SLA variance differences at crossovers between SLA datasets computed using the WTC from GPD1 and that from ECMWF (blue) and between GPD2 and ECMWF (green), using all points with valid SLA.

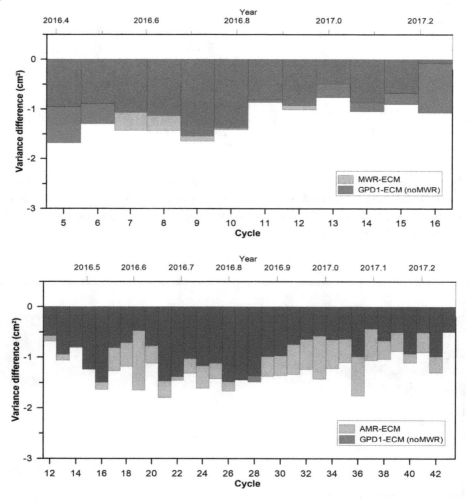

Figure 19. Temporal evolution of weighted along-track SLA variance differences for SLA datasets computed using the WTC from the on-board MWR and from ECMWF (orange) and those from GPD1 and ECMWF (blue) for S3A (**Top**) and J3 (**Bottom**). Only points with a valid MWR have been used.

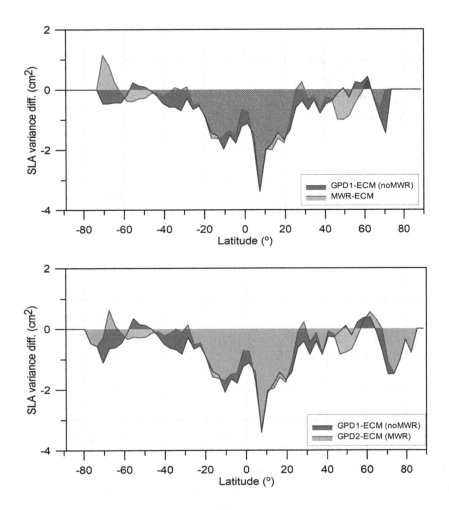

Figure 20. (**Top panel**): variance differences of SLA versus latitude, for SLA datasets computed using the WTC from the MWR and ECMWF (orange) and those from GPD1 and ECMWF (blue), over the period of S3A cycles 05 to 16 and using only points with a valid MWR; (**Bottom panel**): variance differences of SLA versus latitude between GPD1 and ECMWF (blue) and between GPD2 and ECMWF (green), over the period of S3A cycles 05 to 16 using all points with valid SLA.

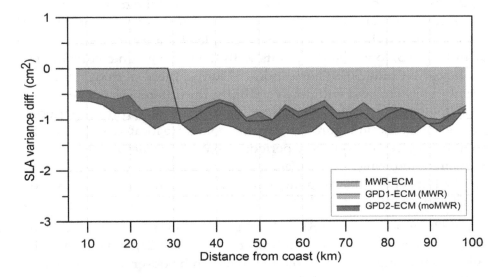

Figure 21. Variance differences of SLA versus distance from coast for SLA datasets computed using the WTC from MWR and ECMWF (orange), those from GPD1 and ECMWF (blue) and those from GPD2 and ECMWF (green) over the period of S3A cycles 05 to 16. In the first case only valid MWR points were selected while in the last two cases all points with valid SLA were considered

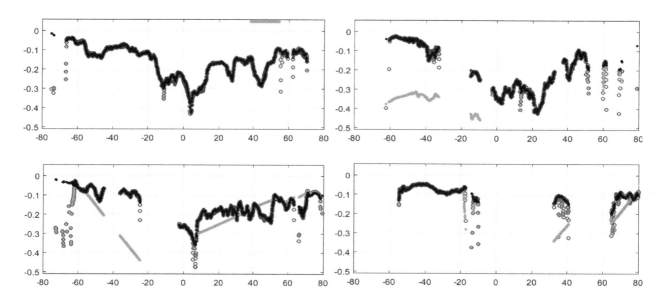

Figure 22. From top to bottom and left to right: WTC (in metres) for S3A cycle 06 passes 041, 646, 660, and 670, as a function of latitude (degrees). Shown are: ECMWF Operational (blue), MWR (red), GPD2 (black), and composite (green) WTC.

4. Discussion

This section presents a thorough discussion of the results presented in Section 3.

In terms of the global statistical parameters, GMI and Sentinel-3A agree very well, with mean and RMS of WPD differences (WPD(GMI)–WPD(S3A)) of 0.17 cm and 0.95 cm, respectively. The overall pattern of the WPD differences at match points shown in Figure 3 seems to indicate that at the low latitudes, where the WPD is larger, GMI measures wetter than S3A, while at the extreme southern latitudes, corresponding to regions of low WPD values, there is a tendency for GMI to measure drier than S3A. However, the scale factor (1.004) and small offset (0.12 cm) reveal a very good overall match between the two sensors, also demonstrated by the very small scatter of the scatter plot of the two WPDs illustrated in Figure 4.

The most curious feature of the comparison between GMI and S3A is the observed strong periodic pattern of 41 days. When analysing the way the spatial pattern of the match points changes with time (Figures 5 and 6), it can be observed that there are periods for which all match points between the two sensors are located only at high latitudes. An example of these periods is the one corresponding to the green bar in Figure 5, to which the points with the same green colour in Figure 7 (all above latitude 50°N or below latitude 50°S) correspond. During these periods, since the WPD at high latitudes is low (only a few cm) and with low variability, the differences between the two sensors are very small, with absolute values of less than 1 cm. On the contrary, during periods such as the one corresponding to the orange bar in Figure 5, the match points are located at the low latitudes, between ±50° (orange points in Figure 6). Over these periods, the WPD reaches higher values (up to 40–50 cm) and has larger variability, inducing larger differences between the two sensors. These results are a clear demonstration of the impact of data sampling in this type of study, which, if not properly accounted for, can lead to misinterpretations.

The authors of [26] report that the S3A MWR L1 brightness temperatures exhibit a difference of up to 1 K between ascending and descending tracks for the 23.8-GHz channel, with unknown origin. With a view to investigate if this potential bias can be observed in the current L2 products, the differences WPD (GMI)—WPD(S3A) have been represented in the bottom panel of Figure 5 separately for ascending (in blue) and descending (in red) tracks. No significant differences can be observed between the ascending and descending track differences. In terms of overall statistics, the S3A WPD differences with respect to GMI for the ascending tracks are about 1 mm larger than those for the descending tracks (compare dark blue with dark red lines in the bottom panel of Figure 5). This difference is too small, insignificant and cannot be related with the reported 1 K difference since the latter can lead to WPD differences ranging from 0.2 cm to 1 cm.

In spite of the small analysis period, the RMS differences (between GMI and S3A) computed for periods of the S3A cycles (27-day) evidence a long-term stable signal with an 82-day periodic pattern. This pattern occurs due to the fact that S3A orbit is Sun-synchronous, while GPM orbit is not. Due to its orbit characteristics, the GPM orbital plane takes 82 days to complete a full rotation with respect to the Sun and therefore with respect to the S3A orbital plane, explaining the observed periodic signal. In spite of these periodicities observed in the differences between GMI and the S3A MWR, due to the fact that GMI is on a non-Sun-synchronous orbit with a 65° inclination, the number of match points with S3A is larger than for example with any SSM/IS sensor, as these are on Sun-synchronous orbits, out of phase with respect to the S3A orbit by 3–6 h.

The comparison with the radiometers on board J2 and J3 indicate that on average these sensors measure drier than S3A by about 0.6–0.7 cm, while the RMS of the differences is within 1.2–1.3 cm. The scale factor (1.01), the negative offset of −0.7 and −0.8 cm and the blue predominant colour in the spatial pattern of the WPD differences shown in the top and medium panels of Figure 8 confirm this tendency.

For SARAL, the results are a bit different. The RMS of the differences is larger (1.5 cm) which can be explained by the fact that the crossover differences span a larger time interval (240 min versus 180 min used for J2 and J3). The scale factor of 0.96, offset of 0.6 cm, and the larger scatter observed in Figure 8 (bottom panel) and 9 (bottom panel) indicate a slightly larger difference between these two sensors, partly explained by the mentioned time difference between the corresponding crossovers.

The statistical parameters of the comparisons between S3A and the MWR aboard the other three altimeter missions are in agreement with the results presented in [24] and further work by these authors, comparing the same sensors with the SSM/I and SSM/IS sensors, which also indicate that J2 and J3 measure drier than the latter sensors by about 0.6 cm and 1 cm, respectively.

The time evolution of the daily and 27-day RMS differences between J2, J3, SARAL, and S3A illustrated in Figure 10 indicates a stable behaviour of the Sentinel-3A radiometer with respect to these sensors.

The comparisons with WTC derived at coastal GNSS stations show that land contamination can be observed in the S3A MWR observations up to 20–25 km from the coast. At this distance, the RMS of the differences between the WTC from S3A and the GNSS-derived ZWD is about 1.8 cm, increasing rapidly towards the coast, a clear indication that in the band of 0 km up to 25 km the MWR observations are not valid. On the contrary, no land contamination can be observed in the GPD1 WTC, derived solely from third party data. This stresses the need for robust criteria to remove the land-contaminated MWR observations and the important role of solutions such as GPD+ in providing valid WTC in the coastal regions.

Section 3.3 presents an extensive comparison between the S3A WTC and various corrections: GPD1 (from the GPD+ algorithm using only third party data), GPD2 (from the GPD+ algorithm, preserving the valid S3A MWR observations), the ECMWF operational model, and the composite WTC. In Figures 12 and 13, it can be observed that GPD1 and GPD2 are very similar, in spite of the fact that GPD1 does not use any observations from the S3A on-board MWR. It can also be concluded that, in regions with a valid MWR, the ECMWF operational model values are also very close, the main difference being that some, though important, small scale features are missing. These figures are clear illustration of typical S3A MWR-derived WTCs, where invalid observations can be observed near the coast, at high latitudes and at low latitudes, associated with heavy rain events.

The spatial distribution of the RMS of WPD differences between GPD1 and S3A MWR for the whole period of study (cycles 05 to 16), shown in Figure 14, indicates that the largest differences between these two WTCs are associated with regions with the largest WTC variability and ocean circulation patterns. Large differences also occur in some coastal regions, a possible indicator that some noisy MWR observations may still be present, in spite of the fact that all MWR measurements at distances from coast less than 30 km have been removed.

Both the statistical parameters in Table 1 and Figure 16 evidence that the S3A WPD is closer to GPD1 (RMS difference less than 1 cm) than to the ECMWF model (RMS difference 1.3 cm). The mean values of these differences are very small (0.1 cm and 0.2 cm for the comparison with GPD1 and ECMWF, respectively). The scale factors are both 1.01 and the offsets are also very small (0.2 cm, −0.1 cm). The time evolution of these differences, shown as 27-day RMS values in Figure 15, is stable.

The better agreement of S3A MWR-derived WTC with GPD1 in comparison with the ECMWF model is also observed in the scatterplots of Figure 16, where a smaller scatter can be observed for the former WTC pair. The better agreement of the S3A MWR with GPD1 than with the ECMWF model is explained by the fact that GPD1 WTC is also based on observations and the model has a poor temporal sampling of 6 h.

Figures 17–21 present the assessment of the S3A MWR-based WTC by means of SLA variance analysis. In this analysis, the WTC from S3A MWR, GPD1, and GPD2 are compared against the ECMWF model. In this way, both the performance of each WTC with respect to this model and their relative performance can be analysed.

The SLA variance analysis at crossovers presented in Figures 17 and 18 indicates that the S3A MWR reduces the variance both with respect to ECMWF and GPD1 (Figure 18, top panel), the first reduction being significant (mean cycle values of 1.1–1.5 cm^2), while the latter (not shown, inferred from Figure 18, top panel) is quite small (mean cycle values of 0.1–0.3 cm^2). The same is observed when comparing the performance of GPD1 and GPD2 against ECMWF (Figure 18, bottom panel). GPD1 reduces the mean cycle variance by 0.8–1.4 cm^2 while GPD2 reduces by 1.0–1.6 cm^2. Note that the comparisons involving the MWR only use the valid MWR points, while the comparisons involving only GPD1 and GPD2 use all points with valid SLA.

While the results from the variance analysis at crossovers indicate that the S3A MWR consistently reduces the SLA variance with respect to the WTC that only incorporates non-collocated radiometers (GPD1) and that the latter increases the variance with respect to the GPD+ WTC that makes use of the S3A MWR (GPD2), the same does not happen in along-track SLA variance analysis. In the along-track SLA variance differences, computed cycle by cycle (Figure 19), in most cycles GPD1 leads to a slightly larger SLA variance reduction with respect to ECMWF than the GPD WTC that incorporates the S3A MWR (GPD2). This is also demonstrated in Figures 20 and 21, which present the along-track variance differences in function of latitude and distance from coast, respectively. The top panel of Figure 20 shows that, in comparison with the MWR, GPD1 conducts to a slightly larger variance reduction with respect to ECMWF, except for some high-latitude bands. The bottom panel of the same figure shows that, in comparison with GPD2, with the exception of some high latitude bands, GPD1 also leads to a slightly larger SLA variance reduction with respect to ECMWF. In the first comparison only valid MWR points have been used, while in the latter all points with valid SLA have been analysed. This result is more evident in the SLA variance analysis function of distance from coast (Figure 21) where GPD1 consistently reduces the variance with respect to both MWR and GPD2.

This result is not expected since, in all previous analyses performed by the authors, the WTC derived from observations of the on-board MWR always reduced the SLA variance with respect to the WTC that only uses non-collocated observations, in all types of diagnoses and when only valid MWR points are selected. As an example, the bottom panel of Figure 19 illustrates for J3 the same variance analysis shown in the top panel for S3A, for the same period, proving that, contrary to S3A, the J3 AMR-2 WTC is consistently better than that of GPD1. Similar results have been obtained for Envisat, confirming that the GPD+ WTC that incorporates the valid observations from the Envisat radiometer consistently reduces the SLA variance with respect to the correction that only uses third party observations. These results indicate that although the overall performance of the S3A MWR seems good, improvements are still required to retrieve a WTC with the quality of that for J3.

In the latitude plots (Figure 20) some ice contamination can still be observed in the MWR-derived WTC and GPD2, suggesting that more robust editing is still desirable in the polar regions. We recall that a strong editing has already been performed by rejecting all MWR observations above latitude 70°N and below 70°S. Consequently, the preliminary version of the GPD+ WTC that makes use of the on-board MWR (GPD2) can still be improved, even with the present MWR dataset.

The examination of the composite correction present on the analysed S3A products shows that the implementation of this WTC still has serious problems, depicted in Figure 22. Moreover, the correction is not present in a large percentage of points, e.g., 21% of the points with valid SLA for cycle 06. Similar to the GPD+ WTC, the composite WTC aims at generating a continuous correction, valid everywhere, including the coastal zones and high latitudes. Also, similar to the GPD2 type of correction described in this paper, it also preserves the valid MWR values. The main difference between GPD+ and the composite WTC is that, on every invalid point, the former retrieves a new estimate by data combination of all available observations, while the latter uses the model values, adjusted to the closest valid MWR points. This requires robust criteria to detect valid/invalid MWR values, otherwise the model will be adjusted to spurious MWR observations, resulting in large biases as shown e.g., in the bottom left panel of Figure 22 or straight lines as shown in the right panels of the same figure. Similar behaviour has been observed, though in only a few occurrences, in the composite correction present in the Archiving, Validation and Interpretation of Satellite Oceanographic data (AVISO) Corrected Sea Surface Heights (CORSSH) products of TOPEX/Poseidon [23]. Due to potential implementation problems and the fact that GPD+ use observations while the composite is solely based on model values, it has been shown that the GPD type of WTC is a significant improvement with respect to the composite WTC, particularly in the coastal zone [9,23].

5. Conclusions

This study presents an independent assessment of the wet tropospheric correction derived from the two-band microwave radiometer deployed on Sentinel-3A, present on the L2 NTC products from the "Spring 2017" Reprocessing Campaign", Processing Baseline 2.15.

Studies such as this one play an important role in the improvement of altimeter-derived products, as satellite-derived data improvement is a long and continuous process. Sentinel-3 is a challenging mission in many aspects, with the research community having a long way to go to fully exploit the capabilities of its instruments, in particular SRAL. Six years after its end of mission, Envisat data are still being reprocessed and the reprocessing of TOPEX/Poseidon data has been delayed for several years.

Considering the relatively short time span of the dataset (10 months), the overall performance of the S3A MWR seems stable, with no drifts or irregular behaviour having been observed. A pronounced periodic signal has been observed in the differences with respect to GMI, due to the orbit configurations of the spacecraft housing these sensors.

When considering the linear adjustment of S3A MWR-derived WPD to those from four different sensors, scale factors very close to 1 have been obtained: GMI (1.00); J2 and J3 (1.11); and SARAL (0.96). The corresponding values for the WTC are the same since WTC is symmetric to WPD. All WPD offsets have absolute values of less than 1 cm: GMI (0.1 cm), J2 (-0.7 cm), J3 (-0.8 cm), and SARAL (0.6 cm). The corresponding offsets for the WTC are symmetric to those for WPD. The mean WPD differences have absolute values below 1 cm: GMI (0.2 cm), J2 (-0.6 cm), J3 (-0.7 cm) and SARAL (0.0 cm), the mean WTC differences being the corresponding symmetric values. RMS WPD (or WTC) differences (cm) of 0.95 (GMI), 1.3 (J2), 1.3 (J3), and 1.5 (SARAL) demonstrate a good agreement with all these sensors, in particular with GMI.

The comparison with GNSS shows land contamination in the S3A MWR observations up to 20–25 km and RMS differences between S3A MWR WTC and GNSS-derived ZWD of 1.8 cm at these distances. These results are in agreement with those found for the radiometers of other altimetric missions, for which the corresponding RMSs of differences are 1.7–1.8 cm [43].

Extensive comparison has been performed with the ECMWF model-derived WTC and two WTC versions from the GPD+ algorithm: one using only third party data (GPD1) and another preserving the valid S3A MWR points (GPD2). Direct WTC comparisons show small statistical parameters for the differences between all these corrections. As expected, smaller RMS values have been observed for the differences with respect to the GPD+ WTC (1 cm for GPD1) than for the differences with respect to ECMWF (1.3 cm). The scale factors with respect to both GPD1 and ECMWF are 1.01 and both offsets and mean differences have absolute values smaller than or equal to 0.2 cm.

As expected, all SLA variance analyses evidence a reduction of 1–2 cm^2 of the SLA computed with the S3A MWR WTC with respect to the ECMWF model, when mean cycle values are considered. The SLA variance analysis at crossover points indicates a small (mean cycle values less than 0.5 cm^2) but consistent variance reduction with respect to GPD1, however, the opposite is observed in the along-track variance analysis. This seems to suggest that improvements in both the retrieval algorithm and the criteria used to detect valid/invalid MWR observations are required to achieve a WTC with accuracy similar to the AMR-2 WTC for Jason-3. These results also highlight that the along-track comparison with the GPD1 type of WTC provides complementary and very important insight into the quality of the S3A MWR-derived WTC, not evident in the assessment against the other.

Similar to other dual-frequency radiometers on board the previous ESA missions, strong ice and land contamination is observed, the former in particular making the establishment of validation criteria for the MWR observations difficult at the high latitudes. Consequently, all observations with latitude absolute values larger than 70° have been rejected.

The composite WTC present in the analysed products was revealed to have implementation problems and was unsuitable for use, improvements being required in future versions of S3A data.

As a whole, this study contributes to a better knowledge of the wet path delay affecting satellite altimeter observations. In spite of the good overall performance of the S3A MWR when compared against other state-of-the-art radiometers, current limitations are also identified. Once a tuned retrieval algorithm is achieved, meeting the state-of-the art performances of current and past similar instruments, the inability of this type of radiometer to measure over non-ocean surfaces such as e.g., land, ice, or wetlands, prevent its use in these regions. Therefore, in regions where accurate satellite altimetry measurements are of crucial importance such as coastal or inland water zones, and accurate MWR-derived WTC retrieval is not possible, alternative methods such as the GPD+ type of corrections must be used.

The GPD+ WTC solely based on third-party observations, together with microwave radiometers on board other altimeter missions and GMI, were revealed to be very useful and independent tools to validate the Sentinel-3 radiometer-based wet tropospheric correction. Future work includes the monitoring of upcoming versions of this dataset and generation of updated versions of the GPD+ WTC making use of ameliorated MWR data, which will contribute to the generation of an accurate version of this important range correction for S3A SRAL measurements over the whole ocean, including coastal zones and at high latitudes.

In regions such as coastal and continental water zones, radiometers with additional high-frequency channels, such as the one being built for Jason-CS/Sentinel-6 [51] will provide smaller footprints, of great relevance for resolution of smaller scales of variability of the WTC.

Over inland water regions and to some extent in coastal regions, the availability of tropospheric corrections at high rate (frequency higher than 1 Hz), either from radiometers with smaller footprints or from high-resolution atmospheric models is also of great relevance, in particular in the exploitation of Sentinel-3 high-rate altimeter data. The advantage of having range corrections at a high rate applies not only to the WTC but also to the dry component of the tropospheric path delay, the dry tropospheric correction (DTC). In regions of sharp topography variations such as some steep coastal regions, lakes or rivers, one of the factors with most impact on the retrieval of accurate tropospheric corrections is their height dependence, particularly large for the DTC (1 cm per 40 m of height variation) [13].

The future of tropospheric corrections for satellite altimetry includes: WTC from MWRs with additional high-frequency channels and improved retrieval algorithms (over the open and coastal ocean and over lakes); and WTC and DTC from high-resolution atmospheric models, both computed at surface height using accurate Digital Elevation Models (DEM) or the altimeter-measured surface height (over inland water regions). In all cases, high-rate (e.g., 20 Hz) corrections are desirable to better account for spatial variability of km order or less, in particular the variability associated with the surface height variations.

Acknowledgments: This work was supported by the European Space Agency in the scope of the SAR Altimetry Coastal and Open Ocean–Performance Exploitation and Roadmap Study (SCOOP) project, Subcontract to SCOOP Contract N. 4000115385/15/I-BG. It is also a contribution to the Validation of Coastal ALtimetry from Sentinel-3 (VOCALS3) project. The authors would like to thank RADS for providing the S3A data, Remote Sensing Systems for providing the GMI products, and the European Centre for Medium-Range Weather Forecasts (ECMWF) for making the ECMWF operational model available.

Author Contributions: M.J.F. conceived, designed and performed the experiments and wrote the paper; C.L. contributed to data analysis editing and review of the paper.

References

1. Benveniste, J. Radar altimetry: Past, present and future. In *Coastal Altimetry*; Vignudelli, S., Kostianoy, A., Cipollini, P., Benvensite, J., Eds.; Springer: Berlin/Heidelberg, Germany, 2011.

2. Stammer, D.; Cazenave, A. *Satellite Altimetry over Oceans and Land Surfaces*, 1st ed.; CRC Press: Boca Raton, FL, USA, 2017.

3. Le Traon, P.Y. From satellite altimetry to argo and operational oceanography: Three revolutions in oceanography. *Ocean Sci.* **2013**, *9*, 901–915. [CrossRef]

4. Le Traon, P.Y.; Dibarboure, G.; Jacobs, G.; Martin, M.; Remy, E.; Schiller, A. Use of satellite altimetry for operational oceanography. In *Satellite Altimetry over Oceans and Land Surfaces*; Stammer, D., Cazenave, A., Eds.; CRC Press: Boca Raton, FL, USA, 2017.

5. Ablain, M.; Cazenave, A.; Larnicol, G.; Balmaseda, M.; Cipollini, P.; Faugere, Y.; Fernandes, M.J.; Henry, O.; Johannessen, J.A.; Knudsen, P.; et al. Improved sea level record over the satellite altimetry ERA (1993–2010) from the climate change initiative project. *Ocean Sci.* **2015**, *11*, 67–82. [CrossRef]

6. Ablain, M.; Legeais, J.F.; Prandi, P.; Marcos, M.; Fenoglio-Marc, L.; Dieng, H.B.; Benveniste, J.; Cazenave, A. Satellite altimetry-based sea level at global and regional scales. *Surv. Geophys.* **2017**, *38*, 7–31. [CrossRef]

7. Quartly, G.D.; Legeais, J.F.; Ablain, M.; Zawadzki, L.; Fernandes, M.J.; Rudenko, S.; Carrère, L.; García, P.N.; Cipollini, P.; Andersen, O.B.; et al. A new phase in the production of quality-controlled sea level data. *Earth Syst. Sci. Data* **2017**, *9*, 557–572. [CrossRef]

8. Nerem, R.S.; Ablain, M.; Cazenave, A.; Church, J.; Leuliette, E. A 25-year satellite altimetry-based global mean sea level record: Closure of the sea level budget and missing components. In *Satellite Altimetry over Oceans and Land Surfaces*; Stammer, D., Cazenave, A., Eds.; CRC Press: Boca Raton, FL, USA, 2017.

9. Legeais, J.F.; Ablain, M.; Zawadzki, L.; Zuo, H.; Johannessen, J.A.; Scharffenberg, M.G.; Fenoglio-Marc, L.; Fernandes, M.J.; Andersen, O.B.; Rudenko, S.; et al. An accurate and homogeneous altimeter sea level record from the esa climate change initiative. *Earth Syst. Sci. Data* **2017**, *2017*, 1–35. [CrossRef]

10. Birkett, C.; Reynolds, C.; Beckley, B.; Doorn, B. From research to operations: The usda global reservoir and lake monitor. In *Coastal Altimetry*; Vignudelli, S., Kostianoy, A.G., Cipollini, P.J.B., Eds.; Springer-Verlag: Berlin/Heidelberg, Germany, 2011.

11. Cretaux, J.-F.; Nielsen, K.; Frappart, F.; Papa, F.; Calmant, S.; Benveniste, J. Hydrological applications of satellite altimetry: Rivers, lakes, man-made reservoirs, inundated areas. In *Satellite Altimetry in Coastal Regions*; CRC Press: Boca Raton, FL, USA, 2017.

12. Remy, F.; Memin, A.; Velicogna, I. Applications of satellite altimetry to study the antarctic ice sheet, satellite altimetry in coastal regions. In *Satellite Altimetry over Oceans and Land Surfaces*, 1st ed.; Stammer, D., Cazenave, A., Eds.; CRC Press: Boca Raton, FL, USA, 2017; p. 670.

13. Fernandes, M.J.; Lazaro, C.; Nunes, A.L.; Scharroo, R. Atmospheric corrections for altimetry studies over inland water. *Remote Sens.* **2014**, *6*, 4952–4997. [CrossRef]

14. Chelton, D.B.; Ries, J.C.; Haines, B.J.; Fu, L.L.; Callahan, P.S. Satellite altimetry. In *Satellite Altimetry and Earth Sciences: A Handbook of Techniques and Applications*; Fu, L.L., Cazenave, A., Eds.; Academic Press: San Diego, CA, USA, 2001.

15.   Vieira, T.; Fernandes, M.J.; Lázaro, C. Analysis and retrieval of tropospheric corrections for cryosat-2 over inland waters. *Adv. Sp. Res.* **2017**, *46*. [CrossRef]

16.   Cipollini, P.; Benveniste, J.; Birol, F.; Fernandes, M.J.; Obligis, E.; Passaro, M.; Strub, P.T.; Valladeau, G.; Vignudelli, S.; Wilkin, J. Satellite altimetry in coastal regions. In *Satellite Altimetry over Oceans and Land Surfaces*, 1st ed.; Stammer, D., Cazenave, A., Eds.; CRC Press: Boca Raton, FL, USA, 2017.

17.   Keihm, S.J.; Janssen, M.A.; Ruf, C.S. Topex/poseidon microwave radiometer (TMR). III. Wet troposphere range correction algorithm and pre-launch error budget. *IEEE Trans. Geosci. Remote Sens.* **1995**, *33*, 147–161. [CrossRef]

18.   Eymard, L.; Obligis, E. The Altimetric Wet Troposheric Correction: Progress Since the ERS-1 Mission. In Proceedings of the15 Years of Progress in Radar Altimetry, Venice, Italy, 13–18 March 2006.

19.   Tournadre, J.; Lambin-Artru, J.; Steunou, N. Cloud and rain effects on altika/saral ka-band radar altimeter-part i: Modeling and mean annual data availability. *IEEE Trans. Geosci. Remote Sens.* **2009**, *47*, 1806–1817. [CrossRef]

20.   Sentinel-3 Team. *Sentinel-3 User Handbook*, 2nd ed.; GMES-S3OP-EOPG-TN-13-0001; European Space Agency: Paris, France, 2013.

21.   Dinardo, S.; Fenoglio-Marc, L.; Buchhaupt, C.; Becker, M.; Scharroo, R.; Joana Fernandes, M.; Benveniste, J. Coastal sar and plrm altimetry in german bight and west baltic sea. *Adv. Sp. Res.* **2017**. [CrossRef]

22.   Tournadre, J. Improved level-3 oceanic rainfall retrieval from dual-frequency spaceborne radar altimeter systems. *J. Atmos. Ocean. Technol.* **2006**, *23*, 1131–1149. [CrossRef]

23.   Fernandes, M.J.; Lazaro, C.; Ablain, M.; Pires, N. Improved wet path delays for all esa and reference altimetric missions. *Remote Sens. Environ.* **2015**, *169*, 50–74. [CrossRef]

24.   Fernandes, M.J.; Lazaro, C. Gpd+ wet tropospheric corrections for cryosat-2 and gfo altimetry missions. *Remote Sens.* **2016**, *8*, 851. [CrossRef]

25.   EUMETSAT, S3a STM Reprocessing—"Spring 2017" (Level 0 to Level 2). Available online: eum/ops-sen3/rep/17/940906 (accessed on 31 August 2017).

26.   EUMETSAT, Sentinel-3a Product Notice—STM l2 Marine ("Spring Reprocessing Campaign"). Available online: um/ops-sen3/doc/17/944329 (accessed on 15 September 2017).

27.   Scharroo, R.; Leuliette, E.; Naeije, M.; Martin-Puig, C.; Pires, N. Rads Version 4: An efficient way to analyse the Multi-Mission altimeter database. In Proceedings of the ESA Living Planet Symposium, Prague, Czech Republic, 9–13 May 2016; ESA: Prague, Czech Republic, 2016.

28.   Obligis, E.; Eymard, L.; Tran, N.; Labroue, S.; Femenias, P. First three years of the microwave radiometer aboard envisat: In-flight calibration, processing, and validation of the geophysical products. *J. Atmos. Ocean. Technol.* **2006**, *23*, 802–814. [CrossRef]

29.   Thao, S.; Eymard, L.; Obligis, E.; Picard, B. Comparison of regression algorithms for the retrieval of the wet tropospheric path. *IEEE J. Sel. Top. Appl. Earth Obs. Remote Sens.* **2015**, *8*, 4302–4314. [CrossRef]

30.   Collecte Localisation Satellites (CLS). *Surface Topography Mission (STM) Sral/Mwr L2 Algorithms Definition, Accuracy and Specification*; S3PAD-RS-CLS-SD03-00017; CLS: Ramonville St-Agne, France, 2011.

31.   Mercier, F.; Rosmorduc, V.; Carrère, L.; Thibaut, P. Coastal and Hydrology Altimetry Product (PISTACH) Handbook. 2010. Available online: https://www.aviso.altimetry.fr/fileadmin/documents/data/tools/hdbk_Pistach.pdf (accessed on 19 March 2018).

32.   Meissner, T.; Wentz, F.J.; Draper, D. *Gmi Calibration Algorithm and Analysis Theoretical Basis Document*; Report Number 041912; Remote Sensing Systems: Santa Rosa, CA, USA, 2012; 124p.

33.   Draper, D.W.; Newell, D.A.; Wentz, F.J.; Krimchansky, S.; Skofronick-Jackson, G.M. The global precipitation measurement (GPM) microwave imager (GMI): Instrument overview and early on-orbit performance. *IEEE J. Sel. Top. Appl. Earth Obs. Remote Sens.* **2015**, *8*, 3452–3462. [CrossRef]

34.   Fernandes, M.J.; Nunes, A.L.; Lazaro, C. Analysis and inter-calibration of wet path delay datasets to compute the wet tropospheric correction for cryosat-2 over ocean. *Remote Sens.* **2013**, *5*, 4977–5005. [CrossRef]

35.   Stum, J.; Sicard, P.; Carrere, L.; Lambin, J. Using objective analysis of scanning radiometer measurements to compute the water vapor path delay for altimetry. *IEEE Trans. Geosci. Remote Sens.* **2011**, *49*, 3211–3224. [CrossRef]

36.   Wentz, F.J. A well-calibrated ocean algorithm for special sensor microwave/imager. *J. Geophys. Res. Oceans* **1997**, *102*, 8703–8718. [CrossRef]

37.   Wentz, F.J. *SSM/I Version-7 Calibration Report*; RSS Technical Report 011012; Remote Sensing Systems: Santa Rosa, CA, USA, 2013; p. 46.

38. Brown, S. A novel near-land radiometer wet path-delay retrieval algorithm: Application to the *Jason-2/OSTM* advanced microwave radiometer. *IEEE Trans. Geosci. Remote Sens.* **2010**, *48*, 1986–1992. [CrossRef]

39. Dumont, J.P.; Rosmorduc, V.; Picot, N.; Bronner, E.; Desai, S.; Bonekamp, H.; Figa, J.; Lillibridge, J.; Scharroo, R. OSTM/Jason-2 Products Handbook. 2011. Available online: http://www.ospo.noaa.gov/Products/documents/J2_handbook_v1-8_no_rev.pdf (accessed on 19 March 2018).

40. Dumont, J.P.; Rosmorduc, V.; Carrere, L.; Picot, N.; Bronner, E.; Couhert, A.; Guillot, A.; Desai, S.; Bonekamp, H.; Figa, J.; et al. Jason-3 Products Handbook. 2016. Available online: https://www.aviso.altimetry.fr/fileadmin/documents/data/tools/hdbk_j3.pdf (accessed on 19 March 2018).

41. Brown, S.; Islam, T. Jason-3 GDR calibration stability enabled by the cold sky maneuvers. In Proceedings of the Ocean Surface Topography Science Team Meeting, Miami, FL, USA, 23–27 October 2017.

42. Bronner, E.; Guillot, A.; Picot, N. SARAL/Altika Products Handbook. SALP-MU-M-OP-15984-CN. 2013. Available online: https://www.aviso.altimetry.fr/fileadmin/documents/data/tools/SARAL_Altika_products_handbook.pdf (accessed on 19 March 2018).

43. Vieira, T.; Fernandes, M.J.; Lázaro, C. Independent assessment of on-board microwave radiometer measurements in coastal zones using tropospheric delays from gnss. *IEEE Trans. Geosci. Remote Sens.* **2018**, under review.

44. Davis, J.L.; Herring, T.A.; Shapiro, I.I.; Rogers, A.E.E.; Elgered, G. Geodesy by radio interferometry—Effects of atmospheric modeling errors on estimates of baseline length. *Radio Sci.* **1985**, *20*, 1593–1607. [CrossRef]

45. Kouba, J. Implementation and testing of the gridded vienna mapping function 1 (VMF1). *J. Geod.* **2008**, *82*, 193–205. [CrossRef]

46. Fernandes, M.J.; Pires, N.; Lazaro, C.; Nunes, A.L. Tropospheric delays from gnss for application in coastal altimetry. *Adv. Space Res.* **2013**, *51*, 1352–1368. [CrossRef]

47. Dee, D.P.; Uppala, S.M.; Simmons, A.J.; Berrisford, P.; Poli, P.; Kobayashi, S.; Andrae, U.; Balmaseda, M.A.; Balsamo, G.; Bauer, P.; et al. The era-interim reanalysis: Configuration and performance of the data assimilation system. *Q. J. R. Meteorol. Soc.* **2011**, *137*, 553–597. [CrossRef]

48. Fernandes, M.J.; Lazaro, C.; Nunes, A.L.; Pires, N.; Bastos, L.; Mendes, V.B. Gnss-derived path delay: An approach to compute the wet tropospheric correction for coastal altimetry. *IEEE Geosci. Remote Sens. Lett.* **2010**, *7*, 596–600. [CrossRef]

49. Miller, M.; Buizza, R.; Haseler, J.; Hortal, M.; Janssen, P.; Untch, A. Increased resolution in the ecmwf deterministic and ensemble prediction systems. In *ECMWF Newsletter*; ECMWF: Reading, UK, 2010; Volume 124, pp. 10–16.

50. Thao, S.; Eymard, L.; Obligis, E.; Picard, B. Trend and variability of the atmospheric water vapor: A mean sea level issue. *J. Atmos. Ocean. Technol.* **2014**, *31*, 1881–1901. [CrossRef]

51. eoPortal:Satellite Missions Directory, Copernicus: Sentinel-6/Jason-CS (Jason Continuity of Service) Mission. Available online: https://directory.eoportal.org/web/eoportal/satellite-missions/content/-/article/jason-cs (accessed on 20 February 2018).

10

SWOT Spatial Scales in the Western Mediterranean Sea Derived from Pseudo-Observations and an Ad Hoc Filtering

Laura Gómez-Navarro [1,2,*], **Ronan Fablet** [3], **Evan Mason** [1], **Ananda Pascual** [1], **Baptiste Mourre** [4], **Emmanuel Cosme** [2] **and Julien Le Sommer** [2]

[1] Institut Mediterrani d'Estudis Avançats (IMEDEA) (CSIC-UIB), 07190 Esporles, Illes Balears, Spain; evanmason@gmail.com (E.M.); ananda.pascual@imedea.uib-csic.es (A.P.)

[2] Univ. Grenoble Alpes, CNRS, IRD, Grenoble INP, IGE, 38000 Grenoble, France; Emmanuel.Cosme@univ-grenoble-alpes.fr (E.C.); julien.lesommer@univ-grenoble-alpes.fr (J.L.S.)

[3] Institut Mines-Télécom, Telecom-Bretagne, UMR 6285 labSTICC, 29238 Brest, France; ronan.fablet@imt-atlantique.fr

[4] Balearic Islands Coastal Observing and Forecasting System (SOCIB), 07121 Palma de Mallorca, Illes Balears, Spain; bmourre@socib.es

* Correspondence: lauragomnav@gmail.com

Abstract: The aim of this study is to assess the capacity of the Surface Water Ocean Topography (SWOT) satellite to resolve fine scale oceanic surface features in the western Mediterranean. Using as input the Sea Surface Height (SSH) fields from a high-resolution Ocean General Circulation Model (OGCM), the SWOT Simulator for Ocean Science generates SWOT-like outputs along a swath and the nadir following the orbit ground tracks. Given the characteristic temporal and spatial scales of fine scale features in the region, we examine temporal and spatial resolution of the SWOT outputs by comparing them with the original model data which are interpolated onto the SWOT grid. To further assess the satellite's performance, we derive the absolute geostrophic velocity and relative vorticity. We find that instrument noise and geophysical error mask the whole signal of the pseudo-SWOT derived dynamical variables. We therefore address the impact of removal of satellite noise from the pseudo-SWOT data using a Laplacian diffusion filter, and then focus on the spatial scales that are resolved within a swath after this filtering. To investigate sensitivity to different filtering parameters, we calculate spatial spectra and root mean square errors. Our numerical experiments show that noise patterns dominate the spectral content of the pseudo-SWOT fields at wavelengths below 60 km. Application of the Laplacian diffusion filter allows recovery of the spectral signature within a swath down to the 40–60 km wavelength range. Consequently, with the help of this filter, we are able to improve the observation of fine scale oceanic features in pseudo-SWOT data, and in the estimation of associated derived variables such as velocity and vorticity.

Keywords: satellite altimetry; SWOT; western Mediterranean Sea; fine scale; SWOT simulator; ROMS model; filtering

1. Introduction

The Surface Water and Ocean Topography (SWOT) satellite mission is a joint mission by the National Aeronautics and Space Administration (NASA) and the *Centre National d'Études Spatiales* (CNES), with contributions from the UK and Canadian Space Agencies [1]. Presently, the satellite's launch is planned for 2021 [2]. It will provide water elevation maps for oceanographic and hydrological purposes [3,4]. The novelty of this satellite is that it carries a wide-swath altimeter with unprecedented horizontal resolution and global coverage. On the other hand, the associated irregular temporal sampling will constitute a challenge for the exploitation of the data. SWOT will have a 21-day repeat cycle and the revisit time will vary from approximately 10 days at the equator to two days at the poles [5,6]. This implies temporal variability in spatial coverage as the number of observations per repeat cycle will increase with latitude. Moreover, there will also be a temporal variability within a cycle. During each cycle, there are periods of time with a higher temporal sampling. This is due to a longer revisit time so that SWOT also fulfills its hydrological objectives by providing coverage of the bulk of the global land surface [7]. In satellite measurements, there is always a compromise between spatial and temporal resolution. As SWOT aims for global coverage, i.e., high spatial resolution, we lose in temporal resolution (SWOT's repeat cycle will be longer than, for example, the 10-day repeat cycle of the Jason altimeter satellites [8]).

One of the primary oceanographic objectives of the SWOT mission is to characterize the ocean meso- and submesoscale circulation [9] determined from ocean surface topography at spatial resolutions of 15 km (spatial resolution is defined to be perturbation wavelength in the oceanographic context). The resolution capacity of current along-track one-dimensional altimeter data, depending on the altimeter, has been found to be between 40 and 50 km at western boundary currents and between 70 and 110 km at the eastern basins [10]. Two-dimensional gridded products based on the altimetric constellation allow for mapping wavelengths down to 200 km [11]. The SWOT mission is expected to allow to capture wavelengths down to 15 km on its two-dimensional swaths [12], therefore increasing substantially the resolution capacity of present-day altimeter data. The possibility of characterizing the submesoscale is a major breakthrough. While the mesoscale has historically received a lot of attention [13], the submesoscale has previously been out of reach. Theoretical calculations and advanced modeling suggest that submesoscale processes are key to understanding ocean fluxes [14–16]. A pertinent example is the occurrence of mid-ocean plankton blooms [17].

In the Mediterranean Sea, intense mesoscale and submesoscale variability interact across sub-basin and basin scales [18–20]. This variability has an indirect impact on the Atlantic Ocean circulation due to exchange through the Strait of Gibraltar and, subsequently, influence on the great ocean conveyor belt [21–23]. Three scales of motion are therefore overlaid, making an amalgam of intricate processes that require high resolution and can help assess the potential impact that SWOT will have on the study of processes occurring at different scales.

Understanding small scale variability in the Mediterranean Sea is important as it is a region with intrinsically smaller spatial scales than those found in other parts of the world ocean at similar latitudes. Ref. [24] showed that the grid resolution necessary to resolve the first baroclinic deformation radius in the Mediterranean is around 1/16°, whilst in the Atlantic Ocean at the same latitude it is only 1/6°. This implies that smaller structures need to be resolved in the Mediterranean Sea compared to the mid-latitudes of the Atlantic Ocean. This is further demonstrated by [25] who show that lower values of the first baroclinic Rossby radius of deformation are present in the western Mediterranean Sea. These values are approximately between 2 and 16 km, in comparison with a 20–30 km range found at mid-latitudes of the Atlantic Ocean [26]. The Mediterranean values are actually closer to the values found in the Arctic Ocean [27].

The western Mediterranean Sea is one of the areas of the global ocean that will be sampled during the SWOT fast-sampling phase [28]. This phase covers the first 60–90 days after launch, during which the satellite will provide daily high resolution Sea Surface Height (SSH) measurements over a limited repeated orbit for purposes of calibration/validation of the SWOT sensor/instrument.

The goal of this study is to assess the capacity of SWOT to resolve the fine scales in the western Mediterranean. Our first objective is to generate pseudo-SWOT data from numerical model outputs in this region in order to understand its temporal and spatial sampling pattern in this area. We then apply a noise-reduction processing technique to pseudo-SWOT data to find out the spatial scales that SWOT may ultimately be able to resolve. Given SWOT's irregular time sampling and consequent variable spatial coverage, in this paper, we focus on the spatial scales resolved within a swath.

2. Data and Methods

2.1. The SWOT Simulator

With a view to characterizing the potential of SWOT-derived SSH data, we consider a simulation-based framework using the SWOT Simulator for Ocean Science (version 1). This simulator accounts for both SWOT space-time sampling patterns and noise processes. Using as input the SSH fields from an Oceanic General Circulation Model (OGCM), the SWOT simulator generates SWOT-like outputs along a ground swath and the nadir following the orbit ground tracks [29]. Hereinafter, we refer to these outputs as SSH outputs. Note that these simulated fields correspond to Absolute Dynamic Topography (ADT) values in altimetric terminology.

A flowchart of the simulator workflow is provided in Figure 1. Two features should be pointed out. Grid files, generated in the first step of the flowchart, account for the planned orbit of the satellite and the specified domain. Instrument noise and geophysical errors are added during the last step of the flowchart, following recent technical characteristics established by the SWOT project team [29]. Instrument noise is composed of Ka-band Radar Interferometer (KaRIN) noise, roll, phase, baseline dilation and timing errors (see [30]). In this version of the simulator, the only geophysical error is associated with the wet troposphere. Therefore, it is important to keep in mind that additional noise patterns, such as sea state bias [29] or the effects of internal waves [2] are not accounted for in the generated pseudo-SWOT data.

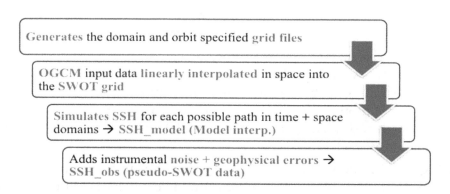

Figure 1. Flowchart of the SWOT simulator procedure.

For more details on the instrument noise and geophysical errors added by the SWOT simulator, see Appendix B.

2.2. Input Data: The Western Mediterranean OPerational (WMOP) Model

A high resolution OGCM of the western Mediterranean region provides input data for the SWOT simulator. We used the WMOP model [31] developed at SOCIB (Balearic Islands Coastal Observing and Forecasting System). More specifically, we consider a 7-year free run simulation of the model spanning the period 2009 – 2015, with spatial coverage from the Strait of Gibraltar to the Sardinia Channel (Figure 2). WMOP is a regional configuration of the Regional Oceanic Model System (ROMS) model [32] with a spatial resolution of approximately 2 km. WMOP is forced with high resolution atmospheric forcing (HIRLAM model from the Spanish Meteorological Agency AEMET), with temporal resolution of 3 h and spatial resolution of 5 km. These features make WMOP a suitable choice to evaluate the potential of SWOT-derived SSH data to resolve mesoscale processes in the western Mediterranean Sea. The presence of fine scale features of a few kilometers is illustrated in Figure 2. In Figure 2, we show snapshots of model relative vorticity (normalized by f) for days corresponding to pass 15 (Figure 2, left) and pass 168 (Figure 2, right) of cycle 2 of the SWOT orbit (see Figure 3).

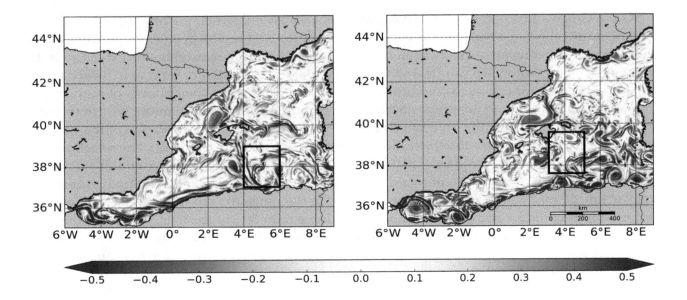

Figure 2. WMOP relative vorticity normalized by f on 23 January 2009 (**left**) and 3 February 2009 (**right**). Black boxes indicate the two regions studied in Section 3: box 1, pass 15 (**left**) and box 2, pass 168 (**right**).

2.3. Analysis and Processing of SWOT-Derived SSH Data

2.3.1. Geostrophic Velocity and Vorticity

Zonal (u_g) and meridional (v_g) (with respect to the SWOT grid) surface geostrophic velocity components are calculated as:

$$u_g = -\frac{g}{f}\frac{\partial \eta}{\partial y}, \tag{1}$$

$$v_g = \frac{g}{f}\frac{\partial \eta}{\partial x}, \tag{2}$$

where g is the gravitational acceleration, f the Coriolis parameter and η the sea level elevation.

The absolute geostrophic velocity (V_g) is obtained with:

$$V_g = \sqrt{u_g + v_g{}^2}. \tag{3}$$

Geostrophic relative vorticity, ζ, is calculated from the zonal and meridional velocities:

$$\zeta = \frac{\partial v_g}{\partial x} - \frac{\partial u_g}{\partial y}. \tag{4}$$

2.3.2. Noise Filtering

As illustrated in Section 3, noise greatly affects the computation of the velocities derived from the pseudo-SWOT data. We therefore investigate filtering procedures for noise removal. The geometry of the SWOT data prevents us from using classical Fourier and convolution-based low-pass filters [33]. Fourier-based filters impose circularity constraints, which cannot be fulfilled; the masks associated with convolution-based filters should be significantly smaller than the width of the SWOT swath, which greatly limits low-pass filtering capabilities. We then considered a Partial Derivative Equation (PDE)-based formulation, such that the low-pass filtering results from an iterated Laplacian diffusion:

$$\partial_t a\,(t,y,x) - \triangle a\,(t,y,x) = 0 \iff \frac{\partial a}{\partial t} = \frac{\partial^2 a}{\partial y^2} + \frac{\partial^2 a}{\partial x^2}. \tag{5}$$

As the Green's function for the heat equation is a Gaussian kernel, the implementation of this PDE-based diffusion is equivalent to a Gaussian convolution and results in an isotropic filtering, that is to say that the filtering acts equally in all directions [34]. Using a four-neighbourhood discretization of the Laplacian operator, we can deal with missing data (e.g., nadir) or land (e.g., island) pixels. The Laplacian operator comes to compute a local mean over the four neighbours of a given pixel. Withdrawing land pixels and missing data from the computation of this local mean, we can iterate the Laplacian diffusion to reach the expected filtering level for all pixels. Each iteration of the Laplacian diffusion can be regarded as a low-pass filtering with a high cut-off frequency. The selection of the number of iterations of the Laplacian diffusion then allows us to reach lower cut-off frequencies. By contrast, the direct application of two-dimensional low-pass filters for cut-off frequencies in the range [30 km, 60 km] would result in filter supports in the range [60 km, 120 km], meaning that no filtering output could be computed for any pixel closer than 30 km (rest. 60 km) from the swath boundaries or a missing data or land pixel. Overall, the filtering level is set by the number of iterations of the Laplacian diffusion and the parameter lambda. This is shown in the following equation, which shows the implementation that we use:

$$a^{k+1} = a^k - \lambda \triangle a^k. \tag{6}$$

With this being an iterative method, in contrast to a traditional Gaussian filter, we can apply cut-off wavelengths greater than the width of a half-swath. In Appendix A, we apply the filter to white noise to show how different combinations of the filter's parameters (lambda and number of iterations) are associated with different cut-off wavelengths ($\lambda_c s$).

2.3.3. Filter Evaluation

To evaluate the performance of the filter and its different parameterizations, the following variables are calculated:

- The radial power spectral density: This variable was calculated to obtain the SWOT spatial spectra. The radially averaged power spectral density (power spectrum) of an image (in our case, the SWOT swath data) is computed.
- The Root Mean Squared Error (RMSE): The RMSE was calculated for the SSH, velocity and vorticity variables as follows:

$$RMSE = \sqrt{\frac{\sum(data - estimate)^2}{N}}, \tag{7}$$

where N is the number of points. Data is taken to be SSH_{model} (or its derived variables, i.e., velocity and vorticity) without filtering. An estimate is taken to be the simulated noisy SSH_{obs} fields (or its derived variables) without filtering, and filtered with different $\lambda_c s$. RMSE values are therefore calculated for different estimates.

3. Results

3.1. Spatial and Temporal Sampling

Pseudo-SWOT data were generated for the full WMOP time period (1 January 2009 to 11 September 2015). This was done for the Science orbit and corresponds to a total of 123 cycles. In a complete cycle, 292 passes are available over the globe, 12 of them crossing our study region (Figure 3). Data were ingested and processed by the SWOT simulator at an across and along track resolution of 2 km. As mentioned in Section 1, one of the specificities of future SWOT data will be their irregular temporal sampling. To better illustrate this, the passes of cycle 2 are plotted in Figure 3. During each cycle, there are periods of time with a higher temporal sampling. This is due to a longer revisit time so that SWOT also fulfills its hydrological objectives as described in Section 1. For instance, the temporal sampling during cycle 2 is as follows: from day 21.3 to 23.9 during which five passes are made; and from day 31.3 to 33.8 during which six passes are made. Then, from day 23.9 to 31.3 and from day 33.8 to 41.2, there are no measurements. Consequently, during each day within a cycle in this study region, there can be two, one or no passes at all. Even with this irregular sampling and without any processing of the data, the final SSH_{obs} map (subplot of day 41.2) allows us to observe some features such as, for example, the signal of the Algerian Current following the north African coastline and several cyclonic and anticyclonic mesoscale eddies.

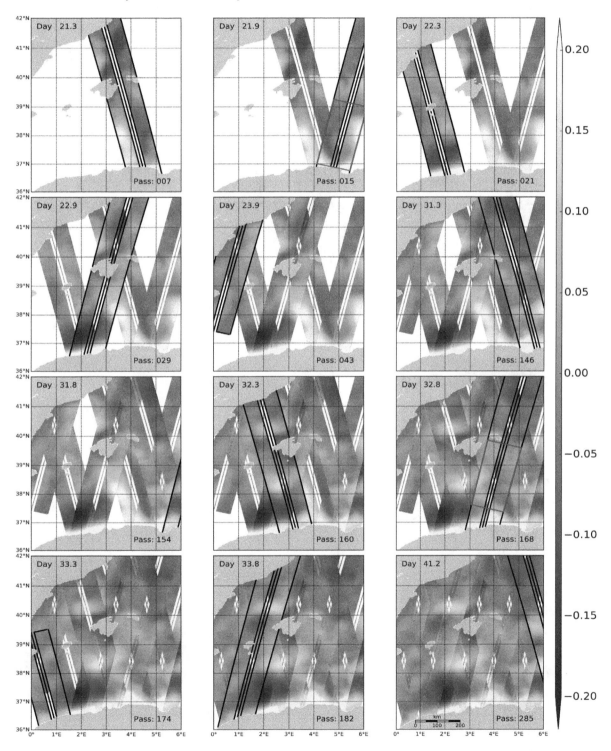

Figure 3. SSH_{obs} (m) obtained for cycle 2. Time increasing from left to right, top to bottom. Days from the beginning of the simulation are shown at the top left corner and the corresponding pass number at the bottom right corner. Outline of the active pass is shown in black. The red boxes show box 1 (pass 15) and box 2 (pass 168).

3.2. Pre-Filtering Analysis of Simulator Outputs

In this study, we focus on the analysis of spatial scales of individual passes. Due to the irregular time sampling of the SWOT data, future studies will be devoted to temporal interpolation of passes. Moreover, prior swath filtering is necessary to determine the quality of the dynamical variables that can be derived from SWOT data, and how it can be improved before combining different swaths for temporal interpolation. As an illustration, we focus on the treatment of two $2° \times 2°$ boxes. Box 1 is within pass 15 and was chosen close to the north African coast as it is a region where anticyclonic eddies are shed from the Algerian Current [25,35]. For example, in the snapshot shown in Figure 4, part of an anticyclonic eddy is present on the eastern part of the domain. Box 2 is within pass 168, and this subdomain south of the Balearic island of Menorca was chosen because it contains smaller structures than in box 1 (see Figure 2). In Figure 5, filament-like structures and smaller eddies can be observed, especially at the northern part of the domain.

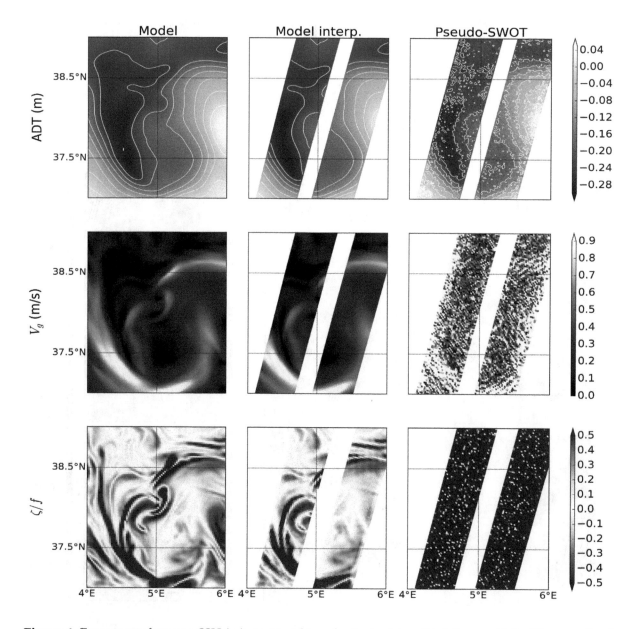

Figure 4. From top to bottom: SSH (m), geostrophic velocity (m/s) and relative vorticity (ζ) normalized by f, on 23 January 2009 corresponding to pass 15 of cycle 2 (box 1). The first, middle and last columns show the data obtained directly from the model (WMOP), from the model interpolated onto the SWOT grid (SSH_{model}), and with added noise (SSH_{obs}), respectively.

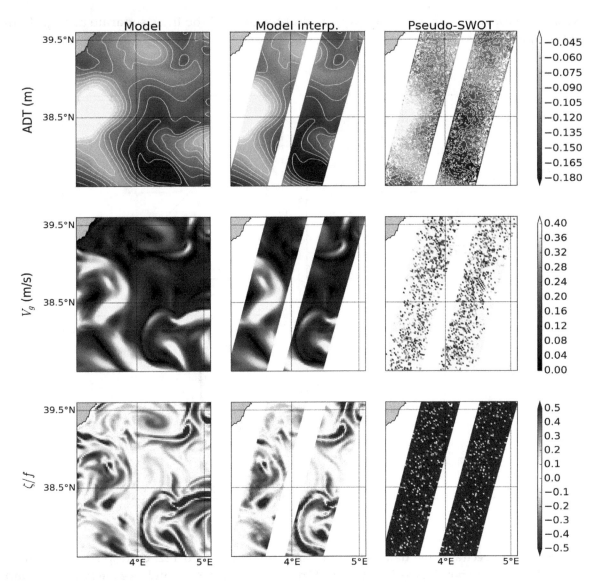

Figure 5. From top to bottom: SSH (m), geostrophic velocity (m/s) and relative vorticity (ζ) normalized by f, on 3 February 2009 corresponding to pass 168 of cycle 2 (box 2). The first, middle and last columns show the data obtained directly from the model (WMOP), from the model interpolated onto the SWOT grid (SSH_{model}), and with added noise (SSH_{obs}), respectively.

The effect of the filter is assessed for SSH and its derived dynamical variables: absolute geostrophic velocity and relative vorticity. These were calculated as explained in Section 2.3.1.

As observed in the first and middle columns (model and model interpolated onto SWOT grid data, respectively) of Figures 4 and 5, SSH and its derived variables reveal fine scale features, but the noise level masks the signal of these features when derived variables are obtained from pseudo-SWOT SSH. We can also see how the effect of the noise is lower in regions with high SSH gradients. If we compare the velocity derived from pseudo-SWOT data of box 1 and 2, for box 1, the region with high values can still be appreciated as they reach 0.9 m/s, but not for box 2 as they only reach 0.4 m/s.

To have information on the spatial scales resolved and the effect of the noise, spatial Fourier power spectra for each filter were calculated as described in Section 2.3.2. The spectra were calculated for each individual cycle, and then averaged over the 122 cycles in which both passes 15 and 168 are available (cycle 123 stops at pass 132). Figure 6 compares the spectra of model data interpolated onto the SWOT grid and the pseudo-SWOT data. The SWOT noise starts to dominate at wavelengths lower than 60 km. In the top panel of Figure 6, the red and blue curves separate at around 60 km for both boxes. If we look at the zoom inset, we see that for pass 15 the lines separate at slightly higher wavelengths than for pass 168.

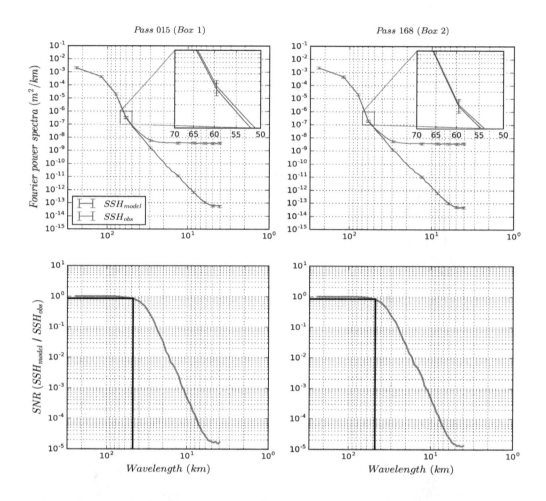

Figure 6. (**Top**) Spectra of the data before filtering, from cycle 1 to 122, corresponding to box 1 (pass 15), (**left**), and to box 2 (pass 168), (**right**). Error bars denote 95% confidence intervals. (**Bottom**) Corresponding Signal to Noise Ratio (SNR), with a horizontal black line indicating where the noise is more than 15% of the signal, and the vertical line the corresponding wavelength.

Nevertheless, there does not seem to be a significant difference between the mean spectra of both passes. Note also how the power spectral energy level of SSH_{obs} at wavelengths lower than 20 km stabilizes around 3.7×10^{-9} for both passes, whilst the energy level of SSH_{model} reduces until it reaches the grid scale. If we look at the signal to noise ratio (SNR), we find that below 50 km wavelength the energy of the noise is significant with respect to that of the signal (SNR values below 15 dB at wavelengths smaller than 47.6 km, i.e., the energy of the noise accounts for more than 15% of the energy of the signal for these scales). Such low SNR values make particularly challenging the denoising issue for scales below 50 km [36]. Consequently, we expect that the best filter parametrization will be one corresponding to λ_c between 47.6 and 60 km.

3.3. SWOT Data Filtering

The Laplacian diffusion filter was applied to remove the noise and, thus, reduce the difference between the spectrum obtained from SWOT estimates with and without noise (Figure 6). Given the results obtained from the non-filtered data spectra, λ_c is first chosen to be 60 km. We then choose smaller λ_cs (50, 40, 30 and 15 km) to see how much lower we can go with this filter. We go down to 15 km, which is the expected wavelength at which SWOT will measure SSH. For comparison, we also choose $\lambda_c = 200$ km, which is the wavelength resolved by present-day altimeter constellation fields [11]. We lastly choose $\lambda_c = 100$ km as an intermediate value between 60 and 200 km.

In Figures 7 and 8, we show the effect of the filter on SSH at different values of λ_cs. For Figure 7, the effect of the filter is mainly seen in the pseudo-SWOT data, especially in the northern part where smaller structures are present. In Figure 8, as there are more, smaller structures, we can see more differences between the filtered outputs with respect to the model interpolated and pseudo-SWOT data. These differences are not only in the shape of the structures that are present, but also in their intensity. On the top row of Figures 7 and 8, the original model is also included to show that differences do also arise from the interpolation onto the SWOT grid. In Figure 7, especially for the 200 km λ_c, we can observe that the original structure present is significantly altered. This emphasizes the importance for development of interpolation techniques to fill the gap between the two swaths with the help of the nadir altimeter data. In Figures 7 and 8, the SSH images for the different filters look very similar, but the differences are amplified when the first derivatives (Figures 9 and 10) and second derivatives (Figures 11 and 12) are calculated.

After applying the Laplacian diffusion filter, we can now retrieve the structures present in the pseudo-SWOT SSH in the absolute geostrophic velocity plots (Figures 9 and 10). With a 15 km λ_c filter, the effect of the noise can be still clearly observed, especially for box 2 where smaller structures are present. As a result, although the main structures are recovered after filtering, their shapes are not accurately retrieved. Even if spurious structures remain, with a 30 km λ_c there is a large improvement with respect to the 15 km λ_c. This improvement seems greater for box 1 than box 2, as the noise seems to have a greater effect within box 2 than box 1. For a 40 km λ_c, in box 1, we can no longer qualitatively see any remaining noise, but we can see some in box 2. For λ_cs greater than or equal to 50 km, the effect of the noise is no longer observed in either box 1 (Figure 9) or box 2 (Figure 10). On the other hand, we observe a large decrease of the magnitude of the velocities from the 15 to the 200 km λ_c. With this filtering method, the intensity of the structures present, and thus the signal, decreases with the increase of λ_c.

In the relative vorticity plots, the loss of signal with the increase of λ_c is even more evident. With no filtering, the relative vorticity of box 1 ranges from $-1.82f$ to $1.66f$ for SSH_{model} and from $-15.22f$ to $18.16f$ for SSH_{obs} (Figure 4). With a 200 km λ_c, this reduces to $-0.23f$ to $0.14f$ for both SSH_{model} and SSH_{obs} (Figure 11). For box 2, with no filtering, the relative vorticity ranges from $-0.71f$ to $1.50f$ for SSH_{model} and from $-17.39f$ to $17.71f$ for SSH_{obs} (Figure 5). For 200 km, it reduces to $-0.07f$ to $0.10f$ for SSH_{model} and from $-0.09f$ to $0.06f$ for SSH_{obs} (Figure 12). There is approximately two orders of magnitude difference between the vorticity calculated from the original data, and that filtered at $\lambda_c = 200$ km. For box 1, the velocity appears to contain no further noise with $\lambda_c = 40$ km, but this filtering is not sufficient to properly reconstruct the relative vorticity. With a 50 km λ_c, some noise is still present, and with 60 km λ_c, there appears to be no remaining noise. For box 2, the velocity appears to have no further noise with a 50 km λ_c, and, similarly to box 1, we use a 60 km λ_c to qualitatively remove remaining noise in the relative vorticity plots. The relative vorticity fields present unrealistic small-scale structures at larger λ_cs values than SSH and velocity. This is expected as the noise effects increase as higher order derivatives are reached. Nevertheless, the larger structures present in the images are recovered from the non-filtered image with a 60 km λ_c filter for both box 1 and 2. Not as much signal is lost with a 60 as with a 200 km λ_c, but some is still lost. For the mesoscale, given the relative vorticity and structures observed in Figures 11 and 12, this does not seem to have a large impact. However, there may be an impact when wanting to observe finer scales as we retrieve normalized relative vorticity much lower than 1.

Spectra were computed for λ_cs of 30, 60 and 200 km to visualize these effects. The corresponding SNR is also calculated in two different ways by using two references. One is by dividing the filtered model-interpolated data by the filtered pseudo-SWOT data, and the other by dividing the non-filtered model-interpolated data by the filtered pseudo-SWOT data. This is shown in Figures 13 and 14.

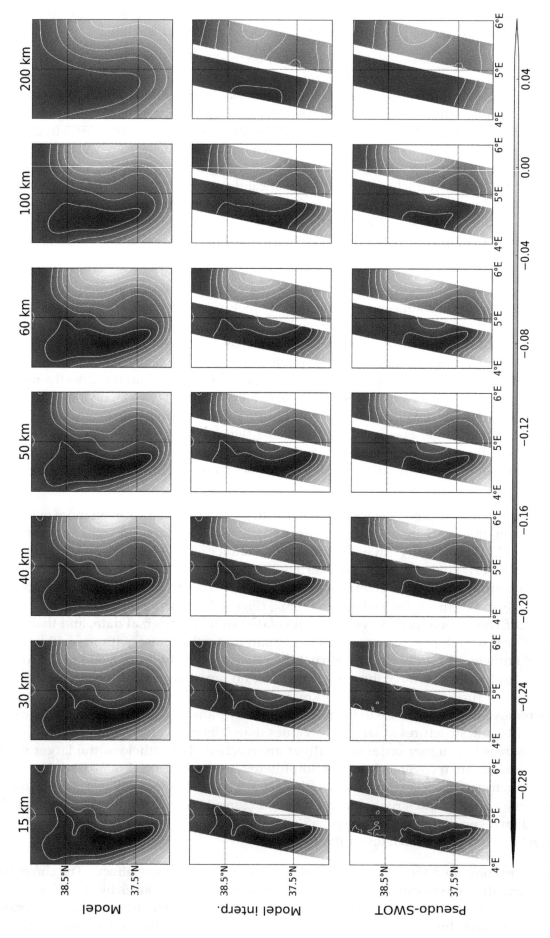

Figure 7. SSH (m) on 23 January 2009 corresponding to pass 15 of cycle 2.

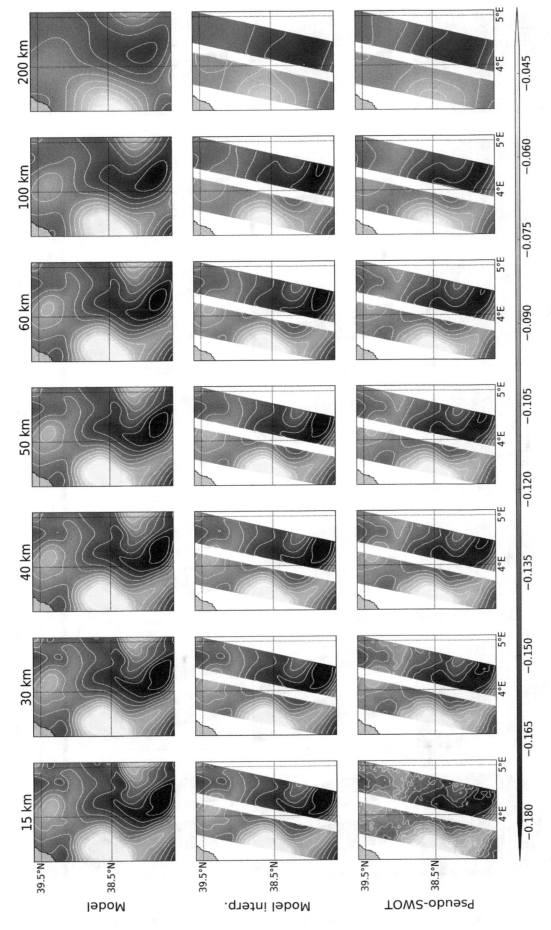

Figure 8. SSH (m) on 3 February 2009 corresponding to pass 168 of cycle 2.

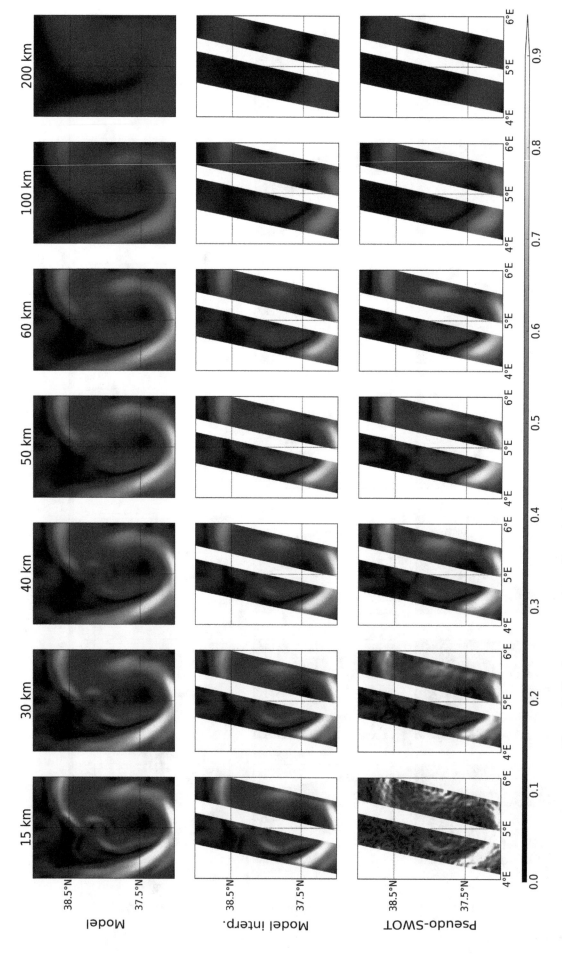

Figure 9. Absolute geostrophic velocity (m/s) on 23 January 2009 corresponding to pass 15 of cycle 2.

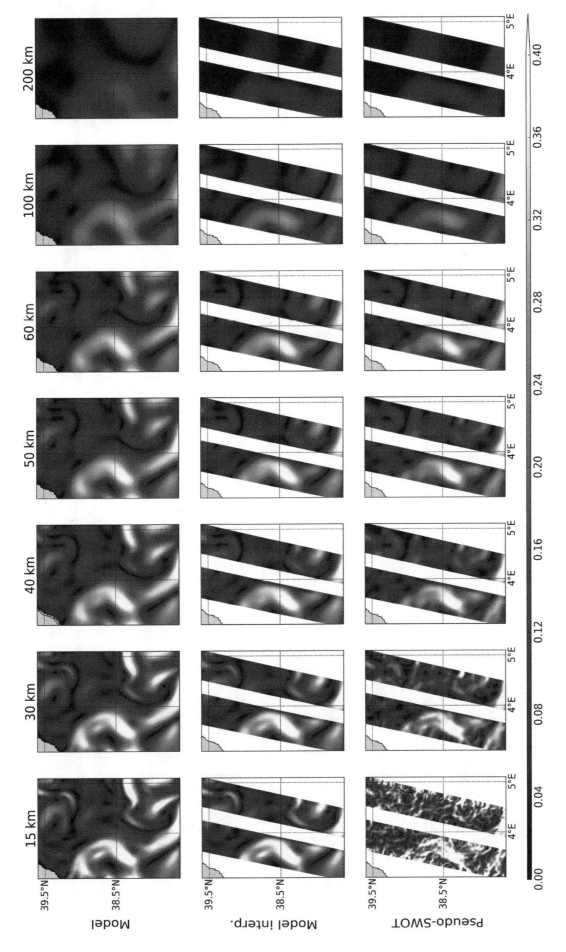

Figure 10. Absolute geostrophic velocity (m/s) on 3 February 2009 corresponding to pass 168 of cycle 2.

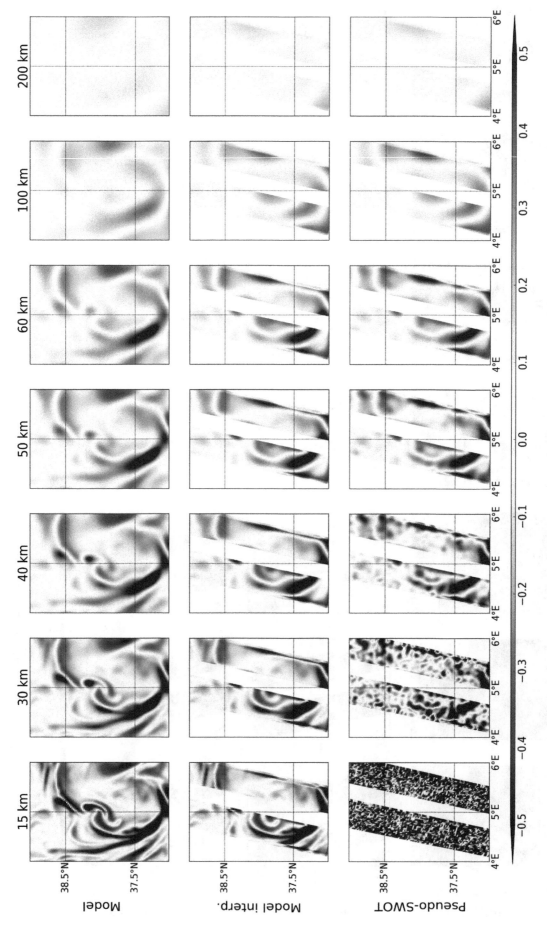

Figure 11. Relative vorticity normalized by f on 23 January 2009 corresponding to pass 15 of cycle 2.

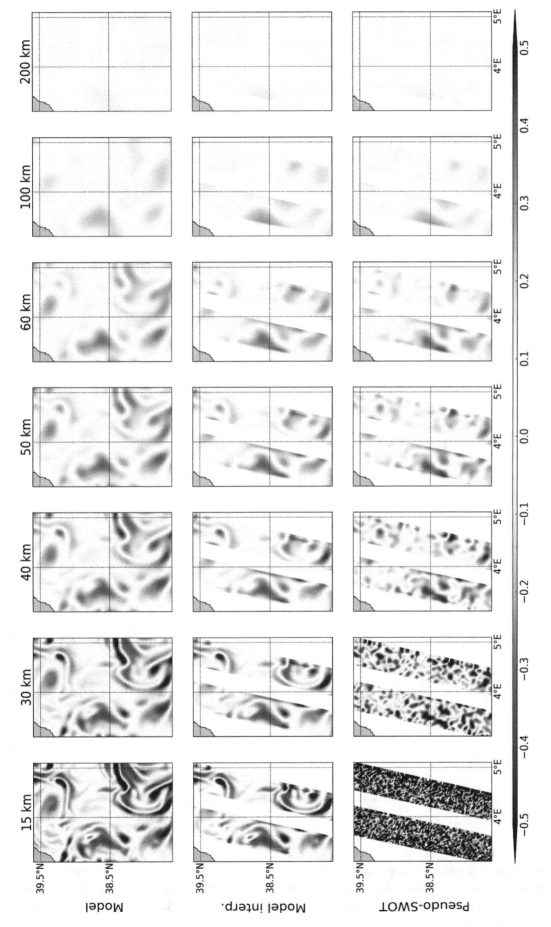

Figure 12. Relative vorticity normalized by f on 3 February 2009 corresponding to pass 168 of cycle 2.

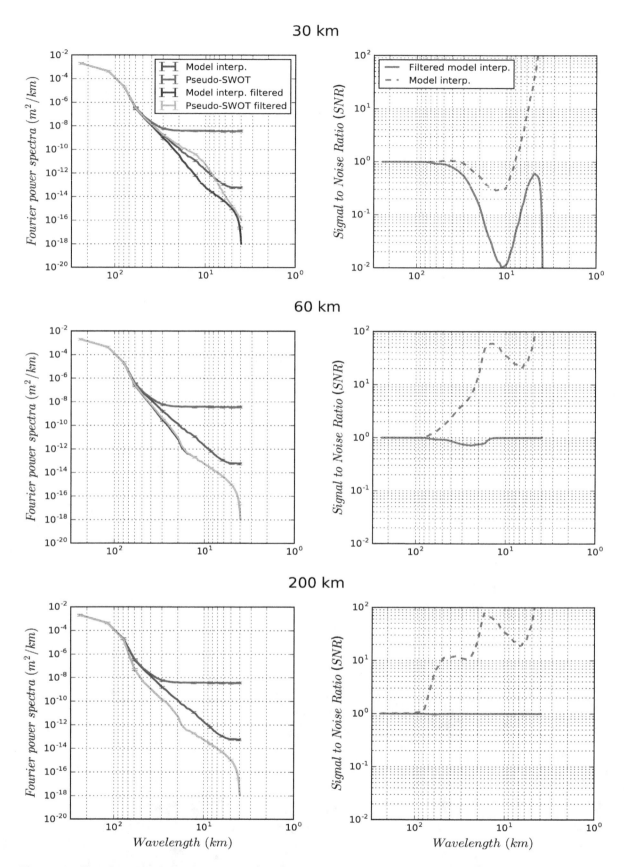

Figure 13. Box 1 region (pass 15) mean of cycles 1 to 122. (**Left**) Spectra of SSH$_{model}$ (blue) and SSH$_{obs}$ (red) before filtering and after applying the different cut-off wavelengths shown in the different rows (30, 60 and 200 km) in purple and orange, respectively. Error bars denote 95% confidence intervals. (**Right**) SNR of SSH$_{model}$ and SSH$_{obs}$, both filtered (solid line) and of SSH$_{model}$ non-filtered and filtered SSH$_{obs}$ (dashed line).

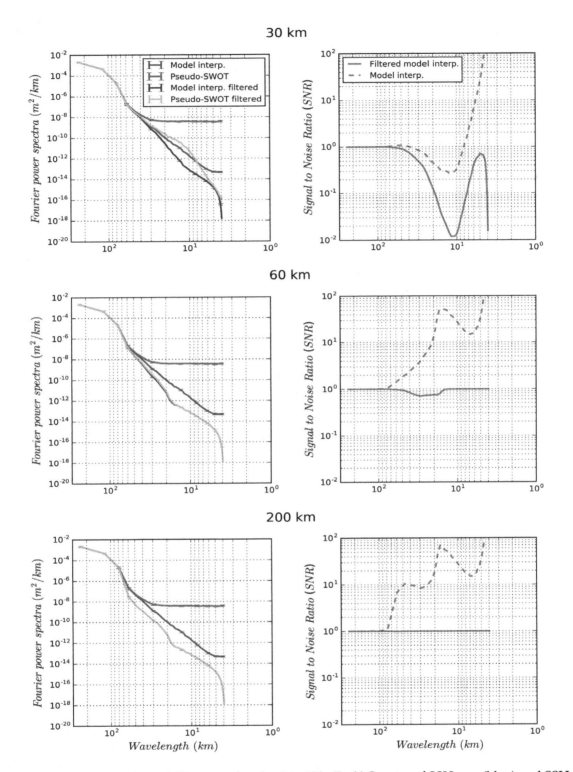

Figure 14. Box 2 region (pass 168) mean of cycles 1 to 122. (**Left**) Spectra of SSH$_{model}$ (blue) and SSH$_{obs}$ (red) before filtering and after applying the different cut-off wavelengths shown in the different rows (30, 60 and 200 km) in purple and orange, respectively. Error bars denote 95% confidence intervals. (**Right**) SNR of SSH$_{model}$ and SSH$_{obs}$, both filtered (solid line) and of SSH$_{model}$ non-filtered and filtered SSH$_{obs}$ (dashed line).

As a consequence of the application of the filter, the separation of the spectral curves of SSH$_{model}$ (model-interp.) and SSH$_{obs}$ (pseudo-SWOT) is reduced. As seen in Figure 6, with no filter, the model-interp. (blue) and pseudo-SWOT (red) curves separate at a wavelength around 60 km. With a 30 km λ_c, the noise level is still high, as observed in Figures 7–12, but the power spectra difference at wavelengths smaller than 60 km between the pseudo-SWOT filtered (yellow) and the model-interp.

(blue) curves is much smaller than it is between the model-interp (blue) and pseudo-SWOT (red) curves (top left panel of Figures 13 and 14). Moreover, the pseudo-SWOT filtered and the model-interp. curves separate at smaller wavelengths. If we look at the dashed line of the top right panel of Figures 13 and 14, we can more accurately determine this wavelength separation value by looking at where the dashed curve starts decreasing. This corresponds to 30 and 40 km wavelengths for box 1 and 2, respectively. Between wavelengths of 60 and 30 km (40 for box 2), the SNR (dashed line) is greater than 1, indicating some over-filtering/smoothing, but this value is very low (1.064 for box 1 and 1.088 for box 2). The pattern of the SNR of the filtered model-interp. over the filtered pseudo-SWOT data (solid line of top right panel of Figures 13 and 14) is similar to that of Figure 6. The noise gains importance as the wavelengths reduce from 60 km, but the SNR values of the 30 km λ_c are lower than the non-filtered ones. On the other hand, this indicates that, given that the filtered model-interp. and pseudo-SWOT spectra are still quite different, a larger λ_c is necessary. In both the left and right panels of the 30 km λ_c of Figures 13 and 14, anomalous patterns are observed below the 10 km wavelength. Not only are these spatial scales very small, but, if we look at the spectra values, they are about 10^{-12} m^2/km or lower. We consider these values to be too low for discussion.

With a 60 km λ_c, the noise is further reduced, but we lose more signal too. If we look at the continuous line of the right panel of Figures 13 and 14, we see that it remains approximately constant at 1. For box 1, it reaches a minimum SNR of 0.715 and for box 2 of 0.697. Looking at the dashed line, the SNR becomes larger than 1 at wavelengths lower than 80 km for both boxes. This means that, although we eliminate all the noise, we are also eliminating part of the signal that was initially present. At wavelengths greater than 10 km, the power spectra of the filtered pseudo-SWOT reach a maximum difference nearly two orders of magnitude smaller than the original model-interp. spectra.

Lastly, with a 200 km λ_c filtering, the model-interp. and pseudo-SWOT spectra curves are identical, and the SNR (solid line) is approximately 1 (Figures 13 and 14, bottom row): however, as observed in Figures 7–12, we lose a lot of signal. In the 20–80 km wavelength range, we can see how the SNR curves (dashed line) rapidly increase and the values are greater with a 200 km λ_c than with a 60 km λ_c. At wavelengths lower than 80 km, on average, there is about one order of magnitude difference between the filtered spectra and the original SSH$_{model}$. It is also interesting to note that the purple and yellow curves separate from the red and blue at 80 km instead of 60, showing that this cut-off exceeds that necessary to remove the noise. This also emphasizes how with SWOT a major advancement could be made as lower cut-off wavelengths will be possible, and thus the observation of smaller scale structures than with contemporary satellites. On the one hand, this result is expected thanks to the 2D swath instead of only 1D nadir data, but, on the other hand, it is important to remember that these are simulated from expected errors and that not all errors are implemented (see Appendix B).

To further quantify the differences observed between the different λ_cs in Figures 7–12, the RMSE between the interpolated model and pseudo-SWOT data shown in these Figures were calculated as described in Section 2.3.3.

In Figure 15, it is interesting to focus on the minimum points of the RMSE curves. Looking at the λ_c corresponding to the different minimum points, for both boxes, the minimum of the curve for SSH and the absolute geostrophic velocity (Vg) is found for a 30 km (29) λ_c. It is slightly higher, 40 km (41), for relative vorticity. This directly relates to the amplification of fine scale structures, and thus the effect of the noise, in the computation of second-order derivatives. It is also in accordance with what is found in the SNR in Figure 6, which shows that we cannot recover the signal at wavelengths lower than 40–50 km. With the qualitative (Figures 7–12) and the spectra (Figures 13 and 14) plots, we saw that, for box 2, as the signal is lower than in box 1, the effect of the noise is greater, and larger λ_cs are necessary. The RMSE plots show us another point of view. As the signal is not as intense in box 2 as in box 1, the over-smoothing (signal lost) due to the Laplacian diffusion filter is lower, and thus we observe lower RMSE values in Figure 15. Moreover, the improvement of the RMSE values is greater for box 2 than box 1. For SSH, the RMSE reduces by 0.05 and 0.07 m from the no filter (0 λ_c) to the minimum RMSE, for box 1 and box 2, respectively. For Vg, the RMSE reduces by 0.42 and 0.455 m/s, and for ζ/f by 3.2 and 3.35.

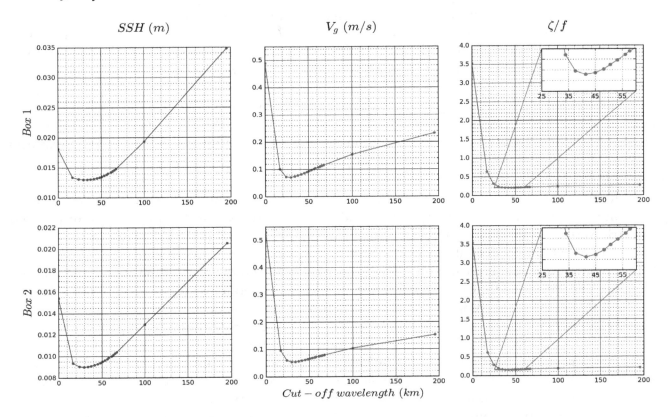

Figure 15. RMSEs of the different variables against the filter's cut-off wavelengths applied to SSH, for pass 15 (**top**) and pass 168 (**bottom**) of cycle 2. Insets show zoom of marked region for better observation of the curves' minimum points.

For SSH, we could say that with a 40 km λ_c it is sufficient, but if we do not want to see the effect of the noise in vorticity, we need a greater λ_c of 60 km. Therefore, together with what is observed in Figures 4–15, we consider that, with this filtering technique applied to this region, SWOT will be able to resolve wavelengths down to a 40–60 km wavelength range. This is the λ_c range where we found that there is a compromise between filtering out the noise of SSH and its derived variables (Vg and ζ/f), and over-smoothing the original image as little as possible. It is important to note that λ_c depends on the signal-to-noise ratio between SSH signals and instrument noise at fine scales. As such, it can be expected to change from region to region (and from season to season) depending on the energy levels at fine scales and on the noise level.

4. Discussion

We find that, by applying our filtering technique to pseudo-SWOT data in the western Mediterranean region, we cannot reach the 15 km wavelength argued for by [12]. We are however able to recover the signal within a swath down to wavelengths of 40–60 km. This wavelength range is close to the one found by [10] at the western boundary currents and nearly half the wavelength values found at the eastern basins by 1D altimeter data. In addition, with the characteristic of SWOT providing 2D SSH data, this will imply a large improvement on the 200 km wavelength resolved by present-day gridded altimetric fields [11]. The SWOT resolved wavelengths found will make it possible to detect structures of 20–30 km in diameter (following [37]) and therefore opens the possibility for observation of fine scales that are unobservable by contemporary altimetric products. This filter is a useful tool for studies comparing the capacities of pseudo-SWOT data with the present altimetric satellite constellation data. As this filter is particularly effective in removing the spatially uncorrelated

KaRIn noise, it may allow the application of already developed techniques that more effectively remove other correlated errors [38].

We find that the presence of structures of different scales and regimes governing—for example, the mesoscale (Ro. order(0.1)) or submesoscale (Ro. order(1)) affects our results on the efficiency of the Laplacian diffusion filter. This filter is therefore sensitive to the presence of different patterns, depending on the region. To reach even smaller scales, it is important to use filtering techniques that conserve and/or retrieve the gradients and, thus, the intensity of the signal present in the observed field. Nonetheless, this sensitivity of the Laplacian diffusion filter could also be due to the effects of noise, depending on the structure in question and its intensity. We also note that the differences between box 1 and 2 found in, for example, the qualitative plots (Figures 7–12) and the minimum RMSE values were not very large; this is not surprising when considering how close in time the two snapshots are.

With this improvement of individual swaths, gridded and possibly daily SWOT SSH maps could be obtained through different interpolation and/or reconstruction techniques. Current gridded altimetric products are obtained using optimal interpolation (OI) [39]. OI may not be the best interpolation method. OI exploits a covariance model of the field to be interpolated. The specification of this covariance model typically relies on a trade-off between the space-time size of the missing data areas and the space-time scales of interest. A covariance model with a large correlation length may lead to an over-smoothing of fine-scale structures, whereas shorter correlation lengths result in filling large missing areas with the background field. From a computational point of view, OI requires a matrix inversion, whose complexity evolves as the cube of the number of observation points. The image-like structure of SWOT data may then be highly computationally-demanding when considering large correlation lengths to fill in large missing areas. Multi-scale OI may be an alternative. However, we expect dynamical interpolation and other data assimilation methods [40–42] to be more adapted both in terms of computational complexity and in their ability to embed relevant dynamic priors to reconstruct horizontal scales down to a few tens of kilometres from SWOT data. On the other hand, SWOT data will greatly improve the present-day OI altimetric products [5]. In addition, SWOT gridded data could be improved in the 40–60 km wavelength range by combining it with the data of a higher temporal resolution.

Dynamic interpolation is an example of a technique that has been investigated by [40] that could help to obtain gridded, daily SSH maps from SWOT. When they apply this method to the Gulf Stream region, they recover the SSH field down to 80 km wavelength. Data-driven schemes recently introduced by [41,42] are also of interest to better reconstruct horizontal scales below 100 km. Overall, for such approaches, it is important to recover the lowest wavelengths possible as spatial resolution loss is likely when producing the gridded maps. Moreover, this spatial resolution loss might be even higher when addressing gridded maps of derived variables. Therefore, the cut-off wavelength should be adjusted to the variables that are to be studied.

Another reconstruction technique that has been investigated in the context of SWOT is a 3D multivariate reconstruction of ocean state. Ref. [43] do this by combining information from SSH and high resolution image structure observations. Once this is achieved, study of the capacity of SWOT to detect fine scale structures could be improved by, for example, better characterizing eddies. As the dataset would be of a higher spatial and temporal resolution than the L2 product, it would then be possible to apply eddy-tracking algorithms like the *py-eddy tracker* [44] or the code developed by [45], which have already been implemented in this region to characterize the western Mediterranean eddy field. A comparison could then be made with the eddies characterized in data from the WMOP model, in the presently available altimetric data and in pseudo-SWOT data.

In future work, the effect of the inter- and intra-annual (or seasonal) variability in the region on the results obtained could be studied too. Although mean spectra were obtained, we focused on two dates in winter. Refs. [46,47] found that there is a strong winter–summer difference in the upper ocean dynamics due to the change of stratification, with the mixed layer depth being deeper in winter.

For example, the reconstruction of mesoscale structures in the upper ocean from pseudo-SWOT data in the Kuroshio Extension region has been studied. They found that the simulated and reconstructed vorticity correlation coefficients varied both inter- and intra-annually [48].

The implementation of filtering techniques that take into account the first and second order SSH derivatives has been started. With this, we hope that in future studies we will be able to recover even smaller wavelengths and to conserve the intensity of the signal after having applied the filter.

New versions of the SWOT simulator will allow the simulation of pseudo-SWOT data during the fast-sampling phase. This makes it possible to start preparing for the calibration/validation phase including the comparison with high resolution in situ data collected during future intensive multi-platform experiments in the western Mediterranean Sea. On the other hand, the only source of geophysical error implemented in the SWOT simulator is still just that related to the wet troposphere. New releases of the SWOT simulator may include the effects of sea state and internal tides [49]. Internal tides and waves are important sources of geophysical errors because, at wavelengths shorter than 50 km, they can affect SWOT data [2]. Therefore, in future work, it would also be interesting to compare our results with an updated version of the SWOT simulator and other OGCMs, especially those that include tides.

5. Conclusions

We have generated simulated pseudo-SWOT data for the western Mediterranean Sea using a SWOT simulator and outputs from an ocean numerical model. To evaluate the output SWOT data, we derived absolute geostrophic velocities and relative vorticities from the pseudo-SWOT SSH data. We find that, due to the satellite's instrumental noise and geophysical errors, the features observed in the pseudo-SWOT SSH are lost in the derived dynamical variables. Looking at the spatial spectra, we find that noise dominates the signal at wavelengths smaller than 60 km. We applied a Laplacian diffusion filtering technique to attempt to remove the noise and hence observe finer scales. We estimated the appropriate cut-off wavelength for each parametrization. To filter out the noise, we applied a series of ascending cut-off wavelengths: 15, 30, 40, 50, 60, 100 and 200 km. We find that in this study region, using this technique, we cannot resolve the expected 15 km wavelength. On the other hand, we are able to recover the signal within a swath down to a 40–60 km wavelength range. This is still an improvement in comparison to wavelengths resolved by present-day 1D altimeters, especially at eastern basins. Robust swath-filtering is an important first step towards meeting our goals for reconstruction techniques that will enable us to combine SWOT and altimetric data in order to produce gridded SSH maps of significantly higher resolution than contemporary products. New versions of the SWOT simulator code include improved representation of instrumental and geophysical errors, and also give us the option to obtain pseudo-data for the SWOT fast-sampling phase. New pseudo-SWOT data will allow us to better refine the results of this study and to examine a wider range of scenarios.

Acknowledgments: The research leading these results has received funding from the Sea Level Thematic Assembly Center (SL-TAC) of the Copernicus Marine and Environment Monitoring Service (CMEMS) and from the Centre National d'Études Spatiales (CNES) through Ocean Surface Topography Science Team (OST/ST) project MANATEE. L. Gómez-Navarro acknowledges CNES and FP-7 PhD funding. This work was supported by a Short Term Scientific Mission (STSM) grant from COST Action ES140. E. Mason was supported by the Copernicus Marine Environment Monitoring Service (CMEMS) MedSUB project. E. Cosme and J. Le Sommer are supported by the CNES through the OST/ST and the SWOT Science Team. The WMOP simulation used in this study was produced in the framework of the MEDCLIC project funded by "La Caixa" Foundation. This study is a contribution to the PRE-SWOT project (CTM2016-78607-P) funded by the Spanish Research Agency and the European Regional Development Fund (AEI/FEDER, UE).

Author Contributions: Ananda Pascual and Laura Gomez-Navarro designed the study; Ronan Fablet, Ananda Pascual and Laura Gomez-Navarro designed the filtering experiments; Baptiste Mourre provided the WMOP simulation data and helped on the analysis of the model data; Laura Gomez-Navarro, Ronan Fablet, Evan Mason, Ananda Pascual, Julien Le Sommer and Emmanuel Cosme contributed to the analysis of the results; Laura Gomez-Navarro wrote the manuscript and all authors contributed to the writing.

Abbreviations

The following abbreviations are used in this manuscript:

ADT	Absolute Dynamic Topography
AEMET	Spanish Meteorological Agency
CNES	Centre National d'Études Spatiales
HIRLAM	HIgh Resolution Limited Area Model
KaRIn	Ka-band Radar Interferometer
NASA	National Aeronautics and Space Administration
OGCM	Oceanic General Circulation Model
PDE	Partial Derivative Equation
RMSE	Root Mean Square Error
ROMS	Regional Oceanic Modeling System
SOCIB	the Balearic Islands Islands Coastal Observing and Forecasting System
SSH	Sea Surface Height
SNR	Signal to Noise Ratio
SWOT	Surface Water Ocean Topography
WMOP	Western Mediterranean OPerational forecasting system

Appendix A

In order to know which number of iterations and lambda to set in the filter's parametrization, the filter was applied to a set of 100 randomly generated white noise fields. Spectra were then obtained and the cut-off wavelength was found by identifying the one that corresponded to where the energy was reduced to a half. An example is shown in Figure A1.

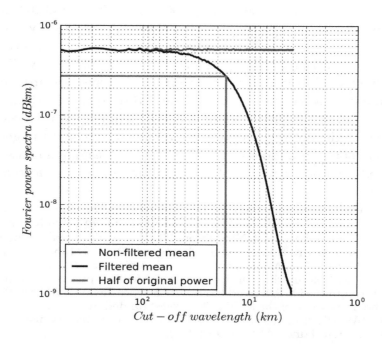

Figure A1. Illustration of how the parameterization corresponding to a 15 (16.72) km cut-off wavelength (λ_c) is estimated. The blue line represents the mean spectra of the 100 non-filtered white noise fields. The black line is the mean spectra of the 100 filtered white noise fields. The horizontal red line shows the half-power spectra of the blue line, and the vertical red line the corresponding wavelength value of the black line, and thus the cut-off wavelength.

In Table A1, we show the different λ_c obtained for a set of lambdas and number of iterations and in Figure A2 a plot of the values shown in Table A1 is presented. As can be observed in Table A1, in most cases, several combinations of lambdas and iterations can give the same cut-off wavelength. We decided to choose the combination corresponding to the smallest lambda, as the smaller the lambda, the smaller the over-smoothing.

Table A1. Cut-off wavelengths (λ_c) values and their corresponding lambda and number of iterations (iter) combinations. The cut-off wavelengths shown in Figures 7–12 are in bold.

Cut-off	Lambda	Iter	Cut-off	Lambda	Iter	Cut-off	Lambda	Iter	Cut-off	Lambda	Iter
16.72	0.05	50	71.88	0.10	450	105.09	0.20	500	141.41	0.35	500
23.95	0.05	100		0.15	300		0.25	400		0.40	450
	0.10	50		0.30	150		0.30	350		0.45	400
29.47	0.05	150		0.45	100		0.35	300		0.50	350
	0.15	50	74.5	0.10	500		0.40	250		0.55	350
33.85	0.05	200		0.20	250		0.50	200		0.60	300
	0.10	100		0.25	200		0.65	150		0.65	300
	0.20	50		0.50	100	110.78	0.25	450		0.70	250
37.58	0.05	250	77.31	0.15	350		0.45	250		0.75	250
	0.25	50		0.35	150		0.55	200	151.91	0.40	500
41.38	0.05	300		0.55	100		0.70	150		0.45	450
	0.10	150	83.63	0.15	400		0.75	150		0.50	400
	0.15	100		0.20	300	117.12	0.25	500		0.55	400
	0.30	50		0.25	250		0.30	400		0.60	350
45.02	0.05	350		0.30	200		0.35	350		0.70	300
	0.35	50		0.40	150		0.40	300		0.80	250
48.19	0.05	400		0.60	100		0.50	250	164.1	0.45	500
	0.10	200	87.19	0.15	450		0.60	200		0.50	450
	0.20	100		0.45	150		0.65	200		0.50	500
	0.40	50		0.65	100		0.80	150		0.55	450
50.58	0.05	450	91.07	0.15	500	124.24	0.30	450		0.60	400
	0.15	150		0.20	350		0.30	500		0.65	350
	0.45	50		0.25	300		0.35	400		0.65	400
53.2	0.05	500		0.30	250		0.40	350		0.70	350
	0.10	250		0.35	200		0.45	300		0.75	300
	0.25	100		0.50	150		0.50	300		0.75	350
	0.50	50		0.70	100		0.55	250		0.80	300
56.12	0.55	50		0.75	100		0.60	250	178.42	0.55	500
57.7	0.10	300	95.31	0.20	400		0.70	200		0.60	450
	0.15	200		0.40	200		0.75	200		0.60	500
	0.20	150		0.55	150	132.27	0.35	450		0.65	450
	0.30	100		0.80	100		0.40	400		0.70	400
	0.60	50	**99.96**	0.20	450		0.45	350		0.70	450
61.15	0.65	50		0.25	350		0.55	300		0.75	400
63.03	0.1	350		0.3	300		0.65	250		0.8	350
	0.35	100		0.35	250		0.80	200		0.80	400
	0.70	50		0.45	200				**195.49**	0.65	500
65.03	0.15	250		0.60	150					0.70	500
	0.25	150								0.75	450
	0.75	50								0.75	500
67.17	0.1	400								0.8	450
	0.20	200								0.8	500
	0.40	100									
	0.80	50									

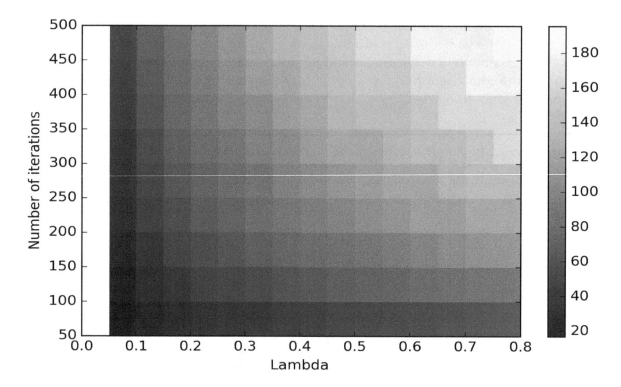

Figure A2. Laplacian diffusion cut-off wavelengths (km) for different combinations of the number of iterations and lambdas.

Appendix B

The noise added by the SWOT simulator can be divided into two types:

- Instrument errors: There are the different types of noise that can affect the signal due to the satellite itself:

  – Ka-Band Radar Interferometer (KaRIn)
  – Roll
  – Timing
  – Phase
  – Baseline dilation

Below are two example cycles of the instrument errors added by the simulator to passes 15 and 168 (Figures A3 and A4). Please note that the color-scale has been adjusted for each error type.

- Geophysical errors: In version 1 of the simulator, only the geophysical error due to the wet troposphere is implemented. Other geophysical errors include those due to the dry troposphere, the ionosphere and the sea state bias (electromagnetic bias). However, the wet troposphere is a major source of geophysical errors and it is implemented via these following two variables:

  – Path delay (pd),
  – Residual path delay (pd_err_1b).

Below, we show two example cycles of the geophysical errors added by the simulator to passes 15 and 168 (Figures A5 and A6). Please note that the color-scale has been adjusted for each error type.

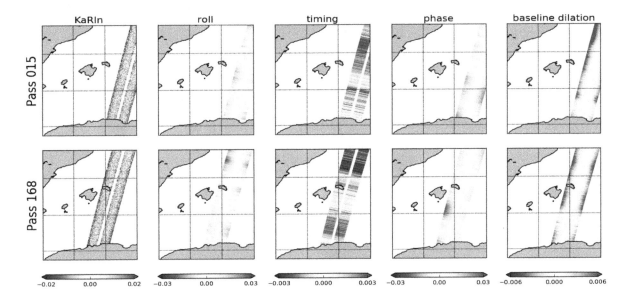

Figure A3. The different instrument noise types (m) for passes 15 pass 168 over cycle 2 are shown.

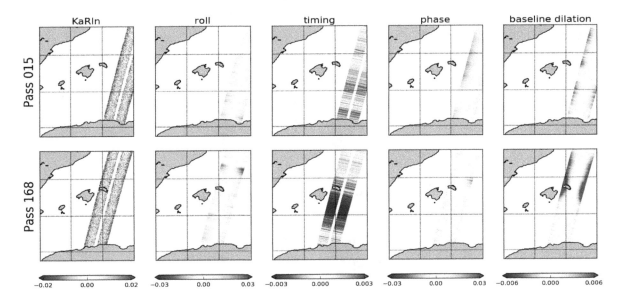

Figure A4. The different instrument noise types (m) for passes 15 and 168 over cycle 30 are shown.

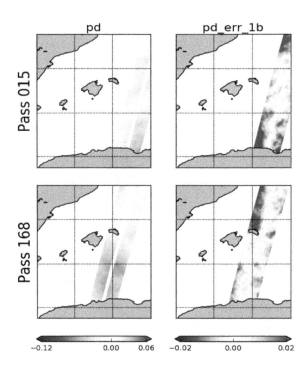

Figure A5. The different geophysical errors (m) for passes 15 and 168 over cycle 2 are shown.

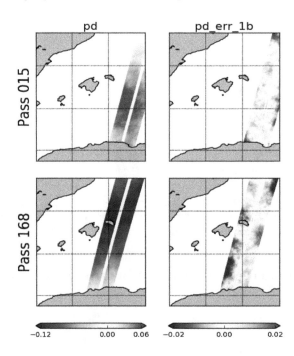

Figure A6. The different geophysical errors (m) for passes 15 and 168 over cycle 30 are shown.

For more details on instrument and geophysical errors, see [30,50].

References

1. Alsdorf, D.; Mognard, N.; Lettenmaier, D. Remote Sensing of Surface Water and Recent Developments in the SWOT Mission. In Proceedings of the Session H21J-06, AGU Fall Meeting, San Francisco, CA, USA, 5–9 December 2011.
2. Fu, L.L.; Morrow, R. A next generation altimeter for mapping the sea surface variability: Opportunities and challenges. In Proceedings of the 48th Liege Colloquium on Ocean Dynamics, Liège, Belgium, 23–27 May 2016.

3. Lee, H.; Biancamaria, S.; Alsdorf, D.; Andreadis, K.; Clark, E.; Durand, M.; Jung, H.C.; Lettenmaier, D.; Mognard, N.; Rodríguez, E.; et al. Capability of SWOT to Measure Surface Water Storage Change. In Proceedings of the Towards high-resolution of oceans dynamics and terrestrial water from space meeting, Lisbon, Portugal, 21–22 October 2010.

4. Rodríguez, E. The Surface Water and Ocean Topography (SWOT) Mission. In Proceedings of the OSTST (Ocean Surface Topography Science Team) meeting, Lisbon, Portugal, 18–20 October 2010.

5. Pujol, M.I.; Dibarboure, G.; Le Traon, P.Y.; Klein, P. Using high-resolution altimetry to observe mesoscale signals. *J. Atmos. Ocean. Technol.* **2012**, *29*, 1409–1416, doi:10.1175/JTECH-D-12-00032.1.

6. Rodríguez, E. Surface Water and Ocean Topography Mission Project Science Requirements Document. Jet Propulsion Laboratory, California Institute of Technology, JPL D-61923; March 2016; 28p. Available online: https://swot.jpl.nasa.gov/documents.htm (accessed on 12 April 2018).

7. Rogé, M.; Morrow, R.; Ubelmann, C.; Dibarboure, G. Using a dynamical advection to reconstruct a part of the SSH evolution in the context of SWOT, application to the Mediterranean Sea. *Ocean Dyn.* **2017**, *67*, 1047–1066.

8. Cipollini, P.; Calafat, F.M.; Jevrejeva, S.; Melet, A.; Prandi, P. Monitoring Sea Level in the Coastal Zone with Satellite Altimetry and Tide Gauges. *Surv. Geophys.* **2017**, *38*, 33–57.

9. Fu, L.L.; Ferrari, R. Observing Oceanic Submesoscale Processes From Space. *Eos Trans. Am. Geophys. Union* **2008**, *89*, 488, doi:10.1109/IGARSS.2011.6049757.

10. Dufau, C.; Orsztynowicz, M.; Dibarboure, G.; Morrow, R.; Le Traon, P.Y. Mesoscale resolution capability of altimetry: Present and future. *J. Geophys. Res. Ocean.* **2016**, doi:10.1002/2015JC010904.

11. Chelton, D.B.; Schlax, M.G.; Samelson, R.M.; de Szoeke, R.A. Global observations of large oceanic eddies. *Geophys. Res. Lett.* **2007**, *34*, doi:10.1029/2007GL030812.

12. Fu, L.L.; Ubelmann, C. On the transition from profile altimeter to swath altimeter for observing global ocean surface topography. *J. Atmos. Ocean. Technol.* **2014**, *31*, 560–568, doi:10.1175/JTECH-D-13-00109.1.

13. Mémery, L.; Olivier, F. Primary production and export fluxes at the Almeria-Oran front: A numerical study. In Proceedings of the EGS-AGU-EUG Joint Assembly, Nice, France, 6–11 April, 2003.

14. Lévy, M.; Klein, P.; Treguier, A.M. Impact of sub-mesoscale physics on production and subduction of phytoplankton in an oligotrophic regime. *J. Mar. Res.* **2001**, *59*, 535–565.

15. Lapeyre, G.; Klein, P. Impact of the small-scale elongated filaments on the oceanic vertical pump. *J. Mar. Res.* **2006**, *64*, 835–851.

16. Omand, M.M.; D'Asaro, E.A.; Lee, C.M.; Perry, M.J.; Briggs, N.; Cetinić, I.; Mahadevan, A. Eddy-driven subduction exports particulate organic carbon from the spring bloom. *Science* **2015**, *348*, 222–225.

17. McGillicuddy, D.J.; Anderson, L.A.; Bates, N.R.; Bibby, T.; Buesseler, K.O.; Carlson, C.A.; Davis, C.S.; Ewart, C.; Falkowski, P.G.; Goldthwait, S.A.; et al. Eddy/wind interactions stimulate extraordinary mid-ocean plankton blooms. *Science* **2007**, *316*, 1021–1026.

18. Allen, J.; Smeed, D.; Tintoré, J.; Ruiz, S. Mesoscale subduction at the almeria–oran front: Part 1: Ageostrophic flow. *J. Mar. Syst.* **2001**, *30*, 263–285.

19. Ruiz, S.; Pascual, A.; Garau, B.; Faugère, Y.; Alvarez, A.; Tintoré, J. Mesoscale dynamics of the Balearic Front, integrating glider, ship and satellite data. *J. Mar. Syst.* **2009**, *78*, S3–S16.

20. Pascual, A.; Ruiz, S.; Olita, A.; Troupin, C.; Claret, M.; Casas, B.; Mourre, B.; Poulain, P.M.; Tovar-Sanchez, A.; Capet, A.; et al. A multiplatform experiment to unravel meso-and submesoscale processes in an intense front (alborex). *Front. Mar. Sci.* **2017**, *4*, 39.

21. Bethoux, J.; Gentili, B.; Morin, P.; Nicolas, E.; Pierre, C.; Ruiz-Pino, D. The Mediterranean Sea: A miniature ocean for climatic and environmental studies and a key for the climatic functioning of the North Atlantic. *Progress Oceanogr.* **1999**, *44*, 131–146.

22. Malanotte-Rizzoli, P.; Artale, V.; Borzelli-Eusebi, G.L.; Brenner, S.; Crise, A.; Gacic, M.; Kress, N.; Marullo, S.; d'Alcalà, M.R.; Sofianos, S.; et al. Physical forcing and physical/biochemical variability of the Mediterranean Sea: A review of unresolved issues and directions for future research. *Ocean Sci.* **2014**, *10*, 281, doi:10.5194/os-10-281-2014.

23. Robinson, A.R.; Leslie, W.G.; Theocharis, A.; Lascaratos, A. Mediterranean sea circulation. In *Ocean Currents: A Derivative of the Encyclopedia of Ocean Sciences*; Elsevier: Amsterdam, The Netherlands, 2001; pp. 1689–1705.

24. Hallberg, R. Using a resolution function to regulate parameterizations of oceanic mesoscale eddy effects. *Ocean Model.* **2013**, *72*, 92–103, doi:10.1016/j.ocemod.2013.08.007.

25. Escudier, R.; Renault, L.; Pascual, A.; Brasseur, P.; Chelton, D.; Beuvier, J. Eddy properties in the Western Mediterranean Sea from satellite altimetry and a numerical simulation. *J. Geophys. Res.* **2016**, *121*, 3990–4006, doi:10.1002/2015JC011371.

26. Chelton, D.B.; Deszoeke, R.A.; Schlax, M.G.; El, K.; And, N.; Siwertz, N. Geographical Variability of the First Baroclinic Rossby Radius of Deformation. *J. Phys. Oceanogr.* **1998**, *28*, 433–460.

27. Nurser, A.J.G.; Bacon, S. Arctic Ocean Rossby radius Eddy length scales and the Rossby radius in the Arctic Ocean Arctic Ocean Rossby radius. *Ocean Sci. Discuss.* **2013**, *10*, 1807–1831, doi:10.5194/osd-10-1807-2013.

28. Wang, J.; Fu, L.L.; Qiu, B.; Menemenlis, D.; Farrar, J.T.; Chao, Y.; Thompson, A.F.; Flexas, M.M. An Observing System Simulation Experiment for the Calibration and Validation of the Surface Water Ocean Topography Sea Surface Height Measurement Using In Situ Platforms. *J. Atmos. Ocean. Technol.* **2018**, *35*, 281–297, doi:10.1175/JTECH-D-17-0076.1.

29. Gaultier, L.; Ubelmann, C.; Fu, L.L. *SWOT Simulator Documentation*; Tech. Rep. 1.0.0, Jet Propulsion Laboratory, California Institute of Technology: Pasadena, CA, USA, 2015.

30. Gaultier, L.; Ubelmann, C.; Fu, L.L. The challenge of using future SWOT data for oceanic field reconstruction. *J. Atmos. Ocean. Technol.* **2016**, *33*, 119–126, doi:10.1175/JTECH-D-15-0160.1.

31. Juza, M.; Mourre, B.; Renault, L.; Gómara, S.; Sebastián, K.; Lora, S.; Beltran, J.; Frontera, B.; Garau, B.; Troupin, C.; et al. SOCIB operational ocean forecasting system and multi-platform validation in the Western Mediterranean Sea. *J. Oper. Oceanogr.* **2016**, *9*, s155–s166.

32. Shchepetkin, A.F.; McWilliams, J.C. The regional oceanic modeling system (ROMS): A split-explicit, free-surface, topography-following-coordinate oceanic model. *Ocean Model.* **2005**, *9*, 347–404.

33. Sonka, M.; Hlavac, V.; Boyle, R. *Image Processing, Analysis, and Machine Vision*; Cengage Learning: Boston, MA, USA, 2014.

34. Aubert, G.; Kornprobst, P. *Mathematical Problems in Image Processing: Partial Differential Equations and the Calculus of Variations*; Springer Science & Business Media: Berlin, Germany, 2006; Volume 147.

35. Escudier, R.; Mourre, B.; Juza, M.; Tintoré, J. Subsurface circulation and mesoscale variability in the Algerian subbasin from altimeter-derived eddy trajectories. *J. Geophys. Res. Ocean.* **2016**, *121*, 6310–6322, doi:10.1002/2016JC011760.

36. Gunturk, B.K.; Li, X. *Image Restoration: Fundamentals and Advances*; CRC Press: Boca Raton, FL, USA, 2012.

37. Klein, P.; Morrow, R.; Samelson, R.; Chelton, D.; Lapeyre, G.; Fu, L.; Qiu, B.; Ubelmann, C.; Le Traon, P.Y.; Capet, X.; et al. Mesoscale/Sub-Mesoscale Dynamics in the Upper Ocean. 2015. NASA Surface Water and Ocean Topography (SWOT). Available online: https://www.aviso.altimetry.fr/fileadmin/documents/missions/Swot/WhitePaperSWOTSubmesoscale.pdf (accessed on 12 April 2018).

38. Ruggiero, G.; Cosme, E.; Brankart, J.; Sommer, J.L.; Ubelmann, C. An efficient way to account for observation error correlations in the assimilation of data from the future SWOT High-Resolution altimeter mission. *J. Atmos. Ocean. Technol.* **2016**, *33*, 2755–2768, doi:10.1175/JTECH-D-16-0048.1.

39. Pujol, M.I.; Faugere, Y.; Taburet, G.; Dupuy, S.; Pelloquin, C.; Ablain, M.; Picot, N. DUACS DT2014: The new multi-mission altimeter data set reprocessed over 20 years. *Ocean Sci.* **2016**, *12*, 1067–1090.

40. Ubelmann, C.; Klein, P.; Fu, L.L. Dynamic interpolation of sea surface height and potential applications for future high-resolution altimetry mapping. *J. Atmos. Ocean. Technol.* **2015**, *32*, 177–184.

41. Fablet, R.; Verron, J.; Mourre, B.; Chapron, B.; Pascual, A. Improving mesoscale altimetric data from a multitracer convolutional processing of standard satellite-derived products. *IEEE Trans. Geosci. Remote Sens.* **2018**, doi:10.1109/TGRS.2017.2750491.

42. Lguensat, R.; Viet, P.H.; Sun, M.; Chen, G.; Fenglin, T.; Chapron, B.; Fablet, R. Data-driven Interpolation of Sea Level Anomalies Using Analog Data Assimilation. 2017. Available online: https://hal.archives-ouvertes.fr/hal-01609851 (accessed on 12 April 2018).

43. Moro, M.D.; Brankart, J.M.; Brasseur, P.; Verron, J. Exploring image data assimilation in the prospect of high-resolution satellite oceanic observations. *Ocean Dyn.* **2017**, *67*, 875–895.

44. Mason, E.; Pascual, A.; McWilliams, J.C. A new sea surface height-based code for oceanic mesoscale eddy tracking. *J. Atmos. Ocean. Technol.* **2014**, *31*, 1181–1188, doi:10.1175/JTECH-D-14-00019.1.

45. Conti, D.; Orfila, A.; Mason, E.; Sayol, J.M.; Simarro, G.; Balle, S. An eddy tracking algorithm based on dynamical systems theory. *Ocean Dyn.* **2016**, *66*, 1415–1427, doi:10.1007/s10236-016-0990-7.

46. D'Ortenzio, F.; Iudicone, D.; de Boyer Montegut, C.; Testor, P.; Antoine, D.; Marullo, S.; Santoleri, R.; Madec, G. Seasonal variability of the mixed layer depth in the Mediterranean Sea as derived from in situ profiles. *Geophys. Res. Lett.* **2005** *32*, doi:10.1029/2005GL022463.

47. Houpert, L.; Testor, P.; de Madron, X.D.; Somot, S.; D'ortenzio, F.; Estournel, C.; Lavigne, H. Seasonal cycle of the mixed layer, the seasonal thermocline and the upper-ocean heat storage rate in the Mediterranean Sea derived from observations. *Progress Oceanogr.* **2015**, *132*, 333–352.

48. Qiu, B.; Chen, S.; Klein, P.; Ubelmann, C.; Fu, L.; Sasaki, H. Reconstructability of 3-Dimensional upper ocean circulation from SWOT Sea Surface Height measurements (early online release). *J. Phys. Oceanogr. J. Phys. Ocean.* **2016**, *46*, 947–963, doi:10.1175/JPO-D-15-0188.1.

49. Lindstrom, E.; Cherchal, S.; Fu, L.L.; Morrow, R.; Pavelsky, T.; Cretaux, J.F.; Vaze, P.; Lafon, T.; Coutin-Faye, S.; Amen, L.; et al. *Summary Report of the 2nd SWOT Science Team Meeting 2017*; Techreport; Meteo-France Conference Centre: Touloue, France, 2017.

50. Esteban-Fernandez, D. SWOT Project: Mission Performance and Error Budget Document. Jet Propulsion Laboratory, California Institute of Technology, JPL D-79084. April 2017; 117p. Available online: https://swot.jpl.nasa.gov/documents.htm (accessed on 12 April 2018).

PERMISSIONS

LIST OF CONTRIBUTORS

Jamon Van Den Hoek
Geography Program, College of Earth, Ocean, and Atmospheric Sciences, Oregon State University, Corvallis, OR 97331, USA

Augusto Getirana
Earth System Science Interdisciplinary Center, University of Maryland, College Park, MD 20740, USA
Hydrological Sciences Laboratory, NASA Goddard Space Flight Center, Greenbelt, MD 20771, USA

Hahn Chul Jung
Hydrological Sciences Laboratory, NASA Goddard Space Flight Center, Greenbelt, MD 20771, USA
Science Systems and Applications, Inc., Lanham, MD 20706, USA

Modurodoluwa A. Okeowo and Hyongki Lee
Department of Civil and Environmental Engineering, University of Houston, Houston, TX 77005, USA

Shenglian Guo and Jun Wang
State Key Laboratory of Water Resources and Hydropower Engineering Science, Wuhan University, Wuhan 430072, China

Youjiang Shen, Dedi Liu and Jiabo Yin
State Key Laboratory of Water Resources and Hydropower Engineering Science, Wuhan University, Wuhan 430072, China
Hubei Province Key Lab of Water System Science for Sponge City Construction, Wuhan University, Wuhan 430072, China

Liguang Jiang and Peter Bauer-Gottwein
Department of Environmental Engineering, Technical University of Denmark, 2800 Kgs. Lyngby, Denmark

Karina Nielsen
DTU Space, National Space Institute, Technical University of Denmark, 2800 Kgs. Lyngby, Denmark

Yingying Chen
School of Marine Science, Nanjing University of Information Science and Technology, Nanjing 210044, China
State Key Laboratory of Tropical Oceanography, South China Sea Institute of Oceanology, Chinese Academy of Sciences, Guangzhou 510000, China

Dongxiao Wang
State Key Laboratory of Tropical Oceanography, South China Sea Institute of Oceanology, Chinese Academy of Sciences, Guangzhou 510000, China

Yunwei Yan
State Key Laboratory of Satellite Ocean Environment Dynamics, Second Institute of Oceanography, State Oceanic Administration, Hangzhou 310000, China

Kai Yu
School of Marine Science, Nanjing University of Information Science and Technology, Nanjing 210044, China
State Key Laboratory of Satellite Ocean Environment Dynamics, Second Institute of Oceanography, State Oceanic Administration, Hangzhou 310000, China

Changming Dong
School of Marine Science, Nanjing University of Information Science and Technology, Nanjing 210044, China
Department of Atmospheric and Oceanic Sciences, University of California, Los Angeles, CA 90095, USA

Zhigang He
College of Ocean and Earth Science, Xiamen University, Xiamen 361000, China

Gaia Piccioni, Denise Dettmering, Marcello Passaro, Christian Schwatke, Wolfgang Bosch and Florian Seitz
Deutsches Geodätisches Forschungsinstitut der Technischen Universität München (DGFI-TUM), Arcisstrasse 21, 80333 München, Germany

Pascal Bonnefond
SYRTE, Observatoire de Paris, PSL Research University, CNRS, Sorbonne Universités, UPMC Univ. Paris 06, LNE, 75014 Paris, France

Jacques Verron
Institut des Géosciences de l'Environnement (IGE)/ CNRS, 38041 Grenoble, France

Jérémie Aublanc, Annabelle Ollivier, Jean-Christophe Poisson, Pierre Prandi and Pierre Thibaut
Collecte Localisation Satellites (CLS), 31520 Ramonville Saint-Agne, France

K. N. Babu, Aditya Chaudhary and Rashmi Sharma
Space Applications Centre (ISRO), Ahmedabad 380015, India

Muriel Bergé-Nguyen and Jean-François Crétaux
Laboratoire d'Etudes en Géophysique et Océanographie Spatiales (LEGOS), 31400 Toulouse, France

Mathilde Cancet
NOVELTIS, 31670 Labège, France

Frédéric Frappart
Geosciences Environment Toulouse (GET), University of Toulouse, National Center for Scientific Reaseach (CNRS), Institute for Research and Development (IRD), UPS. Observatory Midi-Pyrénées (OMP), 14 Av. E. Belin, 31400 Toulouse, France
Laboratory of Studies on Spatial Geophysics and Space Oceanography (LEGOS), University of Toulouse, National Center for Space Studies (CNES), CNRS, IRD, UPS. OMP, 14 Av. E. Belin, 31400 Toulouse, France

Bruce J. Haines
Jet Propulsion Laboratory, California Institute of Technology, Pasadena, CA 91109, USA

Olivier Laurain
Géoazur — Observatoire de la Côte d'Azur, 06905 Sophia-Antipolis, France

Christopher Watson
Surveying and Spatial Science Group, School of Geography and Environmental Studies, University of Tasmania, Hobart 7001, Australia

Florence Birol
Laboratoire d'Etudes en Géophysique et Océanographie Spatiales (LEGOS), Observatoire Midi-Pyrénées, 31400 Toulouse, France

Xi-Yu Xu
The CAS Key Laboratory of Microwave Remote Sensing, National Space Science Center, Chinese Academy of Sciences, Beijing 100190, China
Laboratoire d'Etudes en Géophysique et Océanographie Spatiales (LEGOS), Observatoire Midi-Pyrénées, 31400 Toulouse, France
State Key Laboratory of Remote Sensing Science, Institute of Remote Sensing and Digital Earth, Chinese Academy of Sciences, Beijing 100094, China

Anny Cazenave
Laboratoire d'Etudes en Géophysique et Océanographie Spatiales (LEGOS), Observatoire Midi-Pyrénées, 31400 Toulouse, France
International Space Science Institute, 3102 Bern, Switzerland

Sakaros Bogning
Département de Sciences de la Terre, Université de Douala, BP 24 157 Douala, Cameroun
Jeune Equipe Associée à l'IRD — Réponse du Littoral Camerounais aux Forçages Océaniques Multi-Échelles (JEAI-RELIFOME), Université de Douala, BP 24 157 Douala, Cameroun
LEGOS, Université de Toulouse, CNES, CNRS, IRD, UPS OMP, 14 Av. E. Belin, 31400 Toulouse, France

Jacques Etamé
Département de Sciences de la Terre, Université de Douala, BP 24 157 Douala, Cameroun

Raphaël Onguéné
Jeune Equipe Associée à l'IRD — Réponse du Littoral Camerounais aux Forçages Océaniques Multi-Échelles (JEAI-RELIFOME), Université de Douala, BP 24 157 Douala, Cameroun

Fernando Niño
LEGOS, Université de Toulouse, CNES, CNRS, IRD, UPS OMP, 14 Av. E. Belin, 31400 Toulouse, France

Jean-Jacques Braun
GET, Université de Toulouse, CNRS, IRD, UPS OMP, 14 Av. E. Belin, 31400 Toulouse, France

Gil Mahé and Jean-Pierre Bricquet
HydroSciences Montpellier, Université de Montpellier, CNRS, IRD, 300 Av. Pr E. Jeanbrau, 34090 Montpellier, France

Frédérique Seyler
ESPACE-DEV, Université de Montpellier, IRD, Université des Antilles, Université de Guyane, Université de La Réunion, Maison de la Télédétection, 500 Rue J-F. Breton, 34093 Montpellier, France

Marie-Claire Paiz
The Nature Conservancy Gabon Program Office, Lot 114 Haut de Gué-Gué, 13553 Libreville, Gabon

Cassandra Normandin, Vincent Marieu and Bertrand Lubac
Oceanic and Continental Environments and Paleoenvironments (EPOC), Mixed Research Unit (UMR) 5805, University of Bordeaux, Allée Geoffroy Saint-Hilaire, 33615 Pessac, France

Eric Mougin
Geosciences Environment Toulouse (GET), University of Toulouse, National Center for Scientific Reaseach (CNRS), Institute for Research and Development (IRD), UPS. Observatory Midi-Pyrénées (OMP), 14 Av. E. Belin, 31400 Toulouse, France

Fabien Blarel
Laboratory of Studies on Spatial Geophysics and Space Oceanography (LEGOS), University of Toulouse, National Center for Space Studies (CNES), CNRS, IRD, UPS. OMP, 14 Av. E. Belin, 31400 Toulouse, France

Adama Telly Diepkilé
Department of Education and Research (DER) Math-Informatics, Faculty of Sciences and Technology (FST)/ University of Sciences, Techniques and Technologies of Bomako (USTTB), Bamako 3206, Mali

Nadine Braquet
National Research Institute of Science and Technology for Environment and Agriculture (IRSTEA), IRD, 361 rue Jean-François Breton, 34196 Montpellier, France

Abdramane Ba
Laboratory of Optics, Spectroscopy and Atmospheric Sciences (LOSSA), Department of Education and Research (DER) Physics, Faculty of Sciences and Technology (FST)/University of Sciences, Techniques and Technologies of Bomako (USTTB), Bamako 3206, Mali

Maria Joana Fernandes and Clara Lázaro
Faculdade de Ciências, Universidade do Porto, 4169-007 Porto, Portugal

Centro Interdisciplinar de Investigação Marinha e Ambiental (CIIMAR/CIMAR), Universidade do Porto, 4050-123 Porto, Portugal

Laura Gómez-Navarro
Institut Mediterrani d'Estudis Avançats (IMEDEA) (CSIC-UIB), 07190 Esporles, Illes Balears, Spain
Univ. Grenoble Alpes, CNRS, IRD, Grenoble INP, IGE, 38000 Grenoble, France

Evan Mason and Ananda Pascual
Institut Mediterrani d'Estudis Avançats (IMEDEA) (CSIC-UIB), 07190 Esporles, Illes Balears, Spain

Emmanuel Cosme and Julien Le Sommer
Univ. Grenoble Alpes, CNRS, IRD, Grenoble INP, IGE, 38000 Grenoble, France

Ronan Fablet
Institut Mines-Télécom, Telecom-Bretagne, UMR 6285 labSTICC, 29238 Brest, France

Baptiste Mourre
Balearic Islands Coastal Observing and Forecasting System (SOCIB), 07121 Palma de Mallorca, Illes Balears, Spain

Index

Printed in the USA
CPSIA information can be obtained
at www.ICGtesting.com
JSHW051359091023
49903JS00006B/210

9 781647 404079